MW00988735

OPTIONS

OPTIONS

Essential Concepts and Trading Strategies

THIRD EDITION

EDITED BY THE OPTIONS INSTITUTE

The Educational Division of the Chicago Board Options Exchange

McGraw-Hill

New York San Francisco Washington, D.C. Auckland Bogotá
Caracas Lisbon London Madrid Mexico City Milan
Montreal New Delhi San Juan Singapore
Sydney Tokyo Toronto

Library of Congress Cataloging-in-Publication Data
Options : essential concepts and trading strategies / edited by the
Options Institute, the Educational Division of the Chicago Board Options
Exchange. — 3rd ed.
p. cm.
ISBN 0-07-134169-2
1. Options (Finance) I. Options Institute (Chicago Board Options
Exchange)
HG6024.A3065 1999
332.63'228—dc21 99–18006
CIP

McGraw-Hill

A Division of The McGraw·Hill Companies

1 2 3 4 5 6 7 8 9 0 DOC/DOC 9 0 4 3 2 1 0 9

ISBN 0-07-134169-2

The sponsoring editor for this book was Stephen Isaacs, the editing supervisor was Barry Brown, and the production supervisor was Tina Cameron. It was set in Palatino by Jana Fisher through the services of Barry E. Brown (Broker—Editing, Design and Production).

Printed and bound by R.R. Donnelley & Sons Company.

This book is printed on recycled, acid-free paper containing a minimum of 50% recycled, de-inked fiber.

The Options Institute is proud to dedicate this 3rd edition to two distinguished colleagues:

Harrison Roth (1932–1997)
Harry Roth's 40-plus years in the options industry began long before the advent of exchange-listed and exchange-traded option contracts, as a seller of "over-the-counter" puts and calls in the early 1950s. When the CBOE was created, Harry was there as a pioneer in the listed option business, and he remained a dedicated supporter of options and the CBOE for the rest of his life. His commitment to education was evident through his service as consultant for The Options Institute from its earliest days, where he introduced a great many public investors as well as traders to the benefits and risks of options. He was a founding member of the Committee on Options Proposals (the "CO-OP"), and most recently Senior Options Strategist at Cowen & Company. His book *LEAPS: What They Are and How to Use Them for Profit and Protection* remains the authoritative source on the use of long-term options. Known for his prodigious memory and gift for logic, Harry was an expert bridge player and master puzzler. He will always be remembered as a widely respected and often quoted member of the financial community.

James W. Yates (1940–1996)
Jim Yates had an early vision of what the options industry would become and an enthusiasm about the flexibility options could offer as investing tools. His thoughts often took up where other industry professionals' ended. When at work, Jim lived in a world quantified by distribution curves and trading patterns, but he had a knack for conveying his ideas to even the most mathematically challenged. He launched a career in finance as both a retail and institutional broker with a focus on equity sales. After the CBOE was founded, Jim formed and served as President of DYR & Associates and began distribution of one of the earliest newsletters devoted entirely to options. DYR's unique offerings included a technique for using implied volatility to identify over-bought and over-sold stocks, and an index for option writers—ideas ahead of their time. Jim was instrumental to the creation of The Options Institute for which he continued to design and teach option courses to investors and industry professionals alike. He also worked independently to educate option investors across the nation. His engaging personality and gregarious manner will be missed by all of those lucky enough to have known him.

CONTENTS

If you are an investor seeking clear, practical, and useful information on options, then this book was written for you. The authors are former brokers, traders, option strategists, and professors, all of whom have taught courses for The Options Institute.

The Options Institute is the educational arm of the Chicago Board Options Exchange. It was started in 1985 after a study revealed that the vast majority of investors either knew little or nothing about options trading or had misconceptions about the potentially valuable role that options can play in reducing risk, enhancing income, or increasing the number of alternatives available to investors. The surprising thing is that, after 13 years, the situation is still largely the same. Despite presentations to over 75,000 people including individual and institutional investors, brokers, government regulators, and corporate financial managers, instructors at The Options Institute still encounter misconceptions about options. This book is intended to address those issues.

Although many of the fundamentals of option mechanics, pricing, and textbook strategies remain unchanged, this third edition has many additions and improvements over the second edition. This edition is organized in three parts: Essential Concepts, Investing and Trading Strategies, and Real-Time Applications. Chapter 1, The History of Options, is entertaining and informative, and it has been completely rewritten and expanded since the second edition.

Part 1, Essential Concepts, is largely unchanged from the second edition. Chapter 2, Fundamentals of Options, defines basic options-related terms and presents a non-mathematical explanation of option valuation and price behavior. The final portion of this chapter, The Second Level of Understanding, introduces the four option risk factors, known as the "Greeks," which estimate how option values change when inputs such as price of underlying, time, or volatility change.

Chapter 3, Volatility Explained, discusses a commonsense, non-mathematical way this important subject. Chapter 4, Option Strategies: Analysis and Selection, discusses 15 different option strategies. An expiration profit-and-loss diagram is presented and the nature of the thinking that would justify the selection of the strategy is explained.

Part 2, Investing and Trading Strategies, is comprised of chapters that discuss the use of options from the points of view of different market participants: individual investors and traders, institutional investors, and professional market makers. Chapter 5, Investing and Trading Strategies for the Individual Investor, discusses both investment-oriented strategies and trading, or speculative, strategies. Investment-oriented strategies include basic strategies such as covered writing and protective puts as well as more advanced strategies such as the stock repair strategy. Basic option purchasing and option spreads are two of the trading strategies discussed. Chapter 6, Strategies of Institutional Investors, discusses portfolio management strategies using index options. Both the mechanics and logic for strategy selection are discussed. Chapter 7, How the Trading Floor Operates, explains how a hypothetical order to buy or sell an option could be processed from the time of an investor's instruction to a broker, to when that order is represented in the trading crowd on the exchange floor, to when a trade confirmation is reported back to the investor. Chapter 7 also serves as a prelude to Chapter 8, How Market Makers Trade, which explains how professional market makers make buy and sell decisions. The conclusion of the discussion in this chapter is that off-floor investors and traders are not in competition with full-time professional floor traders. Rather, they are in different businesses; they use different strategies; they manage their trading capital differently; and they have different methods of earning profits.

Part 3, Real-Time Applications, has chapters that are relevant to both investors and traders. Chapter 9, Institutional Case Studies, presents two lifelike situations from different market environments that offer opportunities to apply theory to practice. The first case study is written from the perspective of a portfolio manager and discusses the use of option strategies for risk management. The second case discusses the issues involved when an investment trust seeks to gain approval from their trustees to allow investment in derivative securities. Chapter 10, The Predictive Power of Options, illustrates in depth how one well-known market strategist uses information from the listed options market to make trading decisions. And finally, if you want to get wired, Chapter 11, Electronic Resources, will help you do it. Although the World Wide Web is in a state of constant change, the discussion is intended to help you focus on the important issues of the information you need and assistance you need to make decisions. Lists of available Web sites, software, and media are presented in this chapter, and even if the lists are incomplete or out of date by the time you read them, there are plenty of ideas that will enable you to conduct your own search.

Helping investors and traders learn to help themselves is the real purpose of this chapter.

This edition of *Options: Essential Concepts and Trading Strategies* is concluded with a well-researched and complete Glossary of Option Market Terminology. A valuable resource in itself, the Glossary is one of the many reasons you will want to keep this book readily available as a valuable reference.

ACKNOWLEDGMENTS

This book is the result of a team effort, but the authors are not the only members of the team. We would like to thank William J. Brodsky, Chairman and CEO of the Chicago Board Options Exchange, and John Power, OBOE Vice President and Director of The Options Institute. They supported the effort required to produce this book through the commitment of valuable resources and time allotments. We would also like to thank Debra Peters, Linda Boland, Sue Belling, Michele Kaufman, Dione Rendo, and Niece Atkins of The Options Institute for their energetic and cheerful assistance. Finally, we would like to thank you, the readers and seminar attendees, who ask the questions that this book tries to answer and who participate now or in the future in one of the most significant developments in the modern financial marketplace, the creation and growth of the listed options market.

J. Marc Allaire (Chapter 5, Option Strategies for the Individual Investor) currently holds the position of Options Strategist at Barrington Investment Limited, a Bermuda-based brokerage firm. He is also President of BAM Corp., a money management firm that specializes in options related strategies.

Previously, he traded futures and options for a regional brokerage firm in Montreal before assuming the responsibility of options marketing for the Montreal Exchange. In 1986 he joined Richardson Greenshields of Canada Limited, where he was Vice President and Options Manager. His primary focus was on the retail side of the business, where he concentrated on broker education, strategy development, and sales support.

He joined The Options Institute in 1992 in the position of Senior Staff Instructor, where he remained until taking on his present responsibilities in 1998. He has written extensively about the options market in various publications and textbooks.

Mr. Allaire is a graduate of the University of Ottawa, and he has an MBA from McGill University in Montreal. When not on the trading desk, he enjoys adding to the 600+ bird species he has observed, and the collecting and sampling of fine wines.

James B. Bittman (Chapter 2, Fundamentals of Options, Chapter 3, Volatility Explained; Chapter 6, Strategies for Institutional Investors; and Chapter 8, How Market Makers Trade) is the author of two books, *Options for the Stock Investor*, (McGraw-Hill, 1996), and *Trading Index Options* (McGraw-Hill, 1998). As a Senior Instructor at The Options Institute, he teaches courses for public and institutional investors. He also presents custom courses throughout the United States, Europe, South America, and Southeast Asia. Mr. Bittman is also a member of the faculty of The Illinois Institute of Technology, where he teaches in the masters' level Financial Markets and Trading Program.

In 1980 Mr. Bittman began his trading career as an equity options market maker at the CBOE. From 1983 to 1993, he was a Commodity Options Member of the Chicago Board of Trade, where he traded options on financial futures and agricultural futures.

Mr. Bittman received a BA, *magna cum laude*, from Amherst College and an MBA from Harvard University.

Eric Frait (Chapter 6, Strategies for Institutional Investors) has been in the financial markets and the financial services industry since 1993. His career began as a Futures and Securities Specialist with the Harris Trust and Savings Bank in Chicago, where he provided services for CBOE, CBOT, and CME trading firms. In 1995, he joined CBOE market-maker firm Hull Trading Company, LLC as an Assistant Trader, where he was part of the "Hi-Cap" (Highly Capitalized) options trading team on the floor of the CBOE and the Interest Rate options trading team on the floor of the CBOT. Mr. Frait has been with the CBOE since 1996. Prior to joining The Options Institute as a Senior Instructor, he was a Senior Investigator in the Exchange's Regulatory Division, where he was responsible for market surveillance and for enforcing Exchange and SEC rules and regulations.

He has served as a staff liaison to several Exchange Committees, including the Clearing Procedures Regulatory and Operations Sub-Committees and the Stock System Advisory Committee. Mr. Frait has a BA from the University of Iowa and an MBA, *Beta Gamma Sigma*, from Loyola University, Chicago, with a triple specialization in Finance, Economics, and Financial Derivatives.

Floyd Fulkerson (Chapter 11, Electronic Resources) has 15 years experience in the options business at the CBOE, climbing the ranks from runner on the trading floor to market-maker trading equity options. He began consulting for The Options Institute in 1989, and took the position of Software Project Manager in 1995. He holds a BA degree from Columbia University in New York.

Elliot Katz (Chapter 4, Option Strategies: Analysis and Selection) is First Vice President and Branch Manager for the investment firm Sutro & Co. in Beverly Hills. Previously, he was a branch manager and sales manager for another national brokerage firm. Prior to that, Mr. Katz was an instructor with The Options Institute where he taught options strategies to retail brokers and professional money managers.

Mr. Katz holds a BS in computer science from the State University of New York at Stony Brook. He lives in southern California with his wife, Laurie, and children, Jonathan and Stephanie.

Marshall V. Kearney (Chapter 7, How The Trading Floor Operates; Chapter 8, How Market Makers Trade) is a Senior Instructor at The Options Institute. He began his association with the Chicago Board

Options Exchange as an independent market maker in 1981. In 1992, he became a Registered Options Principal and a founding partner of PTI Securities L. P., a member firm of the CBOE. At PTI Securities, he developed and implemented hedging strategies using listed options for institutional and retail clients. Mr. Kearney authored PTI's weekly strategy letter for over four years. He also composed "Daily Comments" on the PTI web-site, which he helped develop. He has been a regular contributor to many news services including Reuters, WMAQ Radio(Group W), The CBS Radio Network, Derivatives Week, Channel 26 TV, Barron's, and others. Mr. Kearney served on various committees at the CBOE, including the Arbitration Committee from 1984 to 1996.

Andrew B. Lowenthal (Chapter 6, Strategies for Institutional Investors) is a Director of Market Operations at the CBOE. His responsibilities include managing the CBOE FLEX Option product, monitoring and improving efficiency in CBOE's operations, developing and implementing CBOE's screen-based trading system, and administering CBOE's disaster recovery plan.

Mr. Lowenthal has been with the CBOE since 1983. Prior to his current position, which he assumed in June, 1997, Mr. Lowenthal was a Director in the Exchange's Department of Strategic Planning and International Development. Prior to that, he directed the training function within the Exchange's Trading Operations Division, spent a year in the Department of Market Surveillance, and spent three years on the trading floor.

Mr. Lowenthal received both a BA in Finance and an MBA from DePaul University in Chicago. He is a native of the Chicago area, and when not busy with his two young children, he trains for and participates in triathlon and marathon events.

Lawrence G. McMillan (Chapter 10, The Predictive Power of Options) is the author of *Options As a Strategic Investment*, the best-selling work on stock and index option strategies, which has sold over 160,000 copies. The third edition of this work was released in 1993. His second book, *McMillan On Options*, was published in October, 1996. He currently edits and publishes "The Option Strategist," a derivative-products newsletter covering equity, index, and futures options. He also has a unique daily fax service, "Daily Volume Alerts," which selects short-term stock trades by looking for unusual increases in equity option volume. In these capacities, he is the President of McMillan Analysis Corporation. He has spoken on option strategies at many seminars and colloquiums in the United States, Canada, and Europe. In addition, he trades his own account actively in the option markets.

From 1982 to 1989, he was in charge of the Equity Arbitrage Department at Thomson McKinnon Securities, Inc., and then was in charge of the Proprietary Option Trading Department at Prudential-Bache Securities from 1989 through 1990. Before holding those positions, he was the retail option strategist at Thomson McKinnon from 1976 to 1980, trading the firm's proprietary account beginning in 1980.

Mr. McMillan holds a BS in mathematics from Purdue University, and an MS in applied mathematics and computer science from the University of Colorado.

Brian Overby (Chapter 1, The History of Options) is a Senior Instructor at The Options Institute. He has been in the financial industry since 1992 and has experience in both investment banking and retail brokerage. While employed at Charles Schwab & Company, he specialized in options and worked with the firm's most active clients designing strategies and providing trading support. His duties included rewriting the firm's options training manual, developing classes, and instructing courses for internal staff. Mr. Overby also was in charge of organizing and directing options seminars for the firm's retail customers.

Mr. Overby holds a BS in Applied Mathematics from the University of Wisconsin at Steven's Point.

Gary L. Trennepohl (Chapter 9, Institutional Case Studies) is Dean of the College of Business Administration at Oklahoma State University. Previously, he served as Executive Associate Dean and Professor of Finance at Texas A&M University and Head of the Finance Department at the University of Missouri—Columbia.

Dean Trennepohl has co-authored two finance texts, *Investment Management* (1993), and *An Introduction to Financial Management* (1984), and has contributed to The Encyclopedia of Investments. He has also authored or co-authored more than 30 professional journal articles and is active in many professional organizations. During 1993–94 he was president of the Financial Management Association, which has over 10,000 professional and student members worldwide.

For over 15 years he has served as a consultant to pension funds, endowment funds, corporations, and financial institutions. He currently is a visiting faculty member for the Options Institute, and, since 1983, he has conducted yearly financial analysis seminars for business journalists who write for business periodicals and local and national newspapers, including *USA Today*.

Dean Trennepohl holds a BS degree from the University of Tulsa, an MBA from Utah State University, and a PhD in Finance from Texas Tech University.

CHAPTER 1

THE HISTORY OF OPTIONS

Brian Overby

The acceptance of options as versatile investment tools is a recent development. Most of the history of the option contract is one of mystique, but the perception of options has evolved to become an accepted and integral part of modern financial management.

EARLY HISTORY

The first record of an option transaction involving a contractual agreement appears in the Bible. The book of Genesis recounts how Jacob entered into an option-like contract for the purpose of marrying Laban's younger daughter, Rachel. Jacob received permission to marry Rachel when he agreed to work for Laban for seven years. In the language of options, Jacob paid the "premium" of seven years of labor and received "the right, but not the obligation" to marry Rachel.

The earliest recorded speculator in options was Thales, an astronomer and the first eminent Greek philosopher. According to Aristotle, Thales "knew by his skill in the stars while it was yet winter that there would be a great harvest of olives in the coming year; having little money, he gave deposits for the use of all the olive-presses in Chios and Miletus, which he hired at a low price because no one bid against him."[1] An abundant harvest and a great demand for olive presses

[1] Albert I. A. Bookbinder, *Security Options Strategy* (Elmont, NY: Programmed Press, 1976), p. 21.

proved Thales' forecast correct, and, with the contracts in place, Thales sold the rights to the use presses at a great profit.

Although many people associate options only with speculation, the concept of options evolved from a need to control the risk of price fluctuations in agricultural markets. The first documentation of such a use of options occurred in Holland in 1634.[2] Tulips were a status symbol among the 17th century Dutch aristocracy, and, during this time, it was common practice for wholesalers to sell tulips for future delivery. There was great risk, however, in agreeing to sell at a fixed price in the future without knowing, for sure, what the cost would be. To limit this risk and to assure profit margins, many wholesalers bought options from tulip growers. These options entitled wholesalers to buy tulips from growers at a predetermined price for a specific period of time. In other words, the maximum price for the wholesalers was fixed until it was time to deliver to the end customer and receive payment. If tulip prices rose above that maximum price, then wholesalers who owned an option would demand delivery from the grower at that maximum price, thus insuring a profit margin. If, however, prices declined and the option expired worthless, a wholesaler could still secure profits by purchasing tulips at the then current, and lower, market price and fulfill the commitment to the end customer at an even lower cost than anticipated. These options contracts made it possible for many wholesalers to stay in business during periods of extreme volatility in tulip prices. In contrast to the wholesalers who purchased options and had limited risk, wholesalers who obligated themselves to pay a fixed price, regardless of market conditions at the time of harvest, were ruined when tulip prices collapsed.

As demand for tulips increased worldwide, a secondary market developed for options on tulip bulbs. This meant that, in addition to the wholesalers who actually bought and sold tulips and who used options to manage the business risk of a poor harvest, the public began trading tulip bulb option contracts in the hopes of profiting from the flower's new popularity. As tulip prices spiraled upward, many individuals used their savings to speculate in these unregulated option markets. When the Dutch economy entered a recession in 1638, the "tulip craze" came to an end. Tulip bulb prices experienced a staggering drop, and many speculators who had sold put options could not pay for the bulbs

[2] Charles Mackay, L. L. D., *Extraordinary Popular Delusions and the Madness of Crowds*, first published in 1841, Farrar, Straus and Giroux (1932), p. 95.

they were obligated to buy. The Dutch government intervened and attempted to force the sellers of the contracts to make good on their obligations, but the lack of prior regulation made enforcement impossible. These broken contracts further fueled the recession, and, as a result, options developed a terrible reputation throughout Holland and all of Europe. Although option trading on agricultural products continued in Europe among knowledgeable and financially capable parties throughout the 18th century, the trading was on a much smaller scale.

FIRST OPTIONS MARKET IN THE UNITED STATES

In the United States a market for options on stocks developed after the 1791 opening of the New York Stock Exchange.[3] Figure 1-1 shows an advertisement published in 1875 by Tumbridge & Co. who proclaimed to be "Bankers and Brokers." The ad contains the following suggestions for the use of options:

> If you think stocks are going down, secure a Put.
>
> If you think stocks are going up, secure a Call.
>
> A double Privilege pays a profit, no matter which way the market goes.

At this time, a central marketplace for options did not exist. Trading of calls and puts took place in an over-the-counter market and was facilitated by broker-dealers who tried to find a match for an option buyer or seller. The negotiations involved in consummating a trade was a cumbersome process. Not only could there be protracted negotiations over price, every new option contract had its own set of terms and conditions such as length of time, number of shares, and strike price. As a result of every option contract being unique, the buyer and the writer were essentially linked for the term of the option contract. Figure 1-2 is an example of an early call option contracts. In order for either party to close the position, an agreement had to be made with the other party regarding time and price.

In cases where an agreement to close an existing contract could not be reached between the two involved parties, a broker-dealer might advertise the contract in a financial journal under the heading of "Special Options." As illustrated in Figure 1-3, the goal of such an ad was to find another interested party.

---

[3] Herbert J. Filer, *Understanding Put and Call Options* (New York: Crown Publishers, 1959), p. 92.

FIGURE 1-1

An early advertisement for options (1875)

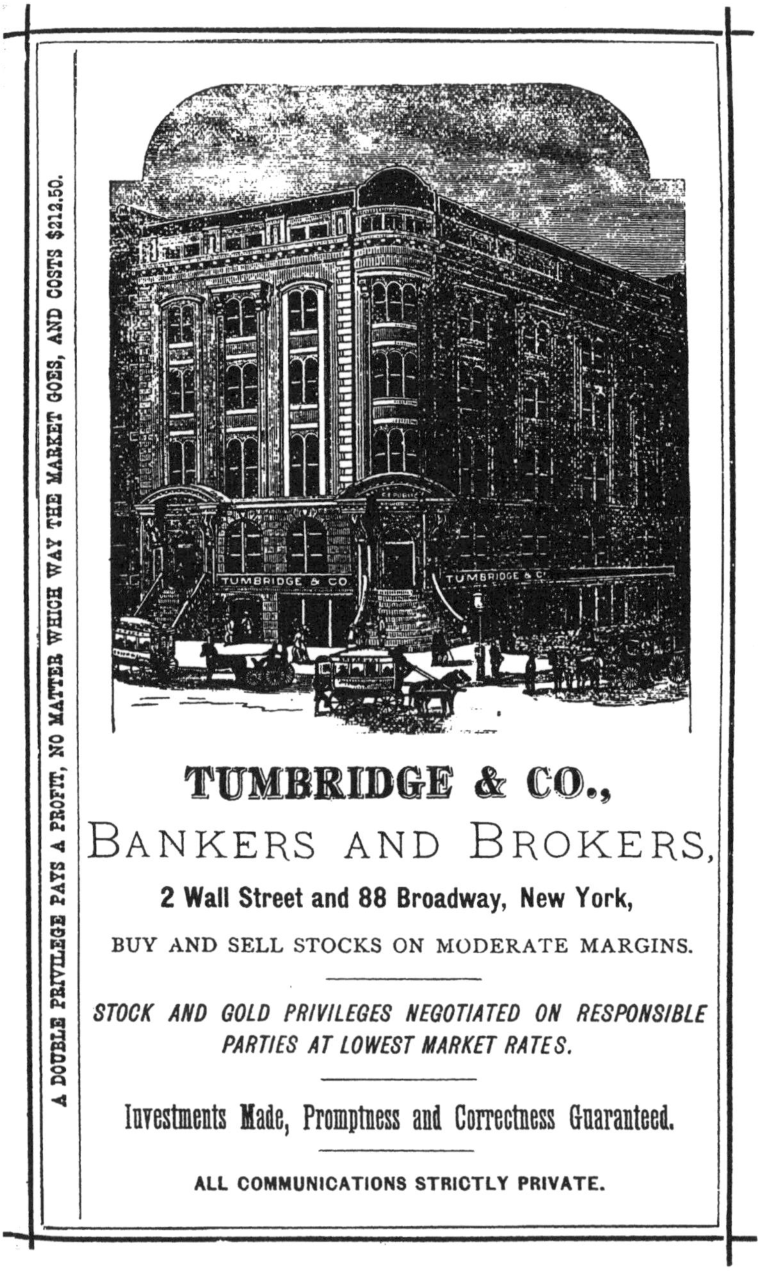

TUMBRIDGE & CO.,

BANKERS AND BROKERS,

2 Wall Street and 88 Broadway, New York,

BUY AND SELL STOCKS ON MODERATE MARGINS.

STOCK AND GOLD PRIVILEGES NEGOTIATED ON RESPONSIBLE PARTIES AT LOWEST MARKET RATES.

Investments Made, Promptness and Correctness Guaranteed.

ALL COMMUNICATIONS STRICTLY PRIVATE.

A DOUBLE PRIVILEGE PAYS A PROFIT, NO MATTER WHICH WAY THE MARKET GOES, AND COSTS $212.50.

FIGURE 1-2

Examples of early option contracts

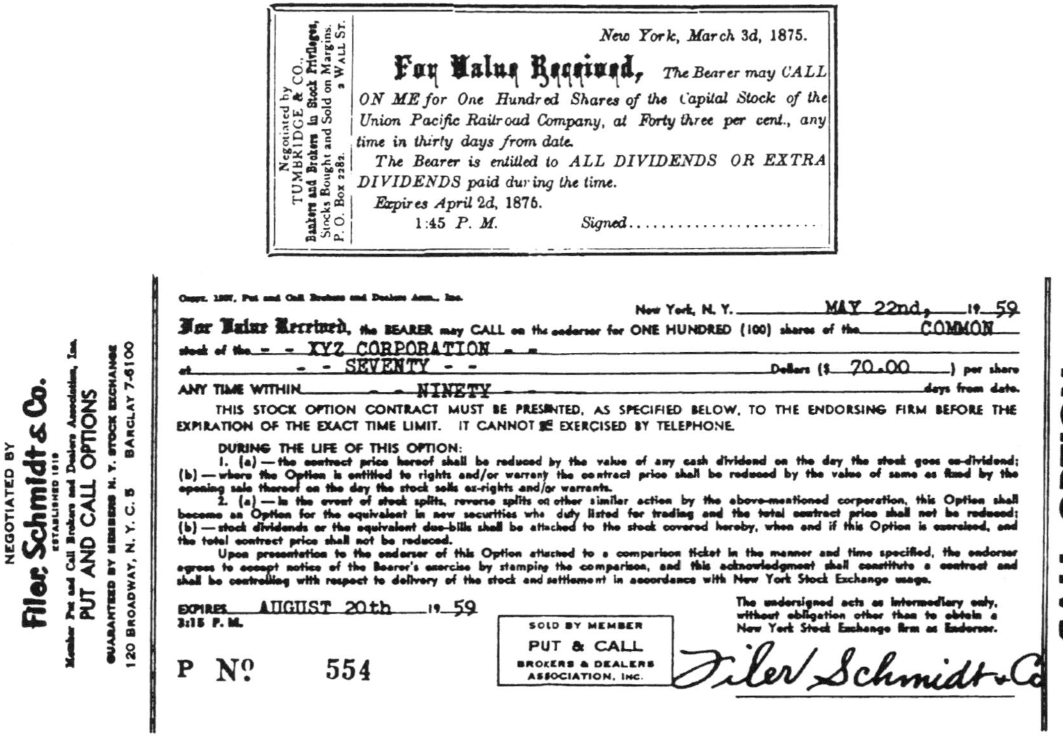

Negotiated by
TUMBRIDGE & CO.,
Bankers and Brokers in Stock Privileges,
Stocks Bought and Sold on Margins.
P. O. Box 2282. 2 WALL ST.

New York, March 3d, 1875.

For Value Received, The Bearer may CALL ON ME for One Hundred Shares of the Capital Stock of the Union Pacific Railroad Company, at Forty three per cent., any time in thirty days from date.

The Bearer is entitled to ALL DIVIDENDS OR EXTRA DIVIDENDS paid during the time.

Expires April 2d, 1876.
1:45 P. M. Signed.......................

NEGOTIATED BY
Filer, Schmidt & Co.
ESTABLISHED 1919
Member Put and Call Brokers and Dealers Association, Inc.
PUT AND CALL OPTIONS
GUARANTEED BY MEMBERS N. Y. STOCK EXCHANGE
120 BROADWAY, N. Y. C. 5 BARCLAY 7-6100

Copyr. 1937, Put and Call Brokers and Dealers Assn., Inc.

New York, N. Y. MAY 22nd, 19 59

For Value Received, the BEARER may CALL on the endorser for ONE HUNDRED (100) shares of the COMMON stock of the - - XYZ CORPORATION - - at - - SEVENTY - - Dollars ($ 70.00) per share ANY TIME WITHIN - - NINETY - - days from date.

THIS STOCK OPTION CONTRACT MUST BE PRESENTED, AS SPECIFIED BELOW, TO THE ENDORSING FIRM BEFORE THE EXPIRATION OF THE EXACT TIME LIMIT. IT CANNOT BE EXERCISED BY TELEPHONE.

DURING THE LIFE OF THIS OPTION:

1. (a) —the contract price hereof shall be reduced by the value of any cash dividend on the day the stock goes ex-dividend; (b) —where the Option is entitled to rights and/or warrants the contract price shall be reduced by the value of same as fixed by the opening sale thereof on the day the stock sells ex-rights and/or warrants.

2. (a) —In the event of stock splits, reverse splits or other similar action by the above-mentioned corporation, this Option shall become an Option for the equivalent in new securities when duly listed for trading and the total contract price shall not be reduced; (b) —stock dividends or the equivalent due-bills shall be attached to the stock covered hereby, when and if this Option is exercised, and the total contract price shall not be reduced.

Upon presentation to the endorser of this Option attached to a comparison ticket in the manner and time specified, the endorser agrees to accept notice of the Bearer's exercise by stamping the comparison, and this acknowledgment shall constitute a contract and shall be controlling with respect to delivery of the stock and settlement in accordance with New York Stock Exchange usage.

EXPIRES AUGUST 20th 19 59
3:15 P. M.

P N? 554

SOLD BY MEMBER
PUT & CALL
BROKERS & DEALERS
ASSOCIATION, INC.

The undersigned acts as intermediary only, without obligation other than to obtain a New York Stock Exchange firm as Endorser.

Filer, Schmidt & Co

CALL OPTION

Source: Herbert Filer, *Understanding Put and Call Options* (New York: Crown Publishers, 1959).

FIGURE 1-3

The Advertising or Offering of Special Options

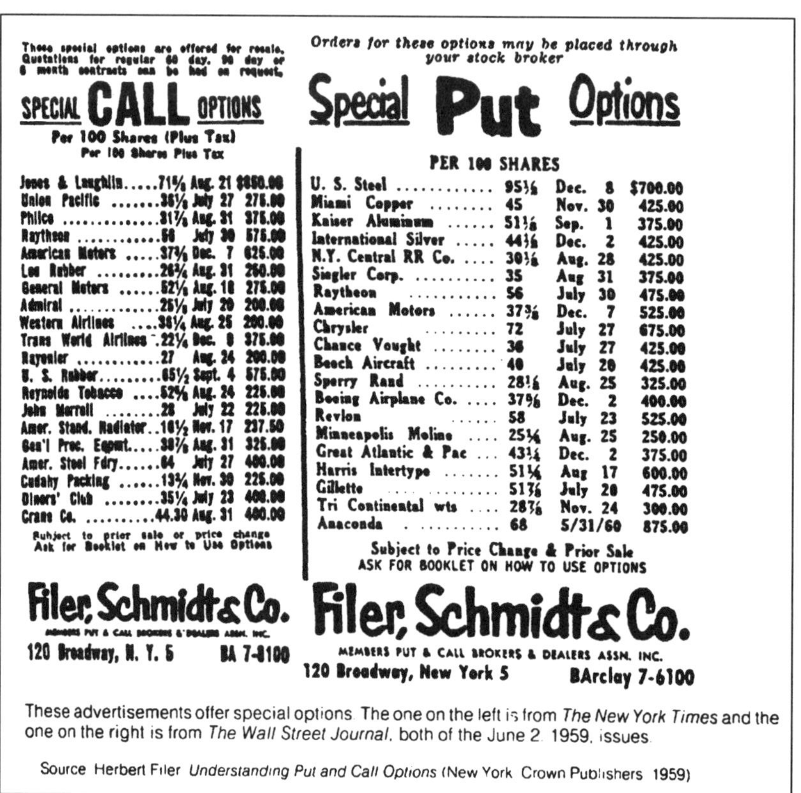

These advertisements offer special options. The one on the left is from *The New York Times* and the one on the right is from *The Wall Street Journal*, both of the June 2, 1959, issues.

Source: Herbert Filer, *Understanding Put and Call Options* (New York: Crown Publishers, 1959).

FORMATION OF THE PUT AND CALL BROKERS AND DEALERS ASSOCIATION

Eventually, out of this loosely knit market, The Put and Call Brokers and Dealers Association, Inc. was formed. The purpose was to establish better networking and to expand business to the extent that common interests allowed.

The difficulties of matching buyers and sellers was not the only problem faced by the infant market. There was, it seemed, an endless struggle with the issue of "fair" pricing. Because no standard existed, buyers and sellers could haggle for days until developments in the market for the underlying stock changed and lessened the interest of one party. Despite these shortcomings of the over-the-counter market, however, trading volume increased in each of the early years of the Association.[4]

ABUSES IN THE 1920s

By the 1920s knowledge of the stock market was expanding ever so slightly into the realm of "the public." At this time, the markets were still unregulated and several unsavory schemes evolved that took advantage of unsuspecting novice investors. Unfortunately, options were sometimes used in these manipulative schemes.

One abusive practice that gained notoriety was the granting of call options to stockbrokers by speculators with large holdings of the underlying stock. In return, the brokers would recommend the stock to their clients. The resulting demand for the stock would benefit both the brokers and the speculators, who would then sell into the rising price. After the brokers' and speculators' positions were liquidated, it was the unsuspecting public customer who was left holding the bag. This unethical practice was employed on a larger scale by "option pools," which acquired options directly from major stockholders of a company, including directors, banks, and the company itself. In some instances, pools owned call options on over 20 percent of the outstanding shares of companies.[5] With control over such a high percentage of the outstanding shares it was easy to manipulate the price of the stock by trading large numbers of shares and simultaneously spreading false rumors. When the prices of stocks traded by the pools began to change, option dealers began following the moves of the pools. As a result, trading in many options was based solely on rumors of whether pools were buying or selling that stock.

In October 1929, both the U. S. stock market and the U.S. financial industry experienced a blow that resulted in changed investor perceptions and the creation of a market regulatory authority. Congressional hearings, held to determine what oversight of the securities industry

---

[4] Albert I. A. Bookbinder, *Security Options Strategy*, p. 110

[5] Filer, *Understanding Put and Call Options*, p. 60.

might prevent future market crashes, resulted in the formation of the Securities and Exchange Commission (SEC).

In addition to its review of the stock and bond markets, the SEC undertook an evaluation of the options market. Because the SEC's review concentrated solely on the manipulative schemes of the 1920s, the fate of the options market appeared dismal.

With the SEC's initial recommendation to outlaw options trading in hand, Congress offered the Put and Call Dealer's Association an opportunity to respond. Recognizing the need to convince Congress of the positive uses of options, the Association asked Herbert Filer, an experienced put–call broker to testify. Filer faced the nearly impossible challenge of saving the options industry, because the initial draft of the bill proposed by the SEC read, in part: "not knowing the difference between good and bad options, for the matter of convenience, we strike them all out."[6]

The almost universal lack of understanding of options and the heightened public awareness of option pools led many in Congress to conclude that all options trading was manipulative. In his response to a Congressional committee, Filer explained the difference between "the options in which [put–call dealers] deal which are primarily offered openly and sold for a consideration, and the manipulative options secretly given, for no fee, but for manipulative purposes."[7]

With that explanation, members of the committee then expressed concern about the number of options that expire worthless. Filer was asked, "If only 12½ percent are exercised, then the other 87½ percent of the people who bought options have thrown money away?" Filer replied, "No sir. If you insured your house against fire and it didn't burn down you would not say that you had thrown away your insurance premium."[8] The committee viewed expired options solely as a monetary loss instead of a means of insuring against market volatility.

Filer was successful in his argument, convincing the committee that options have economic value. The SEC concluded that not all option trading is manipulative and that, when properly used, options are a valuable investment tool. The options business survived post-1929

---

[6] Ibid., p. 78.
[7] Ibid., p. 78.
[8] Ibid., p. 78.

Congressional hearings with the SEC assuming regulating authority under the Securities and Exchange Act of 1934. The SEC still regulates the options industry today.

INITIAL MOTIVATIONS FOR A LISTED OPTIONS MARKET

Shortly after the SEC began regulating the over-the-counter options market, it made a decision that would eventually result in the formation of the listed options market. In 1935 the SEC granted the Chicago Board of Trade (CBOT) a license without expiration to register as a national securities exchange. The CBOT did not take advantage of this license until 1968 when low volume in the commodity futures market caused the organization to explore possible ways of expanding business.

The idea for an open-outcry market in options on stocks came out of the Planning and Marketing Department at CBOT, headed by Joseph W. Sullivan (who eventually became the first president of the Chicago Board Options Exchange (CBOE)). Sullivan believed that the open-outcry method of trading futures contracts could reduce the inefficiencies of the over-the-counter put and call market.

According to William D. Falloon in *Market Maker: A Sesquicentennial Look at the Chicago Board of Trade*, the idea was first discussed in 1968. But when presented to the CBOT Board of Directors by Edmund J. O'Connor, Vice Chairman, in 1969, the concept received little or no support. It was due to the persistence of Ed O'Connor and the eventual support of other CBOT leaders such as Pat Hennessey, Dave Goldberg, and Paul McGuire that the idea of the CBOE was kept alive.

Starting a new exchange required economic justification, financing, and resolution of regulatory issues. The economic justification came, in part, from a report of the CBOT's Advisory Committee, dubbed the Nathan Report, which analyzed how an exchange-based market for calls and puts on stocks would help the U.S. economy and the U.S. securities industry. This committee was comprised of industry leaders, academicians, and members of the institutional investing community.

Because the proposed options were on stocks, the Advisory Committee recognized that it would be essential to coordinate clearing and back office operations with securities brokers. Contacts with the securities industry were developed out of contacts from members of the Committee and from the creation of back-office and other operational functions such as clearing, communication and information technology.

Because the new exchange offered hopes of new business to securities firms, operational contacts were established willingly.

Regulatory oversight was an issue for the new exchange, because the CBOT, which was regulated by the predecessor of today's Commodity Futures Trading Commission, was fearful of having also to report to a second government agency. This concern and the mounting financial costs led to the decision to create the CBOE as a separate entity from the CBOT.

When the projected costs exceeded $2 million, the CBOT Board developed the following financial plan: The new exchange was to finance itself through seat sales, and, initially, 400 seats were sold for $10,000 each. Falloon reported that, after spending $1.5 million on the Advisory Committee report and clearing system review, the CBOT Board agreed to fund an additional $1.7 million in the form of a guaranteed loan. But, Ed O'Connor said, ". . . we had to pay that money back before we opened our doors. In other words, we had to pay them that out of CBOE membership sales. We sold those memberships, and paid the CBOT. When we did that, though, the CBOE was no longer part of the 'family' and was basically on its own. This decision actually broke the organization into two pieces."

FUTURES USED AS MODEL FOR LISTED OPTIONS

The CBOT market for futures contracts was the model for the two most important changes from over-the-counter options, standardization and centralized clearing. Over-the-counter options lacked consistency in terms, but the proposed exchange-traded options would have rules to determine contract size, exercise price, and expiration dates. Over-the counter options also lacked uniform financial backing, because different brokers had different financial conditions. A centralized clearing organization would provide unquestioned financial backing for all market participants as well as facilities and operational integrity. This organization was first known as the CBOE Clearing Corporation until the American Stock Exchange started trading options in 1975. At that time the CBOE and American Stock Exchange combined forces to establish a common clearing facility known as Options Clearing Corporation (OCC). The OCC is a corporation owned now by all the U.S. exchanges that trade listed stock options. It serves as a financially secure counterpart to each participant in an exchange-listed option trade. In effect, the OCC becomes the buyer to every clearing member representing a seller and the seller to every clearing member representing a buyer. Standardization of option contracts and the formation of the OCC set the foundation for a liquid listed options market.

In addition to the innovations that the founding fathers took from the futures market, two professors from the University of Chicago also made a significant contribution that helped the listed options market on its initial footings. In the 1973 issue of the University of Chicago's Journal of Political Economy, Fischer Black and Myron Scholes wrote an article entitled "The Pricing of Options and Corporate Liabilities." The article contained an adaptation of a heat-transfer formula in physics that derived a theoretical price for any financial instrument with a known expiration date. That formula was immediately embraced as the standard to evaluate option prices, because it gave market participants a basis for making judgments about the price relationships of different options on the same underlying and about option classes on different underlyings.

Although option industry participants recognized the importance of the Black-Scholes formula immediately, it took 24 years for its significance to the financial community to gain international acclaim. In 1997, the Nobel Prize for Economics was awarded to the two University of Chicago professors who derived the theoretical option pricing model that bears their names.

CBOE'S OPENING BELL

The Chicago Board Option Exchange (CBOE) opened for business on April 26, 1973.[9] The first trading floor, as shown in Figure 1-4, was a room that was formerly the members' smoking lounge, just off the main trading floor of the Chicago Board of Trade.

On that opening day, the CBOE's future success was far from certain. Many experienced stock market professionals questioned the wisdom of launching a new securities exchange in the midst of one of the worst bear markets on record. Others wondered why "grain traders in Chicago" thought that they could successfully market a new trading instrument that the New York Stock Exchange had judged too complex for the investment public. Nevertheless, CBOE founder and first president Joseph W. Sullivan declared on opening day that this exchange was "long overdue."

---

[9] *Options Update 25th Anniversary*, 1987.

FIGURE 1-4

CBDE opening day, April 26, 1973

Source: CBOE

On CBOE's opening day, when the listed options business was born, only call options were traded on just 16 underlying stocks. On that first day of trading 911 contracts changed hands and it was not long before the doubters were silenced. By the end of the first month of trading, the CBOE's daily volume exceeded the daily volume that was customary for the over-the-counter market, which had been in existence since the late 1800s.

By June 1974, CBOE average daily volume for the month exceeded 20,000 contracts, membership had increased to 477, and the number of stocks on which listed options were traded had risen to 23. Even though miniscule by today's standards, these numbers exceeded the capacity of the former smokers' lounge, and planning began to find more space. On December 2, 1974, only 20 months after it began trading, the CBOE made its first move to new trading facilities, this time to the original, but expanded, Chicago Board of Trade building.

NEW EXCHANGES OPEN

Shortly after this expansion, other exchanges became interested in the opportunities of listing options on their trading floors. In 1975, the

American Stock Exchange and the Philadelphia Stock Exchange (formerly the PBW stock exchange) opened their option trading floors, and, in 1976, the Pacific Stock Exchange opened its option trading floor. However, it was not until 1985 that the New York Exchange joined the rapidly growing listed options industry.

THE MORATORIUM

For the fledgling industry, 1977 was a year that contained both milestones and setbacks. A milestone was reached on June 3, 1977, when put options began trading at the CBOE. Investor acceptance of listed put options came even more quickly than anticipated. Four months later, however, the listed options industry had its first major setback, when the SEC declared a moratorium for all exchanges on additional options listings. No expansion was allowed pending a complete review of the structure and regulatory procedures of all options exchanges and an evaluation of the desirability of developing a central market for options trading. The explosive growth of the listed options industry had been halted. But, even though the moratorium stopped the adding of options on new stocks, it did not slow the increasing trading volume on the existing options, and space was becoming an issue once again.

When the SEC lifted its moratorium three years later, the CBOE responded by adding 25 stock on which options could be traded. Along with this increase in the number of underlying stocks, the CBOE board of directors also approved the purchase of land to build a new trading floor. It was less than three years from that approval to February 21, 1984, when the CBOE moved for the second time, this time to its current 45,000-square-foot trading floor at 400 South LaSalle Street, just south of the original CBOT building. Figure 1-5 shows Exchange members and Exchange officers at the ground-breaking ceremony.

The expansion from under 20,000 square feet to 45,000 square feet facilitated growth, and today the CBOE is the second largest U.S. securities exchange. In 1998 the exchange had 1,943 members, and the price of a membership, or "seat," has risen from the initial cost of $10,000 in 1973 to as high as $735,000. Similarly, the CBOE's first day volume of 911 contracts on only 16 different stocks seems miniscule compared to the current average daily trading volume of over 700,000 contracts on over 1200 stocks and over 60 indexes, with a single-day record trading volume of 2.2 million contracts. Figure 1-6 shows the modern, high-tech CBOE trading floor.

FIGURE 1-5

Ground breaking for CBOE's new building, April, 1982

Source: CBOE

FIGURE 1-6

CBOE's modern, high-tech trading floor, 1998

Source: CBOE

NEW PRODUCTS

Innovative new products have contributed to the success of the CBOE and the listed options industry. Two notable innovations are index options and Long-Term Equity Anticipation Securities (LEAPS). Index options are calls and puts on a group of stocks such as those in the S&P 500 Index and the Dow Jones Industrial Average.

The first index option was the CBOE 100 Index, which began trading on March 11, 1983. In the late 1980s an agreement was made with Standard and Poors Corporation to change the component stocks in the index, and subsequently, the name was changed to the S&P 100 Index or, as it is frequently referred to, the "OEX," which is its ticker symbol. In 1998, OEX index options experienced the highest average daily trading volume of any option class in the world. At the end of 1998, there were over 60 different indexes with listed options trading in the United States.

The other notable invention, LEAPS options, began trading at the CBOE in 1990. LEAPS are options with expiration dates of up to three years when first listed. At the end of 1998, LEAPS options were available on over 150 individual stocks and on over 45 indexes.

INDUSTRY CONSOLIDATION

The 1990s have been an era for mergers and acquisitions for industry in general, and securities exchanges in the U.S. have not been left out of the race to be bigger and better. In April 1997, the CBOE completed the acquisition of the options trading division of the New York Stock Exchange. This acquisition added options on 164 new underlying stocks and on one new underlying index to the CBOE's marketplace.

In June 1998, the National Association of Securities Dealers (NASD) and the American Stock Exchange (AMEX) merged. Only a few weeks later, that merger was expanded to include the Philadelphia Stock Exchange (PHLX). A third event occurred in July, 1998, when the CBOE and the Pacific Exchange (PCX) agreed to consolidate forces.

The explanation for these industry-restructuring events is the explosive growth of computerized trading and the implications of this development on the floor-based open-outcry style of trading. Advances in technology have forced exchanges to combine forces for several reasons. Reduction of operating costs and elimination of duplicated markets are two main reasons.

THE FUTURE

Looking forward, only one thing will happen for certain: the options market will continue to grow and change. In the early 1990s it was estimated that fewer than 5 percent of investors have ever used options. As more and more of the remaining 95 percent learn about the beneficial alternatives that options make possible, growth in option volume is inevitable.

The nature of what the changes will be, however, is less certain. Technology will undoubtedly play an expanding role. But whether trading floors will be replaced by computer screens is still a question. Trading floors offer transparency and competition. Every transaction is open to public scrutiny, and competition between off-floor and on-floor traders offers the opportunity for every market participant to get the best price at any particular moment.

CONCLUSION

The option contract has proved its usefulness to a wide range of investment market participants, and the listed options business has proved its financial strength and ability to transact large numbers of contracts. It can be said that options have gained acceptance as an investment vehicle that can enhance almost every investment philosophy.

PART ONE

ESSENTIAL CONCEPTS

CHAPTER 2

FUNDAMENTALS OF OPTIONS

James B. Bittman

Options are derivative instruments. This means that an option's value and its trading characteristics are tied to the asset that underlies the option. It is this essential defining characteristic that makes options valuable to the knowledgeable investor. A major advantage of options is their versatility. They can be used in accordance with a wide variety of investment strategies. As a result, any investor who understands when and how to use options in pursuit of his or her individual financial objectives can enjoy a clear advantage over other investors. The investor will have an effective means of managing the risk inherent in any investment program. In most investment situations, understanding options gives the investor a wider range of investment choices.

The asset on which the option is traded might be a stock, an equity index, a futures contract, a Treasury security, or another type of security. Although the discussion and examples within this chapter are centered on stock options, the concepts and pricing theories also apply to other kinds of underlying assets.

Whatever the underlying asset, the pricing of an option is commonly thought to be an esoteric and difficult task, certainly not something to be attempted by the mathematically unsophisticated person. At one level, this perception is true—advanced mathematics for the pricing of options have been evident in the past and continue to be utilized. The Black-Scholes option-pricing model, for example, was first developed with stochastic calculus and differential equations. What these techniques are and their manner of application need not concern us here.

The important point is that options pricing can generally be explained using a conceptual approach rather than a highly technical mathematical approach. The discussion of options pricing that follows is directed toward the options investor who seeks an explanation at the intermediate level in accessible terms.

This chapter explains option pricing theory in four steps. First, puts, calls, and related terms are defined. Second, the five elements of an option's theoretical value are explained in a general fashion. Third, each of these elements is examined in greater depth. Fourth, and finally, the concept of *put–call* parity ties together many of this chapter's concepts.

SOME DEFINITIONS

Option

An *option* on an underlying asset is either the right to buy the asset (a *call* option) or the right to sell the asset (a *put* option) at some predetermined price and within some predetermined time in the future.

The key feature here is that the owner of an option has a right, not an obligation. If the owner of the option does not exercise this right prior to the predetermined time, then the option and the opportunity to exercise it cease to exist.

The seller of an option, however, is obligated to fulfill the requirements of the option if the option is exercised. In the case of a call option on stock, the seller has sold the right to buy that stock. The seller of the call option is therefore obligated to sell the stock to the call option owner if the option is exercised. In the case of a put option on a stock, the seller of the put option has sold the right to sell that stock. The seller of the put option is therefore obligated to buy the stock from the put option owner if the option is exercised.

Strike Price and Expiration Date

The predetermined price of the option is known as its *strike price*. When a call option is exercised, the call owner pays the amount of the strike price in exchange for receiving the underlying stock. When a put option is exercised, the put owner receives the amount of the strike price in exchange for delivering the underlying stock. The date after which the option ceases to exist is the *expiration date*. For example, the XYZ SEP 50 call option is the right to buy the stock XYZ at the price of $50 per share until the expiration date in September.

Listed options have clearly defined rules establishing strike prices, contract sizes, and expiration dates. Although rules may vary slightly from exchange to exchange, listed stock options generally have strike prices at intervals of 2.50 from a stock price of $5 to $25. Between stock prices of $25 and $200, option strike prices are generally set at intervals of 5. Above stock prices of $200, strike price intervals are $10.

Stock options in the United States are denominated in quantities of 100 shares each or one round lot of stock. If the XYZ SEP 50 call option in the previous example was quoted at $3, its actual cost would be $300. This is because the $3 quoted price represents the cost on a per-share basis, but the call option contract covers 100 shares. Thus, 100 shares times $3 per share equals the cost of $300.

Expiration Rules

Listed stock options in the United States technically expire on the Saturday following the third Friday of the expiration month. Exceptions are made when legal holidays fall on the Friday or Saturday in question. The Saturday expiration, however, is irrelevant to nonexchange members. The Saturday expiration exists so that brokerage houses and exchange members will have the morning after the last trading day to resolve any errors.

Customers of brokerage firms must concern themselves with two procedures in regard to expiration. First, brokerage firm customers must be aware of their firm's specific rules regarding the deadline for notification for exercise. Second, brokerage firm customers must be aware of the rules for automatic exercise. A call option will be automatically exercised if the stock's last trade in its primary market on expiration Friday is 0.75 or more above the strike price unless the customer has given specific instructions not to exercise. A put option will be automatically exercised if the stock's last trade on expiration Friday is 0.75 or more below the exercise price. Many firms have a final notification deadline of 4:00 PM EST on the expiration Friday, but this rule varies from firm to firm.

Although listed stock options have fairly consistent specifications, listed futures options differ considerably in contract specifications, strike prices, the unit value of price movements, and expiration dates. This is so because the specifications of futures contracts themselves vary. Whereas stock prices are dollar-dominated in 100 share lots and stock option prices move accordingly, a futures contract on corn at the Chicago Board of Trade covers 5,000 bushels, and a futures contract on No. 2 heating oil at the New York Mercantile Exchange covers 42,000 gallons. Even futures contracts on the same underlying asset can vary: the Japanese yen futures contract at the Chicago Mercantile Exchange covers 12,500,000 yen, and the yen contract at the MidAmerica Commodity Exchange covers 6,250,000 yen. As a result of these differences, the futures options trader must be familiar with all the terms of a contract before trading. A trader who does not do this first usually learns very fast, but, unfortunately, it can be an expensive process.

American-Style Options

An *American-style* option has a right (not an obligation) that may be exercised at some predetermined price *at any time until the expiration date.* Sometimes these are referred to as *American* options.

European-Style Options

A *European-style* option has a right that may be exercised *only on the expiration date* of the option. Sometimes these are referred to as *European* options.

The difference between European-style options and American-style options has nothing to do with geography! The distinguishing feature is the right of early exercise that exists with American options and does not exist with European options. Until the CBOE introduced European options on the S&P 500 Index on July 1, 1983, the distinction was not particularly important to investors in options markets because only American-style options had been listed; since then, several other European-style options have been listed.

For the purpose of this discussion, the early exercise feature of American options as it relates to pricing theory need not be considered in detail. It is sufficient to point out that the early exercise privilege of American options is a feature that sometimes has value. As a result, American options sometimes have a higher theoretical value than do European options. With this one distinction in mind, the following discussion of option pricing theory applies to both American and European options.

Price and Strike Price

The relationship of the stock's price to the option's strike price determines whether the option is referred to as in-the-money, at-the-money, or out-of-the-money.

A call option is *in-the-money* when the stock price is above the strike price. A call option is *at-the-money* when the stock price is at the strike price. And a call option is *out-of-the-money* when the stock price is below the strike price. For example, with a stock price of $50, the $45 call is an in-the-money call option, because the call option strike price is below the current market price of the stock. The $50 call is at-the-money, and the $55 call is out-of-the-money.

For a put option, the in-the-money and out-of-the-money designations are opposite those of call options. This is because put options increase in price as the price of the underlying stock decreases.

A put option is *in-the-money* when the stock price is below the strike price. A put option is *at-the-money* when the stock price is at the strike price. And a put option is *out-of-the-money* when the stock price is above the strike price. For example, with a stock price at $50, the $55 put is in-the-money, because the stock price is below the put option's strike price. The $50 put is at-the-money, and the $45 put is out-of-the-money.

Intrinsic Value and Time Value

The price of an option may consist of intrinsic value, time value, or a combination of both. *Intrinsic value* is the in-the-money portion of an option's price. *Time value* is the portion of an option's price that is in excess of the intrinsic value.

If the stock price is above the strike price of a call option, then the stock price minus the strike price represents the intrinsic value of the call option. For example, if the stock price is $53, then the $50 call option has an intrinsic value of $3. Any value above $3 that the market places on this option is time value. Time value exists because the market realizes that the stock may decline below $50, and the stock owner may suffer a loss greater than $3—possibly as much as $53! Because this risk exists, the call option purchaser should be willing to pay more than $3 for the $50 call option because he does not have the same risk if the stock price declines below $50. The call option buyer's risk is limited to the premium paid for the option. The premium paid above $3 for the option—the time value—measures in some sense the market's estimate of the likelihood of the stock price declining below $50. The call buyer who pays $4 for the $50 call option is paying an extra $1 for protection against a stock price decline below $50. If the stock price rises, the call buyer participates in the price rise. The $1 time value paid for the option is the price of the insurance policy against losing money if the stock price were to decline below $50.

FIGURE 2-1

Intrinsic Value and Time Value

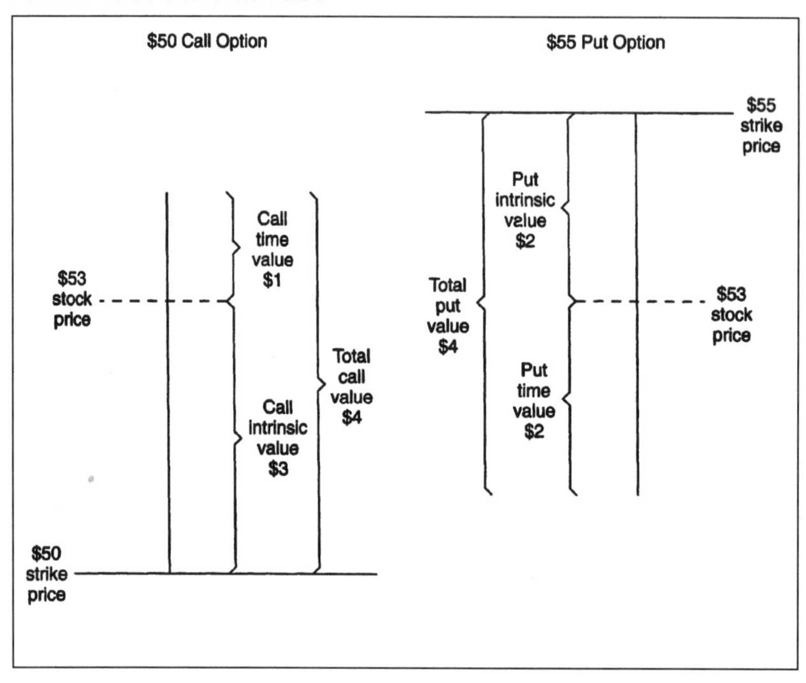

For put options, intrinsic value equals strike price minus stock price because put options are in-the-money when the stock price is below the strike price.

> **Example:** If the stock price is $53, then the $55 put option has an intrinsic value of the $2. Any value above $2 that the market places on this put option is time value.

An out-of-the-money option consists entirely of time value. By definition, the price of an out-of-the-money option has no in-the-money portion; consequently, it has no intrinsic value.

The concepts of intrinsic value and time value for call and put options are illustrated in Figure 2-1.

Parity

An option is trading at *parity* with the stock if it is in-the-money and has no time value. This situation exists when the stock price minus the strike price (for call options) equals the option price. For example, if the $50 call option were trading at $3 when the stock was at $53, then it would be *trading at parity*. In-the-money options, especially deep-in-the-money options, tend to trade at parity when only a few days remain until expiration. This happens when the market perceives that the option is almost certain to be exercised. The logic for this is simple. Because it is only a matter of time until the option is exercised and becomes stock, the option price trades in step with the stock price until exercise occurs.

OPTION PRICE TABLES INTRODUCED

Traditional profit and loss diagrams as presented in Chapter 4 illustrate an option strategy's risk profile *at expiration*. In this chapter, the focus is on how option prices behave prior to expiration, and the important factors that affect an option's theoretical value will be explained. To begin, Table 2-1 shows theoretical call option prices for a $50 call at various times prior to expiration with the underlying stock at different prices. Table 2-2 shows the same information for put options. These tables reveal important characteristics of option price behavior.

First, it should be observed that with the stock at $50 at any time prior to expiration, the call price is greater than the put price. At stock prices other than $50, the call's time value is greater than the put's time value. The reason for this relates to the interest component in call prices, which is discussed later in this chapter under interest rates and put–call parity.

Second, it should be observed that at any time prior to expiration, a $1 move in the underlying stock price will result in an option price change of less than $1. The name of option price change per unit of stock price change is called *delta*, and it is discussed later in this chapter. It is important to note that option price changes per unit of change in the underlying is not constant over changes in stock price or over time.

Third, it should be observed that option prices decline with the passage of time. *Time decay* is, perhaps, the best known aspect about option prices, but there are some misconceptions about time decay. They are discussed thoroughly later in this chapter.

ELEMENTS OF AN OPTION'S VALUE

The five components of an option's theoretical value are:

1. Price of the underlying asset.
2. Strike price of the option.
3. Time remaining until the expiration date.
4. Prevailing interest rates (adjusted for dividends).
5. Volatility of the underlying asset.

TABLE 2-1

$50 Call—Theoretical Values

Stock Price ($)	91 Days	84 Days	77 Days	70 Days	63 Days	56 Days	49 Days	42 Days	35 Days	28 Days	21 Days	14 Days	7 Days	Expiration
55	6½	6⅜	6¼	6⅛	6	5⅞	5¾	5⅝	5½	5⅜	5¼	5⅛	5	5
54	5¾	5⅝	5½	5⅜	5¼	5⅛	5	4⅞	4⅝	4½	4⅜	4¼	4⅛	4
53	5	4⅞	4¾	4⅝	4½	4⅜	4¼	4	3⅞	3⅝	3½	3¼	3⅛	3
52	4⅜	4¼	4⅛	4	3⅞	3⅝	3½	3⅜	3⅛	3	2¾	2½	2¼	2
51	3¾	3⅝	3½	3⅜	3¼	3	2⅞	2⅝	2½	2¼	2	1¾	1½	1
50	3⅛	3	2⅞	2¾	2⅝	2 7/16	2¼	2 1/16	1 15/16	1¾	1½	1¼	⅞	0
49	2⅝	2½	2⅜	2¼	2⅛	1 15/16	1¾	1⅝	1 7/16	1¼	1	¾	7/16	0
48	2 3/16	2 1/16	1 15/16	1¾	1⅝	1½	1⅜	1 3/16	1	13/16	⅝	1 7/16	3/16	0
47	1¾	1⅝	1½	1⅜	1¼	1⅛	1	⅞	¾	9/16	⅜	¼	1/16	0
46	1⅜	1¼	1 3/16	1 1/16	15/16	⅞	¾	⅝	½	⅜	¼	⅛	0	0
45	1⅛	1	⅞	13/16	11/16	⅝	½	⅜	5/16	3/16	⅛	1/16	0	0

TABLE 2-2

$50 Put—Theoretical Values

Stock Price ($)	91 Days	84 Days	77 Days	70 Days	63 Days	56 Days	49 Days	42 Days	35 Days	28 Days	21 Days	14 Days	7 Days	Expiration
55	1 1/8	1 1/16	15/16	7/8	3/4	11/16	9/16	7/16	3/8	1/4	1/8	1/16	0	0
54	1 3/8	1 1/4	1 3/16	1 1/16	1	7/8	3/4	5/8	1/2	3/8	1/4	1/8	0	0
53	1 11/16	1 9/16	1 7/16	1 3/8	1 1/4	1 1/8	1	7/8	3/4	9/16	7/16	1/4	1/16	0
52	2	1 7/8	1 3/4	1 11/16	1 9/16	1 7/16	1 5/16	1 1/8	1	13/16	5/8	7/16	3/16	0
51	2 3/8	2 1/4	2 1/8	2 1/16	1 15/16	1 3/4	1 5/8	1 1/2	1 3/8	1 3/16	15/16	3/4	7/16	0
50	2 13/16	2 11/16	2 9/16	2 7/16	2 5/16	2 3/16	2 1/16	1 15/16	1 3/4	1 9/16	1 3/8	1 1/8	13/16	0
49	3 1/4	3 1/8	3	2 15/16	2 13/16	2 11/16	2 9/16	2 7/16	2 1/4	2 1/8	1 15/16	1 11/16	1 3/8	1
48	3 7/8	3 3/4	3 5/8	3 1/2	3 3/8	3 1/4	3 1/8	3	2 7/8	2 3/4	2 1/2	2 3/8	2 1/8	2
47	4 1/2	4 3/8	4 1/4	4 1/8	4	3 7/8	3 3/4	3 5/8	3 1/2	3 3/8	3 1/4	3 1/8	3	3
46	5	4 7/8	4 7/8	4 3/4	4 3/4	4 5/8	4 1/2	4 3/8	4 1/4	4 1/4	4 1/8	4	4	4
45	5 3/4	5 5/8	5 5/8	5 1/2	5 3/8	5 3/8	5 1/4	5 1/4	5 1/8	5 1/8	5	5	5	5

(As a reminder, the following discussion centers on stock options, but the concepts apply to all types of options.)

Option Price Relative to Stock Price

Column 2 in Table 2-1 shows theoretical values of a $50 call 91 days prior to expiration. As the stock rises from 45 to 46, the theoretical option value rises from 1⅛ to 1⅜. As the stock rises from 50 to 51, the theoretical option value rises from 3⅛ to 3¾. And as the stock rises from 54 to 55, the theoretical option value rises from 5¾ to 6½. The information in this column is presented in the line graph in Figure 2-2. As can be seen, as the underlying stock price rises, the theoretical value of the call rises in a nonlinear fashion at an increasing rate.

As illustrated by the brackets under the graph in Figure 2-2, a call option is in-the-money when the stock price is above the option's strike price. A call option is at-the-money when the stock price is at the option's strike price. And a call option is out-of-the-money when the stock price is below the strike price.

FIGURE 2-2

$50 Call at Various Stock Prices

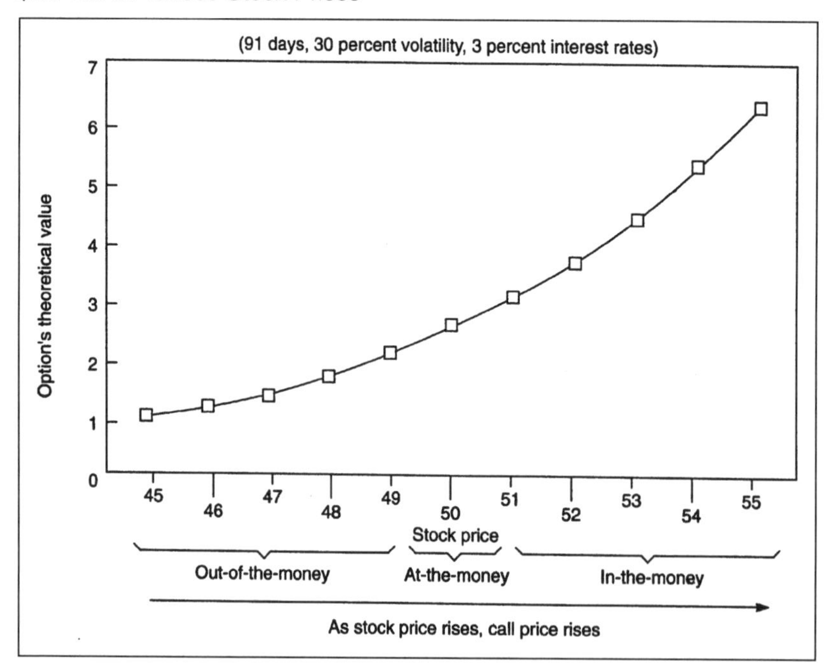

FIGURE 2-3

$50 Put at Various Stock Prices

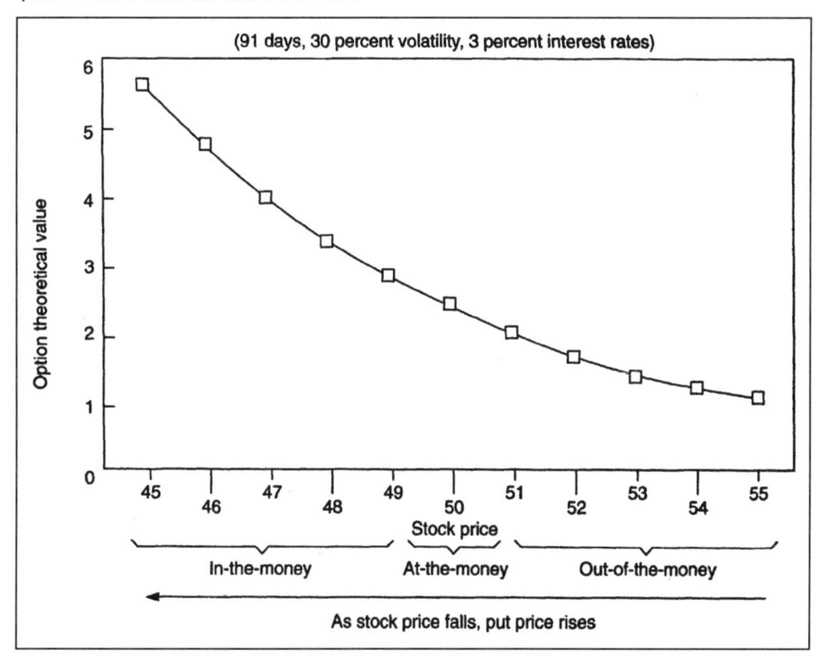

Figure 2-3 shows the information from column 2 in Table 2-2—Put Option Theoretical Values—in line-graph form. The conclusion is that as the underlying stock falls in price, the put option's theoretical value rises in a nonlinear fashion at an increasing rate.

The brackets under the graph in Figure 2-3 illustrate that the designations of in-the-money and out-of-the-money for put options are opposite those for call options. A put option is in-the-money when the stock price is below the strike price, and a put option is out-of-the-money when the stock price is above the strike price.

Option Price Relative to Time

Looking across any row in Table 2-1 illustrates how option prices change with the passage of time. The *time decay* factor in option prices is well known, but there are some misconceptions about how time affects option values.

Figure 2-4 illustrates in line-graph form the call option prices in row 6 of Table 2-1 and the put option prices in row 6 of Table 2-2. With

FIGURE 2-4

Effect of Time—At-the-Money

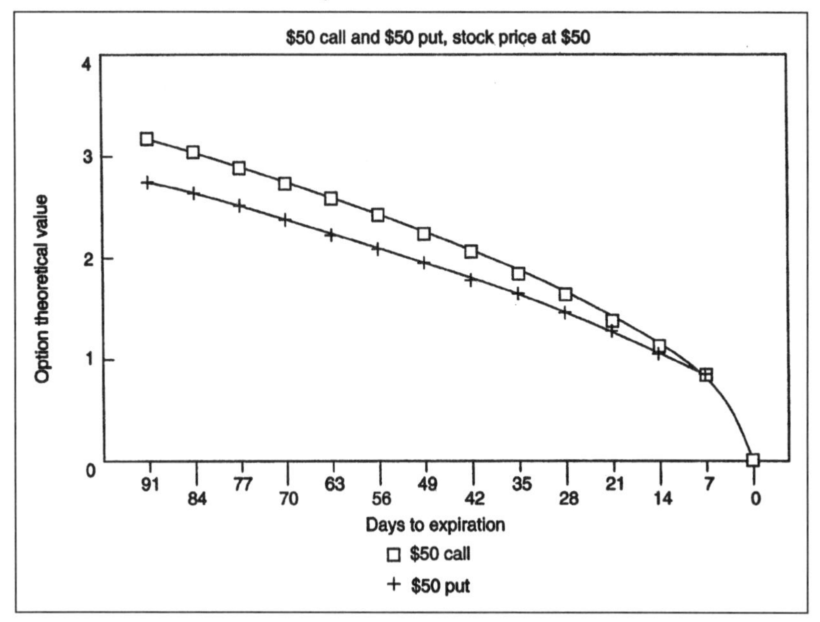

the stock price at $50, the $50 call declines in value from $3\frac{1}{8}$ to 0 over the 91 days to expiration, and the $50 put declines from $2\frac{13}{16}$ to 0. The important observation is that time decay for at-the-money options does not occur in a linear manner. During the week from 91 days to 84 days before expiration, the $50 call declines by $\frac{1}{8}$ from $3\frac{1}{8}$ to 3. During the week 63 days to 56 days, the $50 call declines by $\frac{3}{16}$ from $2\frac{5}{8}$ to $2\frac{7}{16}$. From 28 days to 21 days, the decline is $\frac{1}{4}$ from $1\frac{3}{4}$ to $1\frac{1}{2}$, and the last week, the decline is $\frac{7}{8}$ from $\frac{7}{8}$ to 0.

Time decay for in-the-money and out-of-the-money options, however, is closer to being linear. To illustrate this concept, Figure 2-5 shows how the $50 call declines in value with the stock price at $47, and Figure 2-6 shows how the $50 call declines in value with the stock price at $53. With the stock price at 53, the stock is above the strike price and, therefore, the total option price of $5 at 91 days prior to expiration consists of $3 of intrinsic value and $2 of time value. It must be remembered that only the time value decreases with the passage of time. Consequently, with the stock price 53, the $50 call will decline in value from $5 at 91 days prior to expiration to $3 at expiration—a price equal to its intrinsic value.

FIGURE 2-5

Effect of Time—Out-of-the-Money

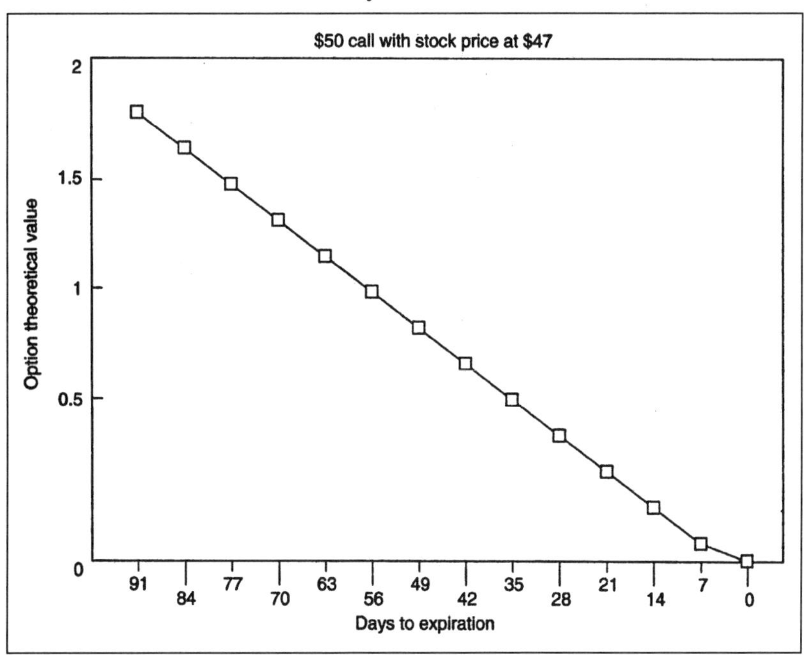

FIGURE 2-6

Effect of Time—In-the-Money

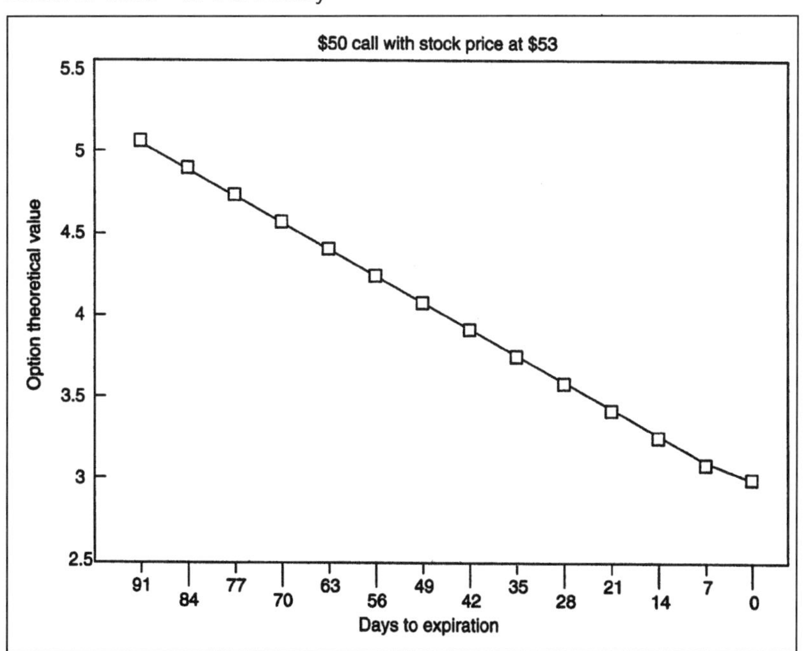

Time decay for put options is similar to time decay for call options. Out-of-the-money and in-the-money options decay in a more linear fashion than do at-the-money puts.

Figure 2-7 consists of the information in Table 2-1, column 2 ($50 call prices 91 days prior to expiration), the prices in column 8 (49 days prior to expiration), column 12 (21 days), and column 15 (expiration). Together, these lines show how call option prices decay over time so that at the expiration date there is no time value in the option's price, and the result is the *hockey-stick* diagram at expiration, which is familiar to many traders of options. Figure 2-8 presents similar time information for put options.

A frequently asked question about options is, "Why does the passage of time affect at-the-money options in a nonlinear fashion?" After all, insurance premiums are almost linear; for example, a one-year house insurance policy costs half of a two-year policy. The answer is that option values are related to the square root of time. Although this sounds esoteric, the relationship is quite simple. If a 30-day at-the-money option has a value of $1, the 60-day option (having twice as

FIGURE 2-7

$50 Call at Various Times Prior to Expiration

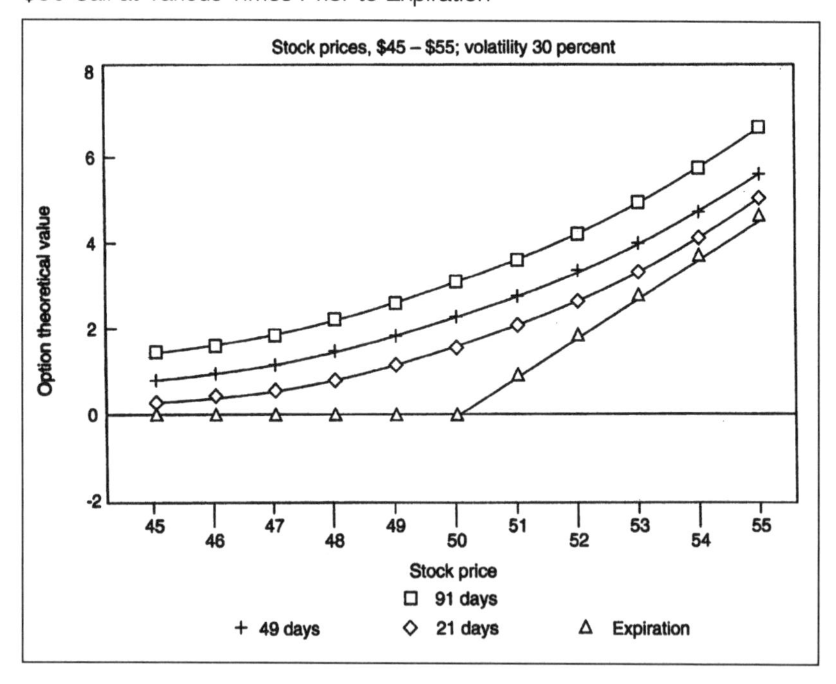

FIGURE 2-8

$50 Put at Various Times Prior to Expiration

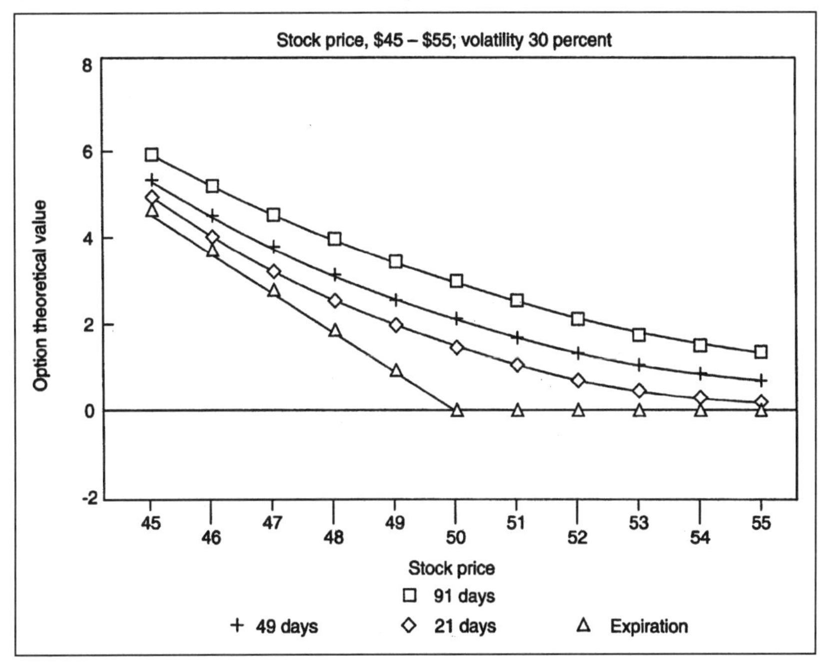

much time until expiration) has a value of approximately $1 \times \sqrt{2}$ or $1 \times 1.414 = 1.414$.

This important concept about the nonlinear impact of time on option prices should give pause to those option traders who resolutely believe in buying only front-month options. Considering time decay alone, option buyers should prefer to buy longer-term options. After all, it is cheaper to buy one six-month option for approximately $2.45 ($1 \times \sqrt{6}$) than it is to buy six 1-month options at $1 each. Conversely, with all else equal, option sellers should have a preference to sell shorter-term options because there is more profit to be made by selling six one-month options than one six-month option. Of course, in the real world, all else is rarely equal. Transaction costs have an impact, and option traders must balance the impact of these real world considerations versus the theoretical.

Interest Rates

The effect of changes in interest rates is most easily understood after having read the section of this chapter concerning put–call parity, in

FIGURE 2-9

Effect of Interest Rates

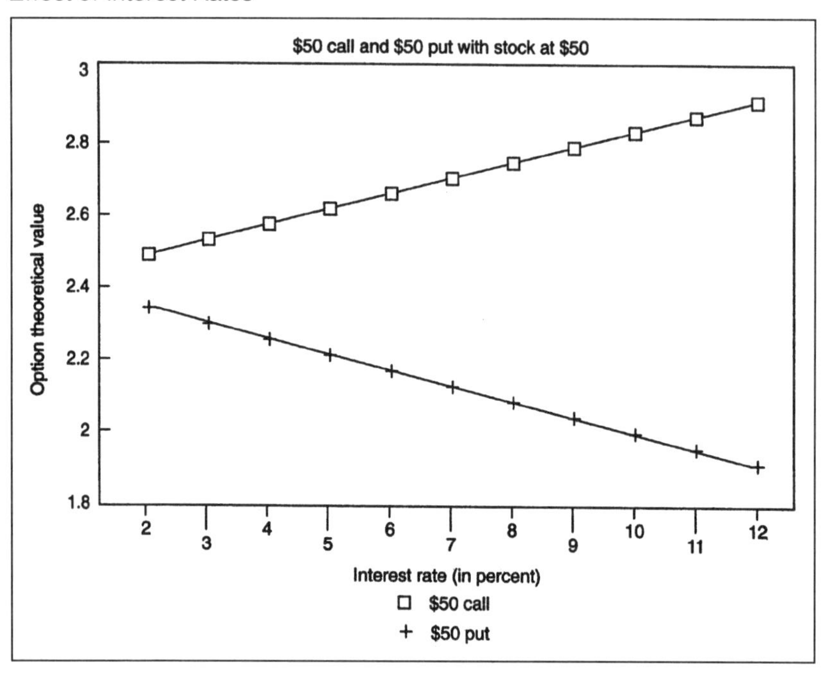

which the arbitrage relationship between options and stock is explained. At this point, it is sufficient to be aware of two points that are illustrated in Figure 2-9.

The first point revealed in Figure 2-9, which is surprising to many option traders, is that rising interest rates cause call prices to rise and put prices to decline. The second point is that the effect of changes in interest rates is small. The figure shows that a rise in interest rates from 2 percent to 12 percent causes call prices to rise from 2½ to 2⅞ and put prices to decline from 2⅜ to 1⅞. Both of these points will be explained more fully in the section on put–call parity.

Option Prices and Volatility

Volatility is discussed in depth in the next chapter. This discussion touches on the major points of how changes in volatility affect option prices. The relationship between volatility and option prices is a direct one—as the volatility percentage increases, so do option prices. Because volatility refers to the price movement of the underlying stock, higher volatility means that greater movement is likely, and greater movement justifies higher option prices.

Figures 2-10, 2-11, and 2-12 show that the relationship between volatility changes and option prices is nearly linear, but that it is a different line for in-the-money options, at-the-money options, and out-of-the-money options.

Figure 2-10 shows that the effect of changes in volatility is linear for both puts and calls and that the difference between the two remains constant when volatility changes. The constant difference is the result of the interest component as discussed above. Figures 2-11 and 2-12 show the effect of changes in volatility on in-the-money and out-of-the-money calls. Although the effect is nearly linear, it should be noted that the slope of the lines is different.

It must be remembered that option prices are based on the expected volatility of the underlying stock. Mathematically, *volatility is nondirectional*! If the market expects greater fluctuation in a stock's price,

FIGURE 2-10

Effect of Volatility—At-the-Money

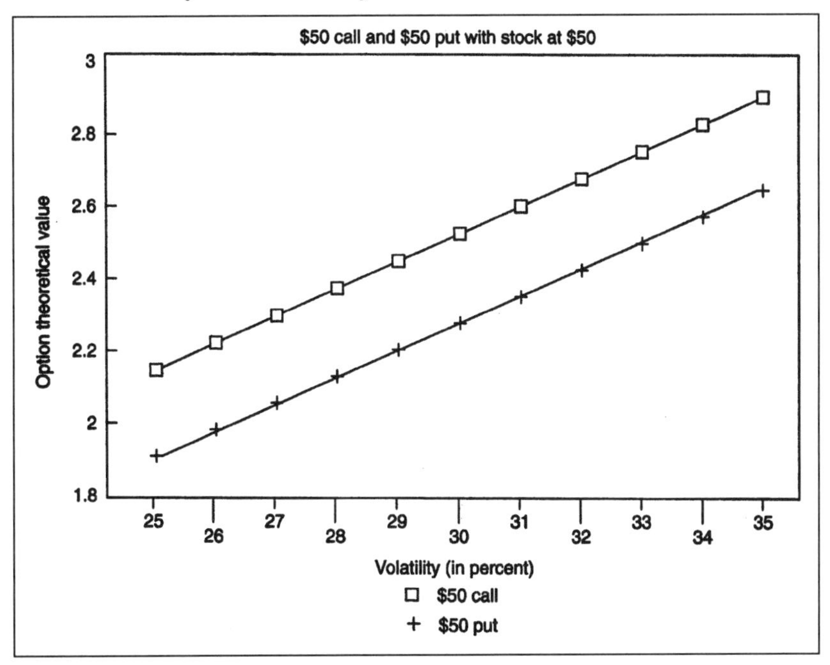

FIGURE 2-11

Effect of Volatility—Out-of-the-Money

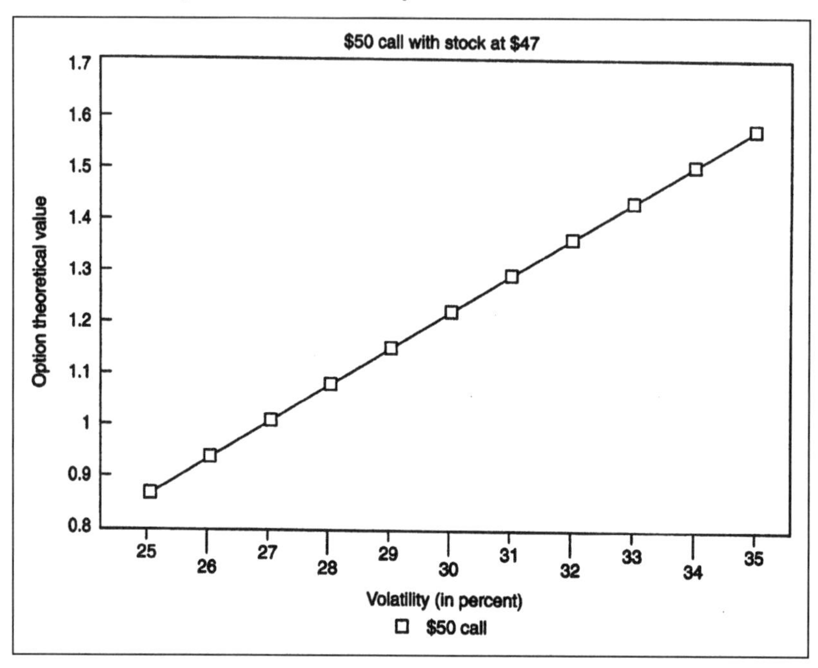

FIGURE 2-12

Effect of Volatility—In-the-Money

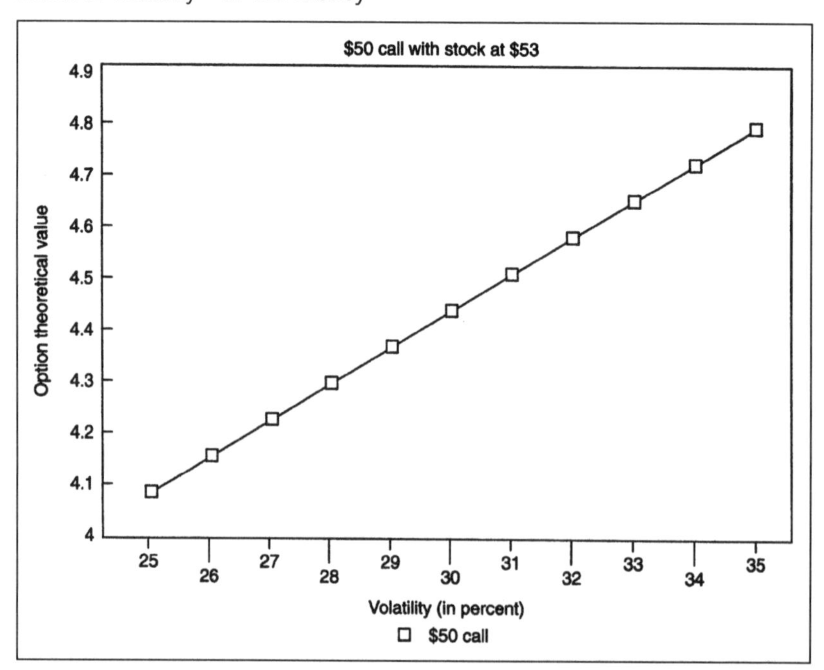

then the fluctuation could be up or down. Consequently, when higher volatility is expected, both call prices and put prices rise.

With an increase in expected volatility, at-the-money options will increase in price more than out-of-the-money options because of how price movements are distributed according to probability theory. Mathematically, a $50 stock always has a greater probability of moving at least $1 than of moving $6. This explains why at-the-money options have higher prices than out-of-the-money options.

Implied Volatility

Implied volatility is the volatility percentage that justifies an option's price. Consequently, the price at which an option is trading tells what volatility level in the stock is implied by the option price. For the professional floor trader who trades a large number of options and who manages large open positions, differences between actual stock price volatility and implied volatility of options may have a significant impact. For the off-floor user of options, however, differences in implied volatility and recent actual volatility rarely are significant. For the off-floor trader, other factors such as stock selection, timing of price movements, and desired rates of return are more important considerations.

Volatility—the Unknown Factor

Of the five components of an option's theoretical value, the stock price, strike price, time until expiration, and prevailing interest rates are readily observable. It is only the volatility of the underlying stock that is unknown.[1] Thus one can conclude that an option's theoretical value is ultimately subjective because the selection of a volatility estimate is subjective. After all, it is the future volatility of a stock that determines an option's true value and the future volatility, of course, cannot be known. These and other concepts about volatility will be developed further in the next chapter.

THE SECOND LEVEL OF UNDERSTANDING

Now that the five elements of an option's theoretical value have been introduced, the second step is to understand how a change in each ele-

[1] Chapter 8 revisits these factors as they pertain to longer-dated options.

ment affects the theoretical value of an option. The questions we address are:

1. Given that changes in option values are not constant with stock price changes, how can changes in option values be measured?
2. How does option price decay change with the passage of time?
3. How do changes in volatility affect option values?

Typically, this is the point where advanced mathematics takes the forefront. This discussion, however, will continue to emphasize concepts. The successful user of options should be aware of the components of option price changes, just as he should be aware of the five elements of value, but a detailed knowledge of the mathematics is not required.

The Effect of Stock Price Change—Delta

During the discussion above under the heading "Option Price Relative to Stock Price," it was stated that "As the underlying stock price rises, the theoretical value of the call rises in a nonlinear fashion at an increasing rate." At first, the option price rise is only a small fraction of the stock price rise. This fraction increases as the stock price rises. When the stock is significantly above the strike price, then the option price movement approaches 100 percent of the stock price movement. It is this "fraction of the stock price movement" that is known as the option's *delta*.

As an example, assume that the underlying stock rises $1 and the option price rises $0.25. In this case, the delta of the option is .25—the option moved 25 percent of the stock price movement.

Delta, however, is not static. The delta of an option changes as the option goes from being an out-of-the-money option to an in-the-money option. The price of an out-of-the-money option changes by a small percentage of the stock price change. The price of an at-the-money option changes by approximately 50 percent of the stock price change. As an option becomes more and more in-the-money, its delta rises and gradually approaches 1.00 or 100 percent. This means that the price of a deep-in-the-money option moves dollar for dollar with the stock price movement. The concept of delta is demonstrated in Table 2-3.

Although the arrow in Table 2-3 points to an instance where the delta equals exactly the option price change for a $1 price rise or fall in the stock, any reader with a calculator can quickly ascertain that the relationship between other option price changes and the corresponding delta are not as exact. This situation occurs because the delta is a theo-

TABLE 2-3

Delta of Call Option

$50 Call Option			
Volatility: 35%		Days to Expiration: 90	
Stock Price	Theoretical Value	Delta	
$56	7¼	.78	
55	6½	.75 ←	The .75 delta implies a .75 value change if the stock price rises or falls by $1.
54	5¾	.71	
53	5⅛	.68	
52	4½	.64	
51	3⅞	.60	
50	3¼	.54	
49	2¾	.50	
48	2⅜	.46	
47	2³⁄₁₆	.41	
46	1⅞	.36	
45	1⁷⁄₁₆	.32	
44	1³⁄₁₆	.27	

retical measure designed for a very small stock price movement, not a full dollar move. For the mathematically sophisticated, there are many books that go into this concept in depth.

A graphical representation of how deltas change with stock price changes is presented in Figure 2-13. As the top half of the graph illustrates, call option deltas are positive and increase as the stock price rises. This is consistent with Table 2-3.

Put options, however, have negative deltas, because put options decrease in value as stock prices rise. This is represented by the bottom half of Figure 2-13. Referring back to Figure 2-3, the $50 put option increases in price at an increasing rate as the stock price moves from 55 down to 45. Consequently, when the stock moves from 55 to 54, the put option increases only slightly in price and at this point has a small negative delta. Subsequently, when the stock price moves from 46 to 45, the option price change is a much larger percentage of the $1 price change in the stock. At this point, the put option has a much larger negative delta.

FIGURE 2-13

Delta: Change in Option Price per 1 Unit Change in Stock

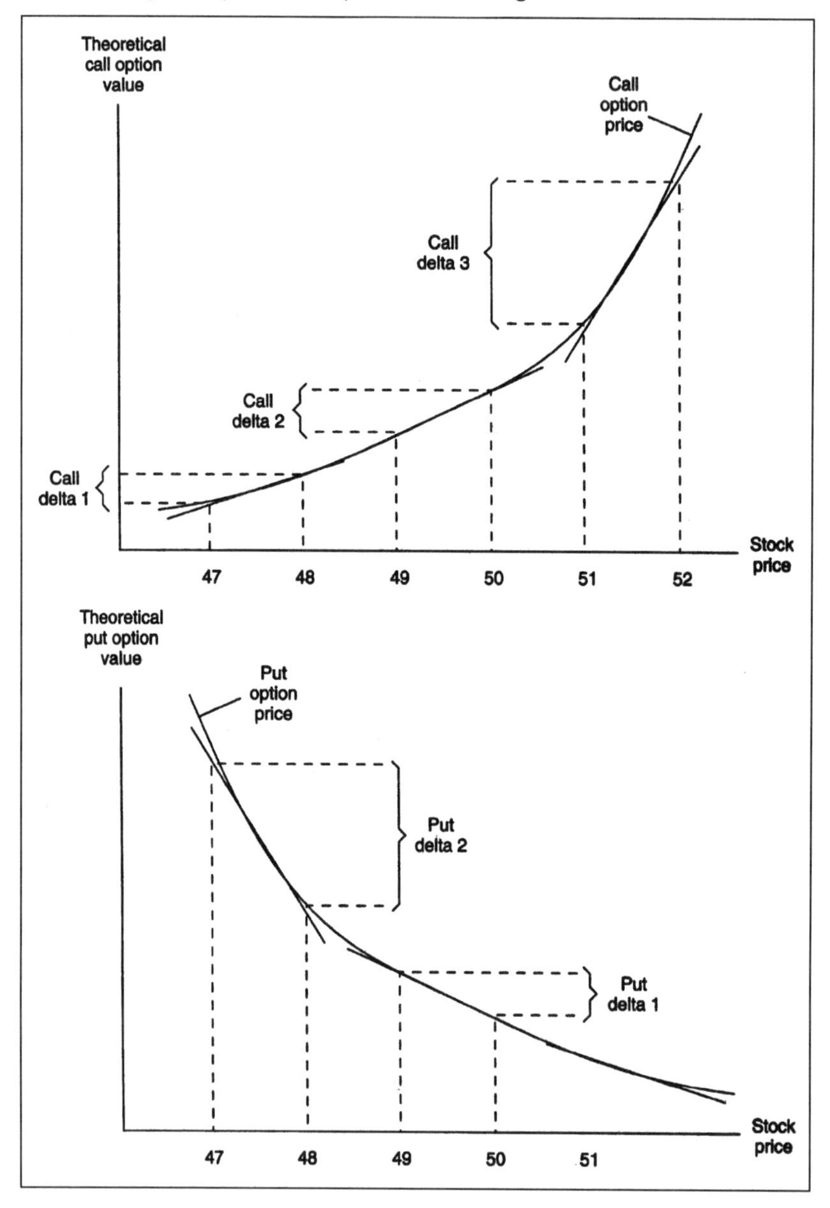

FIGURE 2-14

Effect of Stock Price on Call and Put Deltas[2]

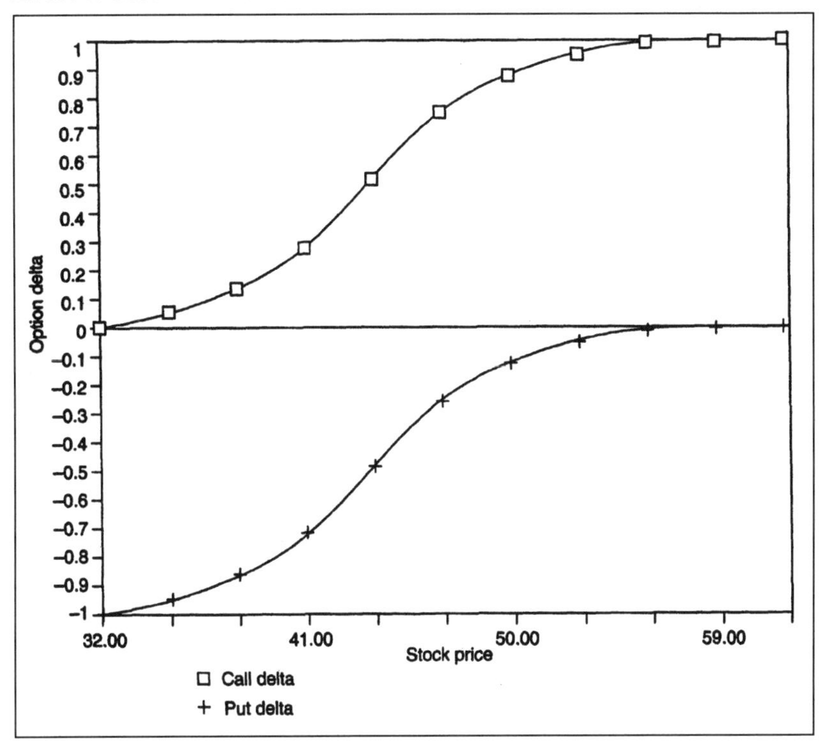

Another method of illustrating the concept of how delta changes with stock price changes is in Figure 2-14. This shows delta on the vertical axis and stock price on the horizontal axis. Call deltas are shown on the top half of the graph, and put deltas are shown on the bottom. The concept is the same as that in Figure 2-13: as stock prices rise, call deltas rise from 0 to 1.00, and put deltas fall from –1.00 to 0.

The delta is important to the user of options because it gives a current estimate of the expected value of an option price change. "Current" is the key word. Too often, option traders think only of what will happen on the date of expiration. Such focus, however, is limiting.

[2] Figures 2-14 through 2-19 in this chapter were produced using The Options Analyst, a set of Lotus 1-2-3 spreadsheets produced by Fin Calc, Inc., Chicago, Illinois.

As an example, assume that you expect XYZ, which is currently at 50, to rally when its earnings are released next week. Rather than buy the stock, you buy the $50 call option to limit your risk in the event the earnings report is unfavorable. For this example, assume that the $50 call option is trading at $2, its delta is .50, and that the stock moves up $1. With a delta of .50 and a stock move of $1, the option can be expected to move by .50 to $2.50, for a profit of .50. The analysis, at expiration, however, yields completely different results. With the stock at $51 on the date of expiration, the $50 call option is worth $1. Because the option was purchased for $2, the result is a loss of $1. This is quite a different result from the profit of $.50 realized on the movement of the stock and the option at the time of the earnings release.

Your trading time frame determines which analysis is correct. Was it your intention to sell the option immediately after the earnings report for a quick profit? Or was it your intention to hold the option until the expiration date with the desire of exercising the option and purchasing the stock if the option was in-the-money?

In the first situation, the trader was counting on short-term upward price movement. The call purchase limited the downside risk, and a .50 profit was realized when the option was sold on the move after the earnings report. In the second situation, the call purchaser wanted to buy the stock, but was using the option as a limited risk alternative during the option's life. Had this purchaser been wrong about the stock and had the stock dropped sharply in price during this time period, then the loss would have been limited to the $2.00 paid for the option.

The Rate of Change in Delta—Gamma

As can be seen from Table 2-3 on, the delta not only changes with price changes of the underlying, it also changes at different rates if the option is in-the-money, at-the-money, or out-of-the-money. For example, when the stock moved from a price of 50 to 51, the delta of the option changed from .54 to .60—a .06 change. However, when the stock rose from 55 to 56, the delta rose from .75 to .78—a .03 change. These rates of change in delta are called *gamma*. By adding a third column to Table 2-3, as in Table 2-4, we can see gamma and how it changes.

In order to demonstrate the concept of gamma clearly, it is necessary to go out at least three decimal points. Such refinements are relevant primarily to professional traders who carry large numbers of options in their portfolios. The typical individual investor should be

TABLE 2-4

Gamma: Rate of Change in Delta

$50 Call and Delta and Gamma			
Stock Price ($)	**91 Days**	**Delta**	**Gamma**
55	6.5153	0.7941	0.0319
54	5.7574	0.7579	0.0362
53	5.0397	0.7177	0.0402
52	4.3660	0.6737	0.0440
51	3.7398	0.6262	0.0475
50	3.1640	0.5758	0.0504
49	2.6411	0.5229	0.0529
48	2.1724	0.4687	0.0542
47	1.7585	0.4139	0.0548
46	1.3989	0.3596	0.0543
45	1.0919	0.3070	0.0526

careful not to place too much emphasis on this concept. For the typical individual investor, stock selection and market forecasting are, by far, the most important concerns. Another way to illustrate how gamma changes is in Figure 2-15.

Figure 2-15 shows that gammas are greatest when the underlying stock is slightly below the option's strike price and smaller when the underlying stock is at a higher or lower price. The exact stock price at which option gammas are greatest is the discounted value of the option's strike price.

Gamma is a sophisticated concept (the second derivative of the price line). Gamma has little importance to the nonprofessional or non-market maker who does not carry large and frequently changing option positions. This concept is used in determining the rate at which an option position changes. Professional traders with large option positions (several hundred long options and short options and thousands of shares of stock—long or short) frequently try to balance their long options against their short options with a goal of being *delta neutral*. This means that their long deltas equal their short deltas. The gamma of the position tells the professional trader how quickly the total position becomes long or short—how fast his or her long deltas get longer versus

FIGURE 2-15

Effect of Stock Price on $50 Call Gammas

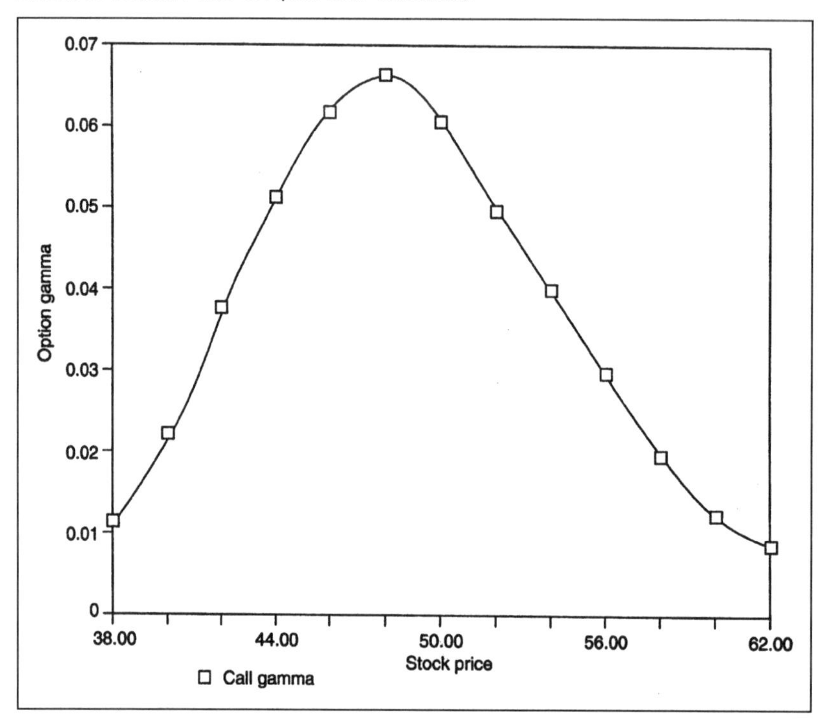

his or her short deltas getting shorter. Again, this concept is relevant primarily to the professional option trader.

How Deltas Change with Time

As time progresses toward the expiration date, the delta of an option changes differently depending on whether an option is at-the-money, in-the-money, or out-of-the-money.

In-the-money options are exercised at expiration. This means that over the time period approaching expiration, the delta of an in-the-money option gradually increases to 1.00 (positive 1.00 for calls and negative 1.00 for puts). Figure 2-16 shows how, with the stock price at 50, the $45 call option delta rises to 1.00 and how the $55 put option delta declines to –1.00.

At the other extreme of delta behavior, the delta of an out-of-the-money option gradually approaches 0 because out-of-the-money options expire worthless. Figure 2-17 illustrates how, with the stock price at 50, the $55 call option delta and the $45 put option delta both gradually decrease to 0.

The delta for an at-the-money option presents an interesting theoretical discussion. The discussion is theoretical because, in reality, rarely is an option exactly at-the-money. That is, a stock price is rarely exactly at a strike price. Normally, a stock price is at least slightly above or below a strike price, thus making either the call or put in-the-money and the other out-of-the-money.

Assume, however, that a stock price is exactly at the strike. Then, as time progresses toward expiration, the delta for both the put and the call remain very close to .50 (positive .50 for call and negative .50 for puts). Although this, at first, may be difficult to comprehend, it is important to

FIGURE 2-16

Delta's Change Over Time for In-the-Money Options

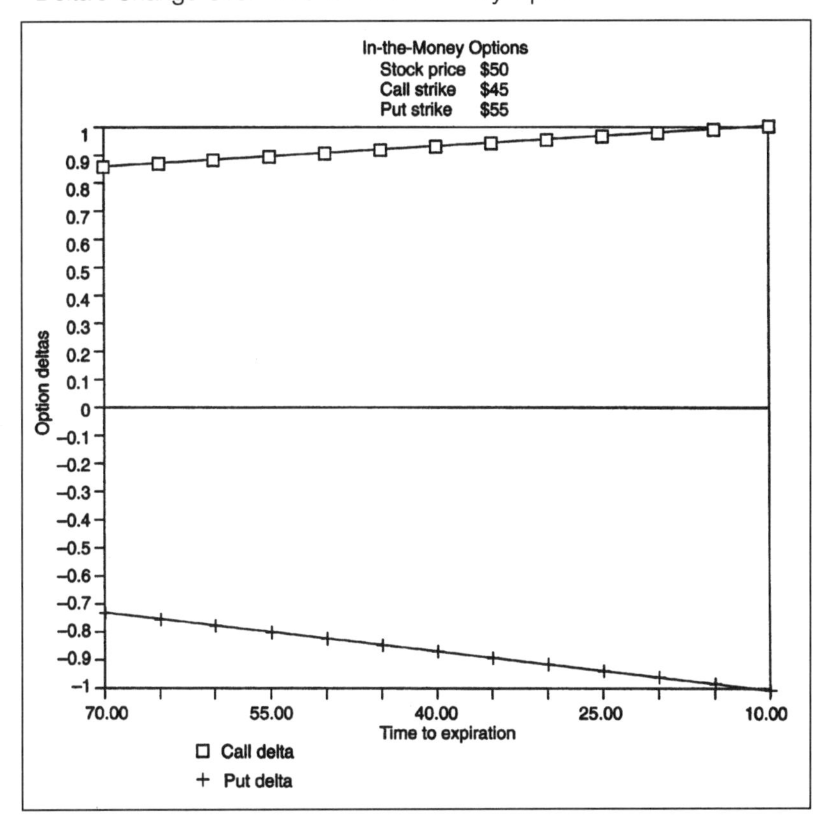

FIGURE 2-17

Delta's Change Over Time for Out-of-the-Money Options

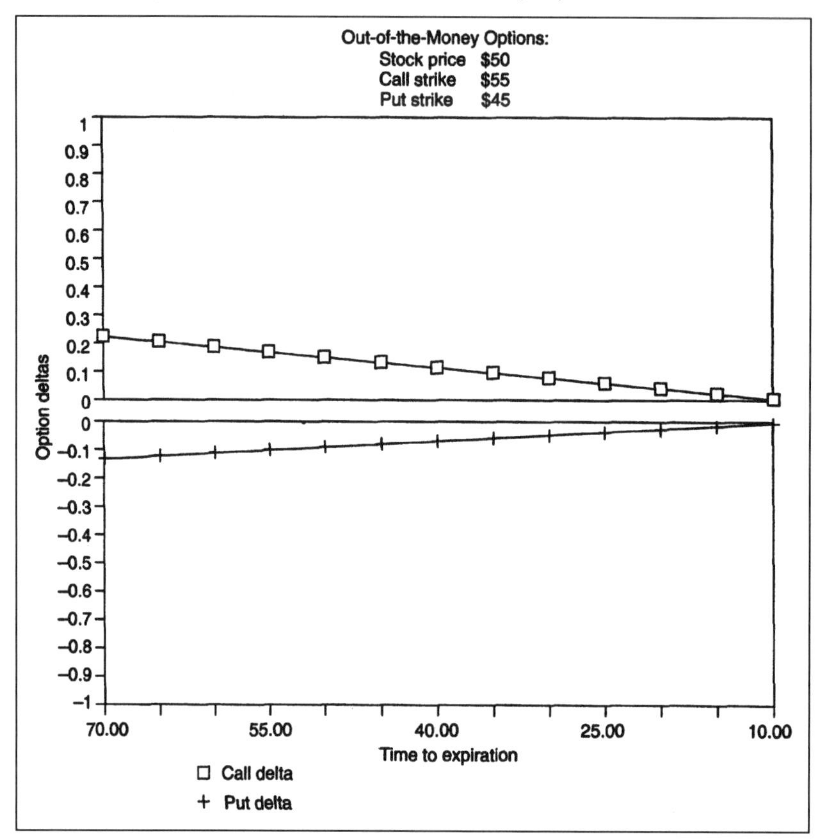

separate in one's mind the effect of time on total option price and the effect of time on delta. The total option price gradually decreases over time, but regardless of the total option price, a 1 unit change in the underlying security will cause about a ½ unit change in the at-the-money option. This remains true at any time right up to the point of the option's expiration. At the instant of expiration, with the stock price exactly at the strike, the delta instantly drops to zero as the option expires worthless. But, in theory, even one second before expiration, the option has a delta of .50. The delta hovering near .50 for at-the-money puts and calls is illustrated in Figure 2-18.

Delta's change over time is important primarily to the professional trader. An in-depth examination is not within the scope of this book.

Time Decay—Theta

Theta is the rate at which an option price erodes per unit of time. As was seen in Tables 2-1 and 2-2 and Figure 2-4, the price of an at-the-money option decays at an increasing rate. During half of an at-the-money option's remaining life, the time value erodes due to the passage of time by approximately 33 percent. The concept of theta is illustrated in Figure 2-19.

Although theta is a sophisticated mathematical concept (it is the first derivative (slope) of the time line in Figure 2-4), it is valuable to all users of options.

For speculators, theta is useful in planning the duration of trades. If a speculator plans to trade out of a purchased option before expiration, then knowledge of time decay helps in deciding when to sell an op-

FIGURE 2-18

Delta's Change Over Time for At-the-Money Options

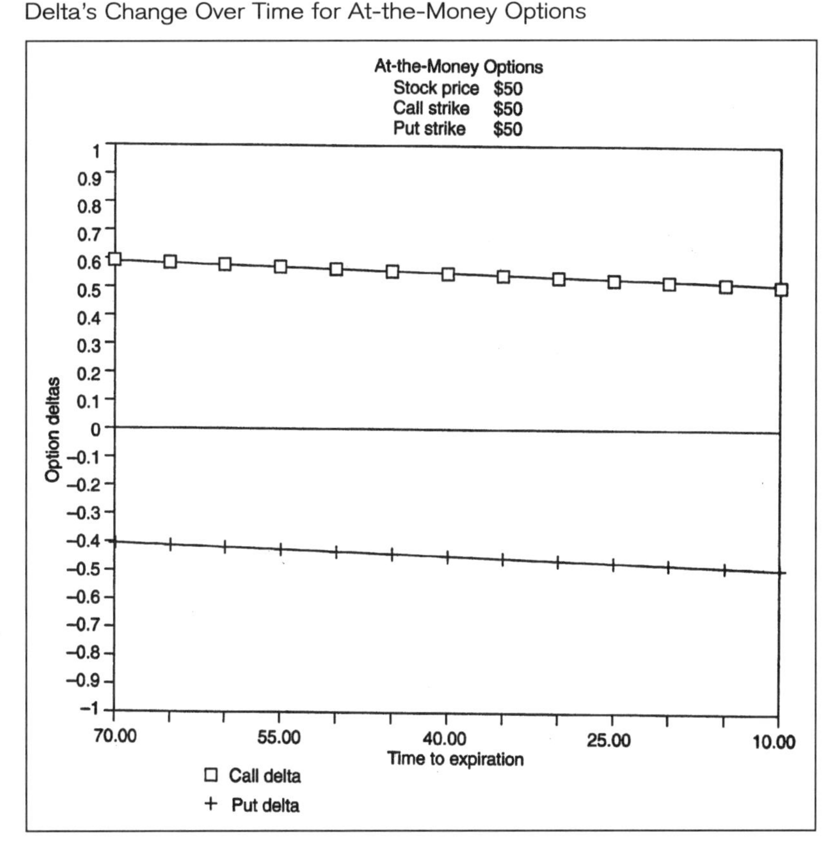

FIGURE 2-19

Theta: Decrease in Option Price per Unit of Time

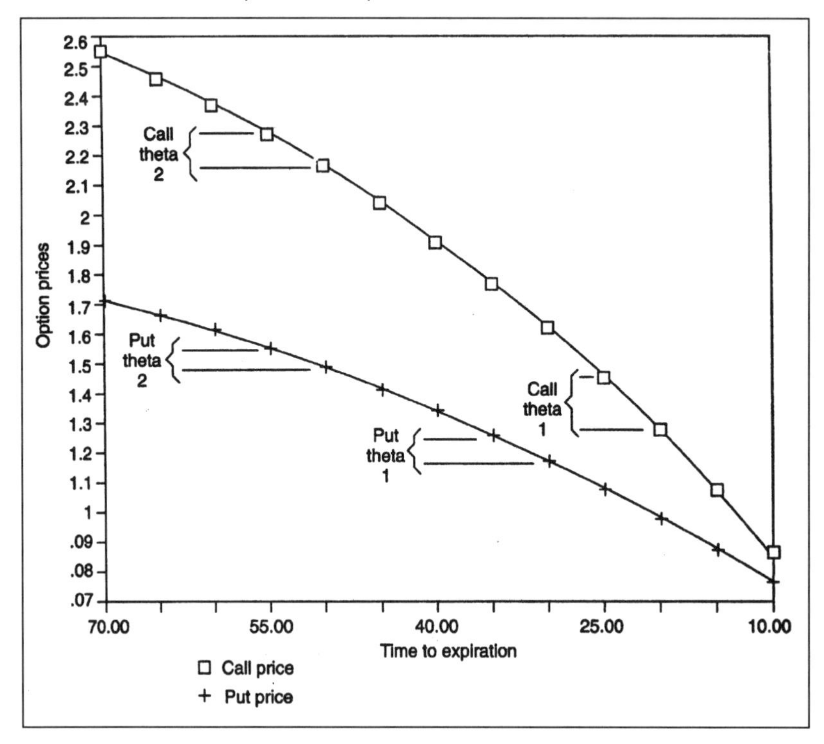

tion. The trader would balance time decay against the delta effect on the option from expected movements in the underlying stock price.

Professional traders use theta as another balancing tool to manage large positions of long and short options.

Changes in Volatility—Vega

The *vega* of an option is the change in an option's value that results from a change in volatility. Vegas are expressed in dollar terms. For example, if a 1 percent change in implied volatility causes a $.25 change in the option's value, then the option is said to have a vega of .25.

The vega is mathematically similar to the delta and theta in that it is a first derivative. As can be seen from Figures 2-10, 2-11, and 2-12, the relationship of volatility to option price is nearly linear. However, the linear relationship for in-the-money, at-the-money, and out-of-the-money options is different for each. The implication is that vega does not change much unless the underlying stock moves considerably in relation to the option's strike price. Again, this is a concept that is most relevant to the professional trader.

THE PUT–CALL PARITY RELATIONSHIP

We have just completed a discussion of the theoretical elements of option value and how changes in each of the elements affect that value. Now our discussion shifts focus—from theoretical value to relative value.

Users of options need some assurance that they are, in some sense, paying a fair price for an option. Traders of stock have the same concern. Fundamental analysis or technical analysis are generally tools used by stock traders to address this concern. In the options market, fair prices are the result of an interplay of participants and a concept known as *put–call parity*. As will be demonstrated, there exists an exact relationship between the prices of calls and puts with the same strike price and expiration and the underlying stock.

Put–call parity is the pricing relationship concept that keeps option prices in line with each other and the underlying stock.

Put–call parity is explained in four steps. First, the maximum risk of a simple stock and option position is defined. Second, it is demonstrated how the use of a second option can eliminate this risk. Third, the role that arbitrage plays in making the market function is discussed. Fourth, real world factors such as dividends, cost of money, and time are factored into the put–call parity relationship.

Step 1: Consider the following stock and option position:

	Per Share	*Total*
Long 100 share XYZ	Cost $52	$5,200
Long 1 $50 XYZ put	Cost $ 3	$ 300

Question: What is the maximum risk of this combined position? For this introductory example, we are assuming no commissions, no dividends, and zero interest rates.

To answer this question, consider what would happen at various stock prices on the expiration date. At a stock price of 50, for example, the put option would be exercised and the stock would be sold at 50 for a $2 loss per share, or $200. Adding the $300 cost of the put option to the

$200 loss on the stock results in a total loss of $500. Similarly, at any stock price from 50 down to 0 a $500 loss would result.

If the stock were to close between 50 and 52 on the option expiration date, the put option would expire worthless for a loss of $300. The per-share loss on the stock, however, would be less than $2, thus making the entire loss less than $500. At 52, of course, the stock position would break even and the total loss would equal $300, the cost of the put option.

At prices above 52, the stock position would show a profit, but that profit would be reduced by the cost of the put option. The break-even stock price on this combined position would be 55, which is where the $300 profit on the long stock would equal the $300 cost of the put option. Above 55, the combined position shows a total profit because the profit on the long stock exceeds the cost of the put option. Figure 2-20 illustrates the range of profit-loss outcomes at various stock prices at the time of option expiration.

FIGURE 2-20

Long Stock and Long Put

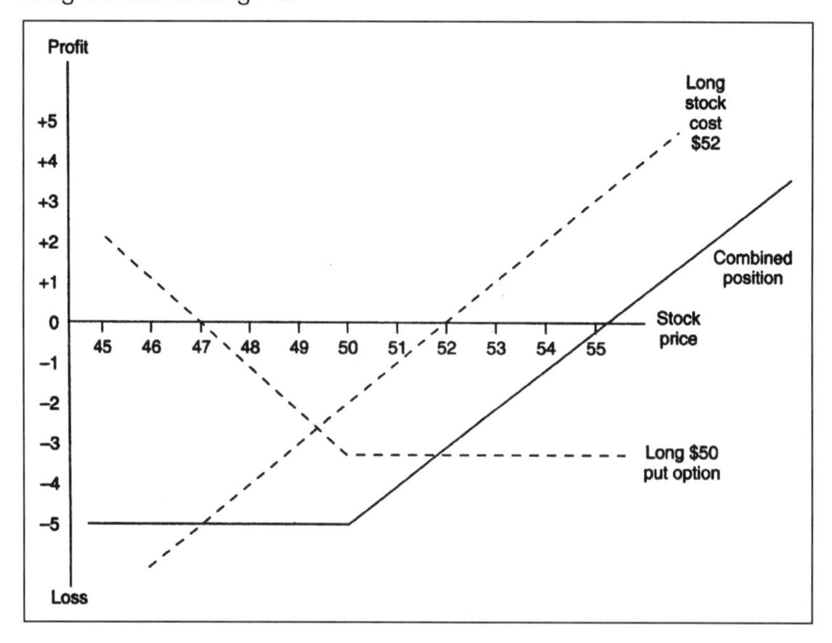

From the above discussion and Figure 2-20, it is clear that the maximum possible loss from the combined position of long stock and long put described above is $500.

Step 2 is eliminating the risk of the long stock and long put position. A second option position—writing a $50 XYZ call at $500—is now added to the original combined position. The new position is as follows:

		Per Share	*Total $*
Short 1 $50	XYZ call	Price $ 5	$ 500
Long 100	XYZ	Cost $52	$5,200
Long 1 $50	XYZ put	Cost $ 3	$ 300

What is the maximum risk of this position?

Again, consider the outcome at various stock prices on the date of option expiration. At 50 or below, the $500 loss on the long stock and long put position is offset by the $500 received from selling the call option. The result is exactly breakeven.

At any price above 50, the put will expire worthless for a $300 loss, but the long stock and short call combination will result in a $300 profit. This is so because at a stock price above 50, the call will be exercised and the stock will be sold at 50 for $200 loss. The $500 received for selling the call, however, is kept. Thus, the $500 profit on the call is reduced by the $200 loss on the stock for a total net profit of $300. This, again, exactly equals the cost of the put option that expired worthless with the stock price above 50.

Net result: At any stock price, the three-way call option—put option—stock position described above achieves breakeven. This is illustrated by Figure 2-21.

Step 3 is arbitrage. The three-sided option—stock combination just described is the basis for how professional traders operate in the marketplace to provide liquidity to other participants. Professional traders, called *arbitrageurs*, are constantly active in the stock and options markets, looking for opportunities to buy stock, buy puts, and sell calls. Their goal, of course, is to make a profit, not to break even as described in the previous example.

In the previous example, with the stock at 52, the $50 call at $5, and the $50 put at $3, the market is said to be *in line* or *at parity*. Professional traders are constantly seeking opportunities where they can sell the call option above $5 or buy the put option below $3. These competitive pressures maintain prices at or near the option's fair value.

FIGURE 2-21

Long Stock, Long Put, Short Call

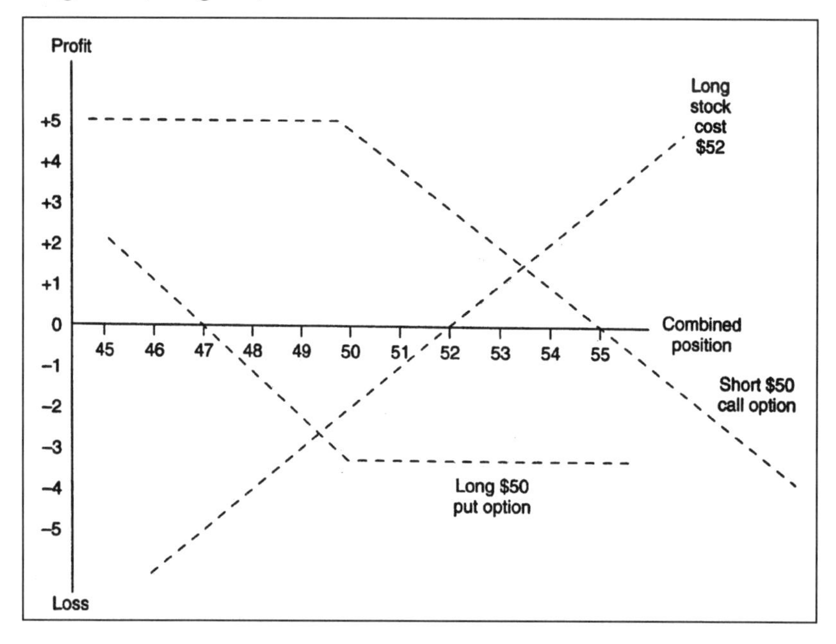

The price is *fair* because the process of arbitrage keeps all prices in line with each other, thus maintaining market equilibrium or parity between calls, puts, and stocks.

For example, if an imbalance of call-buying orders raised call prices, then professional traders would rush to increase their bids for put options. Bidding higher for puts would presumably entice more put sellers into the market. With the increased availability of puts (at the higher prices), the arbitrageurs would be able to buy puts, buy stock, and thus sell the calls being bid for at a higher price.

After this series of transactions occurred, both put and call prices would be higher than previously, but still fair. The new prices are fair because they would be in line with each other under the new market conditions.

Similarly, in another example, if a large order to sell puts at lower prices entered the market, then professional traders would compete against each other to offer call options at lower prices so that they could complete the three-sided option–stock position profitably. After these transactions occurred, both puts and calls would be lower in price than previously. Nevertheless, the prices would still be in line or at parity with each other under the new market conditions.

It should be clear from the preceding discussion that competition among professional traders causes put and call prices to rise and fall together. This concept should put to rest some misconceptions about the options markets expressed in typical statements such as "calls are overpriced" or "puts are cheap." Because of put–call parity, both puts and calls may be at a relatively high price level, or they may both be at a relatively low price level. But, barring transitory distortions, it is almost impossible for one to be "overpriced" or "expensive" relative to the other.

INFLUENCE OF REAL WORLD FACTORS

In the simple three-way example used above, it was assumed that there were no dividends, no cost of money, and no early exercise. In the real world, these factors affect option prices.

Dividends

Reviewing the basic three-way position of long stock, long put, and short call, we now ask: How would this position be affected if the stock paid a $1 dividend?

Without the dividend, the position at the stated prices broke even. The presence of a $1 dividend, therefore, would imply a $1 profit. But what would competitors in the marketplace do when they saw the opportunity to make a $1 profit? Some professional traders or arbitrageurs would be willing to settle for a smaller profit, say 75 cents. Consequently, they would be willing to pay 25 cents more for the put or the stock or sell the call for 25 cents less. Other professional traders would be satisfied with only a 40-cent profit, and their bidding and offering would raise put prices and lower call prices.

Competition in the marketplace would thus raise put prices or lower call prices (or a combination of both) until the basic three-sided stock and option position was back to breakeven. The conclusion to be drawn is that the presence of dividends has the effect of raising put prices and lowering call prices.

Cost of Money

The cost of money is an important consideration for all investors because investment performance is measured against the benchmark riskless investment, usually the short-term Treasury bill. It is no different for the professional option trader who must borrow money to finance large stock and option holdings and who seeks a return to cover those costs and earn a profit.

To illustrate how the cost of money affects option prices, consider the following example and ask this question: At what price must the $50 call option be sold to make this position break even?

	Per Share	*Total*
Short 1 $50 call option	Price ?	?
Long 100 XYZ	Cost $50	$5,000.00
Long 1 $50 put option	Cost $ 3	$ 300.00
9 percent prevailing cost of money		
90 days until option expiration		

The answer to this question is reasoned through as follows:

In 90 days, this three-way position will turn into $50 cash per share because either the stock will be above $50 and an assignment notice will be received for the short $50 call, or the stock will be below $50 and the put will be exercised. In either case, the stock is sold and $50 cash for each share is received. Thus, the question becomes: How much should be invested today to earn 9 percent (annually) if $50 per share will be received in 90 days? Because 90 days is one-fourth of a year, we expect to earn 9 percent × ¼ = 2.25 percent on this three-sided position. Consequently, we must invest 50 × (1 – .0225) = 48.875 or $48⅞ per share. Therefore, the call should be sold for 1⅛ more than the put, or 4⅛; and the final position looks like this:

	Per Share	*Total*
Short 1 $50 call option	Price $ 4⅛	$ 412.50
Long 1 share XYZ	Cost $50	$5,000.00
Long 1 $50 put option	Cost $ 3	$ 300.00
Total invested	$48⅞	$4,887.50

$48⅞ + ($48⅞ × 9% × [¼ of 1 year]) = $50 at option expiration.

Interest Rates

From the above example, if interest rates rise while the stock price, put price, and days until option expiration remain constant, then the call price must rise by the amount of the increased interest cost. This is the necessary result because the cost of carrying the position will increase and the call must be sold for a greater amount to cover that increased cost of carry.

The effect of interest rates on put prices is exactly opposite: as interest rates rise, put prices decline. This can be demonstrated by reasoning through what must happen to put prices when interest rates rise but the other elements (stock price, call price, and days until expiration) remain constant.

Rising interest rates result in an increased cost of carry. To compensate for the higher cost, either the revenue of the position must increase, or the cost of the position must decrease. Because the call price (the revenue side) is assumed to be unchanged, the cost of the position must be reduced. Because the stock price is assumed to be constant, only the put price is left to be reduced to compensate for the increase in cost of carry.

From these examples, it should be clear that the put–call parity relationship will hold at all times. Changes in supply or demand factors for any one or more of the parts of the equation will result in professional traders raising their bids or lowering their offers for other parts of the equation. Consequently, the interaction of different market participants, including competition among professional traders, will result in call, put, and stock prices that are in line with each other, or fair.

SUMMARY

The purpose of this chapter has been to introduce several fundamental concepts about call and put options and the rational nature of option prices. American-style options give the owner a right that may be exercised at any time prior to expiration, and European-style options involve rights that may be exercised only at expiration. Option prices have two components: intrinsic value and time value. In-the-money, at-the-money, and out-of-the-money are designations that refer to the relationship of the current price of the underlying to the option strike price.

The theoretical value of an option depends on five factors: the price of the underlying asset, the strike price of the option, the time remaining to expiration, prevailing cost of carry (interest rates and dividends), and the volatility of the underlying asset. Option values will change by less than the price change of the underlying asset; the ratio of this price change is called delta. Deltas of at-the-money options are approximately .50 regardless of the time to expiration. In-the-money options have deltas greater than .50, and they increase to 1.00 as expiration approaches. Deltas of out-of-the-money options are less than .50 and decrease toward 0 as expiration approaches. A unit of time decay in an option price is called theta. Time decay for at-the-money options is nonlinear, but in-the-money and out-of-the-money options decay in a nearly linear fashion. Volatility is the unknown element in calculating an option's theoretical value; as a result, the calculation of an option's theoretical value is, ultimately, subjective.

Advanced pricing concepts such as how deltas change, how theta changes, and how changes in volatility affect option prices are important primarily to professional option traders who manage large option positions of both long and short options. The typical individual investor and portfolio manager should be most concerned with market prediction, stock selection, and risk management. Options are a valuable investment tool because they offer investors a wider range of risk profiles from which to choose.

CHAPTER 3

VOLATILITY EXPLAINED

James B. Bittman

INTRODUCTION

Volatility is the most used and least understood word in the options business. To most people, the term volatility has an intuitive definition that relates to price movement, and this intuition is correct. On an annual basis, it can be said that a stock that had a 12-month high of 120 and a 12-month low of 80 was more volatile than a stock that traded between 105 and 95. Over a shorter term, it can be said that a stock with an average daily trading range (high price to low price) of $5 is more volatile than another stock that has an average daily trading range of $2 (assuming the average underlying price of both stocks is the same).

Examination of historical stock price movements shows that individual stocks go through periods of high and low volatility. There are many possible explanations for these variations in stock price volatility. One explanation might be general economic factors affecting an industry group of stocks; another might be specific developments for the specific stock. A third possibility could be the psychological state of investors. Regardless of the cause, investors must be aware that the volatility characteristics of a stock can change dramatically at any point in the future.

Another factor that adds to the confusion regarding volatility is that there are four words commonly used in conjunction with volatility. These four words are historical, future, expected, and implied. But what is the concept of volatility that is relevant to the option user? The purpose of the following discussion is to explain when and how each of these four words is used properly. First, we illustrate in a simple example how the historical volatility of a stock is calculated. Second, in conceptual terms, we show how volatility and changes in volatility affect option values. Third, we discuss what expected volatility and implied volatility mean, what overvalued and undervalued mean, and how all the confusing terms associated with volatility can be used properly.

HISTORICAL VOLATILITY

Historical volatility is a measure of actual stock price movement that occurred during a period of time in the past. Specifically, stock price volatility is the annualized standard deviation of a stock's daily returns. A simplified example presented in Table 3-1 illustrates the concept.

Columns 1 and 3 in Table 3-1 show closing prices of two stocks for 10 days. While the prices of both Stock A and Stock B start the 10-day period at 51 and end at 51½, Stock B has larger price changes every day.

TABLE 3-1

Calculation of Historical Volatility

	Stock A		*Stock B*	
	Daily Close	**Daily Return**	**Daily Close**	**Daily Return**
Day 1	$51		$51	
Day 2	$51½	0.0098	$51¾	0.0147
Day 3	$50⅝	–0.0170	$50⅜	–0.0266
Day 4	$50⅞	0.0049	$50¾	0.0074
Day 5	$51⅝	0.0147	$52¼	0.0296
Day 6	$51⅜	–0.0048	$51¾	–0.0096
Day 7	$51⅝	0.0049	$52⅛	0.00725
Day 8	$51½	–0.0024	$51¾	–0.0072
Day 9	$51⅜	–0.0024	$51¼	–0.0097
Day 10	$51½	0.0024	$51½	0.0049
10-day historical volatility (annualized standard deviation)		13.9%		24.7%

Columns 2 and 4 show the daily returns. A daily return is calculated by dividing the daily price change by the starting price. Consequently, from day 1 to day 2, Stock A rose by ½, which was a daily return of positive 0.98 percent on the starting price of 51. From day 7 to day 8, Stock B moved down by ⅜, which was a daily return of negative 0.72 percent on the starting price of 52⅛.

The standard deviation of these daily returns is calculated and annualized, and the result is the 10-day historical volatility presented at the bottom of Columns 2 and 4. As expected, Stock B is shown to have the higher volatility because it experienced greater price fluctuations on very nearly the same base price. It should be noted that there is nothing special about the 10-day period. Historical volatility can be calculated for any time period. The mathematics of calculating standard deviations is beyond the scope of this book. For an extensive mathematical treatment of volatility, there are a number of references available; for example, books on options by Cox and Rubinstein, Jarrow and Rudd, Hull, and Natenburg.

Now that it has been illustrated how the historical volatility of a stock price is calculated, the following discussion presents volatility in a nontechnical way, as a concept that can be understood and used in making investment and trading decisions with options.

VOLATILITY AND OPTION VALUE—A CONCEPTUAL APPROACH

To begin a discussion of volatility, consider the simplistic world presented in Figure 3-1 in which a stock with a starting price of 100 has only two possible outcomes at option expiration: 99 or 101. In this simplistic case, time to expiration is not considered. With the stock price starting at

FIGURE 3-1

Expected Value Calculations (One-Point Volatility, One Time Period, Starting Price 100)

	Probability	Expected Stock Price	Expected Value of 100 Call
100 → 101	50%	.50 × 101	.50 × 1.00
100 → 99	50%	.50 × 99	.50 × -0-
Total		100	.50

100 and only being able to rise one point to 101 or fall one point to 99, the stock price has a one point volatility—a one-point movement without regard to direction. The ending value of the 100 call depends on the ending stock price. If the stock price rises to 101, the 100 call will have a value of 1. If the stock declines to 99, the 100 call will expire unexercised and be worth 0. Knowing the possible final prices for the stock and the probability that each might occur, it is possible to calculate expected values for both the stock price and the 100 call.

Expected value is a statistical concept that means the weighted average outcome. The assumption is that the event is repeated a large number of times so that each possible outcome occurs in accordance with its statistical probability.

In Figure 3-1, there is a 50 percent chance the stock will rise to 101 and a 50 percent chance it will fall to 99. Consequently, the weighted average outcome, or the expected stock price, is 100. The calculation is shown in Figure 3-1 and is as follows:

$$\begin{aligned} \text{Expected value of stock price} &= (50\% \times 101) + (50\% \times 99) \\ &= (.50 \times 101) + (.50 \times 99) \\ &= 50.5 + 49.5 = 100 \end{aligned}$$

Calculating the expected value for the 100 call is accomplished through a similar process. With the stock at a final price of 101, the 100 call is worth 1. With the stock at a final price of 99, the 100 call expires unexercised with a value of 0. Consequently, given a 50 percent probability the 100 call will have a value of 1 and a 50 percent probability it will have a value of 0, the weighted average, or expected value, is 0.50. The calculation is shown in Figure 3-1 and is as follows:

$$\begin{aligned} \text{Expected value of 100 call option} &= (50\% \times 1) + (50\% \times 0) \\ &= (.50 \times 1) + (.50 \times 0) \\ &= .50 + 0 = .50 \end{aligned}$$

CHANGING THE VOLATILITY

The above exercise raises the following question: If the magnitude of the up or down movement is changed, what happens to the expected values?

In Figure 3-2, we assume the stock price, starting at 100, can either rise to 102 or fall to 98. The volatility—movement without regard to direction—has been increased to two points from one point. Figure 3-2

FIGURE 3-2

Expected Value Calculations (Two-Point Volatility, One Time Period, Starting Price 100)

	Probability	Expected Stock Price	Expected Value of 100 Call
100 → 102	50%	.50 × 102	.50 × 2.00
100 → 98	50%	.50 × 98	.50 × –0–
	Totals	100	1.00

shows the expected value calculation for the stock price is still 100. Although this answer may surprise some readers, the stock has a 50 percent probability of ending at 102 and a 50 percent probability of ending at 98. As a result, the expected value of the stock is 100 (the calculation is shown in Figure 3-2), the same as in the previous case, even though the volatility was lower.

In the second case, however, the larger volatility of the underlying stock results in a different expected value for the $100 call. At a final stock price of 102, the 100 call has a value of 2. At a final stock price of 98, the 100 call will expire unexercised and have a value of 0. Because there is a 50 percent chance of either outcome occurring, the expected value of the $100 call in this situation is 1. The calculation is shown in Figure 3-2.

An Analogy to Volatility

There are two differences between the two situations just described in Figures 3-1 and 3-2. First, there is a different size of the up or down movement—the volatility. The two-point movement example can be said to be more volatile than the one-point movement example. The second difference is the expected value for the 100 call: .50 versus 1.00. But it is important to note that there is no difference in the expected value for the underlying stock; both cases had an expected stock price of 100.

This leads to the first important concept about volatility: *an increase in volatility increases the expected value of the 100 call, but does not affect the expected value of the underlying stock.* Although presented in simplified form, this is a general conclusion that holds true in the more sophisticated world of higher mathematics.

A More Advanced Example

The previous example was extremely simplistic in that it assumed only two possible outcomes in one time period. In the real world, of course, many outcomes are possible, and more than one time period exists. Although it is impossible to replicate reality, it is possible to carry the single time period example forward to more time periods so that more outcomes can be generated. The purpose of doing this is to illustrate some important concepts of option price behavior.

The examples just reviewed are one-time-period examples. Figure 3-3 illustrates how a two-time-period example is created, and Figure 3-4 illustrates a three-time-period example. This is called the *binomial process* because in each time period, there are only two possible outcomes. As additional periods are added, more final outcomes become possible.

It is also important to note how the probabilities of final prices change as the number of periods increases. In the single-period case, there are two possible outcomes, each of which has a 50 percent chance of occurring. In the two-period case, however, there are three possible outcomes; and each outcome does not have a 33 percent chance of occurring. If one traces along the branches of the diagram, one can observe that only one out of four ways leads to the final price of 102. Also, only one out of four ways leads to the final stock price of 98. But two out of four ways lead to the final stock price of 100. Consequently, the probabilities in the two-period case are 25 percent of reaching 102, 50 percent of reaching 100, and 25 percent of reaching 98.

FIGURE 3-3

Expanding the Binomial Process to Two Periods (One-Point Volatility, Two Time Periods, and Probabilities of Final Outcomes)

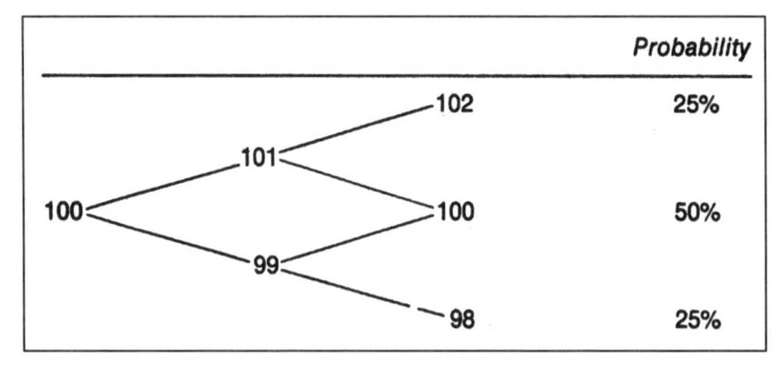

FIGURE 3-4

Expanding the Binomial Process to Three Periods (One-Point Volatility, Three Time Periods, and Probabilities of Final Outcomes)

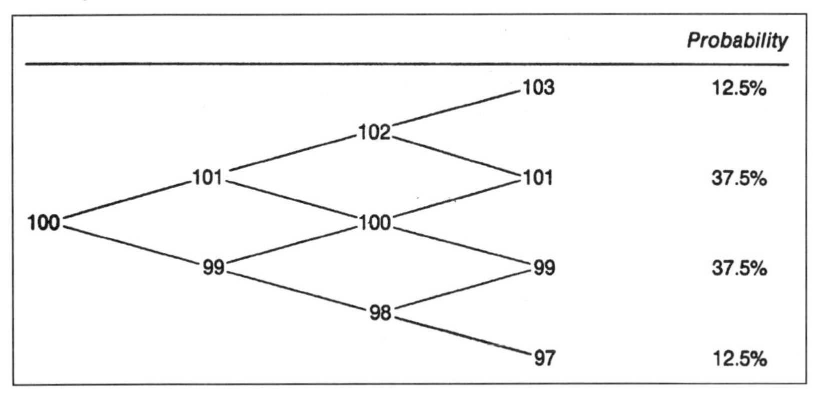

If one traces along the various paths in Figure 3-4 that illustrate a three-time-period binomial example, one sees eight possible ways of arriving at one of the four possible final prices. The probabilities are 12.5 percent of ending at 103, 37.5 percent of ending at 101, 37.5 percent of ending at 99, and 12.5 percent of ending at 97.

The Four-Period Case

We now extend the binomial process to four periods and analyze the implications for option price behavior and volatility. Figure 3-5 assumes a starting price of 100 and an up or down movement of one point per period. Consequently, the range of possible prices at the end of four periods is from up four points to 104 to down four points to 96. The probability of the various final outcomes occurring are as follows: 104—6.25 percent, 102—25 percent, 100—37.5 percent, 98—25 percent, 96—6.25 percent.

Using the expected-value calculations presented earlier, we see that the stock price has an expected value of 100 (see calculations in Figure 3-5) and the 100 call option has an expected value of .75 (calculations are also in Figure 3-5).

Moving Out One Period

The expected value calculations just presented were made from a starting price of 100, which is point A in Figure 3-5. If we assume that during

FIGURE 3-5

Expected Value Calculations (One-Point Volatility, Four Time Periods, Starting Price 100)

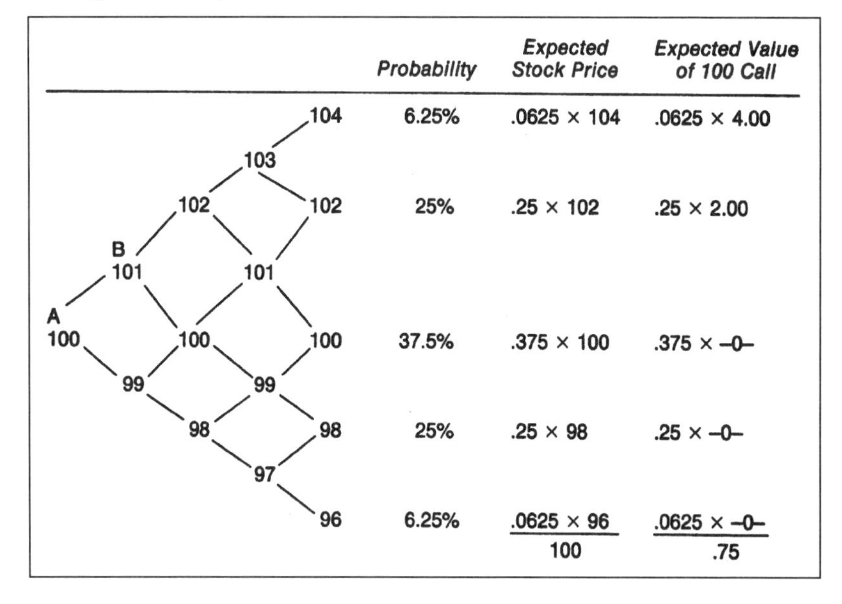

the first period the stock price rises one point to 101 indicated by point B in Figure 3-5, new expected values for both the stock price and the call can be calculated.

Figure 3-6 shows the new situation. With the stock at 101 and only three periods remaining, the range of possible final outcomes has been reduced to 104 at the high end and 98 on the low end. The dashed lines represent the outcomes that are no longer possible. There is also a new set of probabilities for each of these possible outcomes, because now only eight possible ways exist to get to the remaining outcomes. The probabilities are: 104—12.5 percent, 102—37.5 percent, 100—37.5 percent, 98—12.5 percent.

The expected value calculations show that the new expected value of the stock price is 101 and the new expected value of the 100 call is 1.25. The calculations are shown on Figure 3-6.

These new expected value calculations are consistent with the concept of delta presented earlier and illustrated in Tables 2-1 and 2-2 in Chapter 2. When the stock price moves up by one full point, the theoretical value of the 100 call moves up by less than one point; in this case by .50.

Changing the Movement in the Binomial Example

If the size of the up or down movement is increased from one point per period to two points per period in the four-period example, the results are consistent with the concepts just presented. Figure 3-7 shows a four-period example in which the underlying stock price starts at 100 and moves up or down two points during each of four time periods. The range of possible outcomes has expanded to 108 at the high and 92 at the low, with the probabilities of each final outcome as follows: 108—6.25 percent, 104—25 percent, 100—37.5 percent, 96—25 percent, 92—6.25 percent. The expected value of the stock at the end of four periods is 100, and the expected value of the 100 call is 1.50 (the calculations are shown in Figure 3-7).

Figure 3-8 shows a situation in which the starting stock price is 101, there are three time periods to expiration, and the expected volatility is two points per period. (It should be noted that Figure 3-8 does not follow directly from Figure 3-7. However, the numbers calculated in Figure 3-8 are required to complete the upcoming discussion on implied volatility and changes in implied volatility.) Using the price of 101 as the

FIGURE 3-6

Expected Value Calculations (One-Point Volatility, Three Time Periods, Starting Price 101)

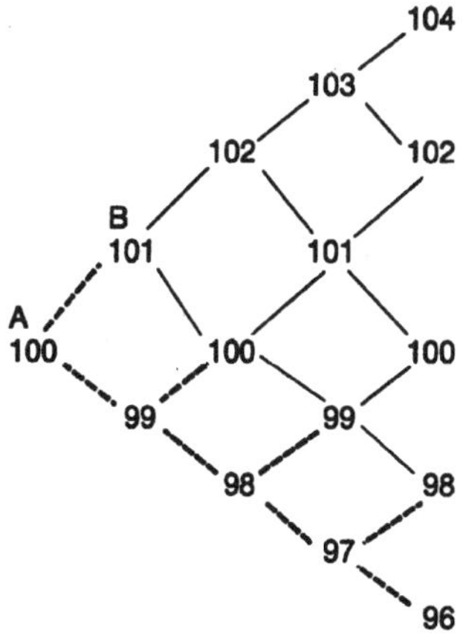

	Probability	Expected Stock Price	Expected Value of 100 Call
104	12.5%	.125 × 104	.125 × 4.00
102	37.5%	.375 × 102	.375 × 2.00
100	37.5%	.375 × 100	.375 × -0-
98	12.5%	.125 × 98	.125 × -0-
96	-----	-------	-------
		101	1.25

FIGURE 3-7

Expected Value Calculations (Two-Point Volatility, Four Time Periods, Starting Price 100)

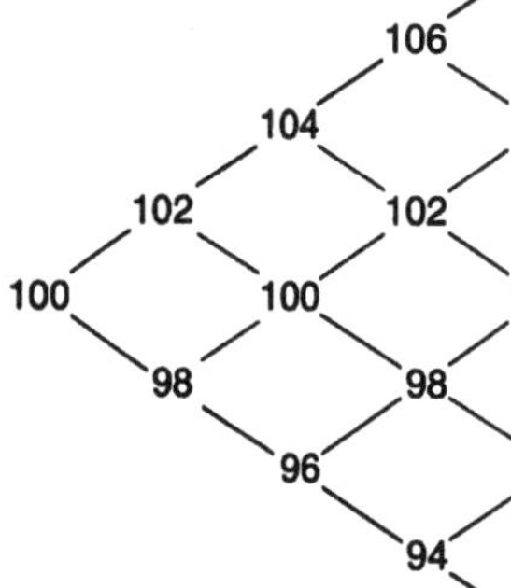

	Probability	*Expected Stock Price*	*Expected Value of 100 Call*
108	6.25%	.0625 × 108	.0625 × 8.00
106			
104 104	25%	.25 × 104	.25 × 4.00
102 102			
100 100 100	37.5%	.375 × 100	.375 × –0–
98 98			
96 96	25%	.25 × 96	.25 × –0–
94			
92	6.25%	.0625 × 92	.0625 × –0–
		100	1.50

FIGURE 3-8

Expected Value Calculations (Two-Point Volatility, Three Time Periods, Starting Price 101)

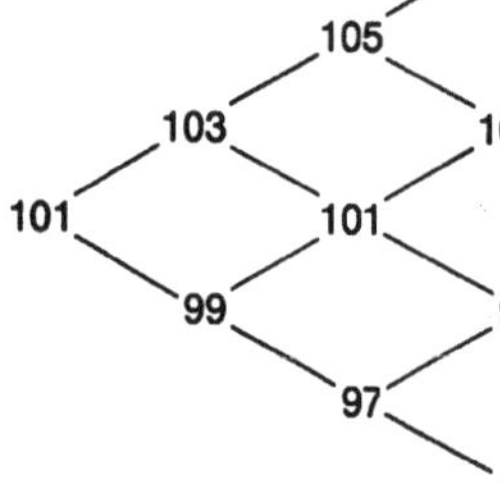

	Probability	*Expected Stock Price*	*Expected Value of 100 Call*
107	12.5%	.125 × 107	.125 × 7.00
105			
103 103	37.5%	.375 × 103	.375 × 3.00
101 101			
99 99	37.5%	.375 × 99	.375 × –0–
97			
95	12.5%	.125 × 95	.125 × –0–
		101	2.00

starting point in Figure 3-8, the new expected value of the stock is 101, and the new expected value of the 100 call is 2.00.

The calculations in Figure 3-8 are consistent with the earlier example in Figure 3-5 and Figure 3-6. Although the stock price is one point higher in Figure 3-8 than Figure 3-7, the expected value of the 100 call is less than one point higher, just as the one-point stock price rise between Figure 3-5 and Figure 3-6 resulted in a less-than-one-point increase in the 100 call.

The comparison of the one-point movement and two-point movement four-period examples is also consistent with the narrow movement and wide movement one-period examples presented earlier. Wider movements per period result in a higher expected value for the 100 call, but not for the underlying stock price.

We now have all the information we need to explain the concepts of expected volatility, implied volatility, and changes in implied volatility.

EXPECTED VOLATILITY

Referring back to Figure 3-5, the expected stock value of 100 is the same as the starting price of 100, and, given the assumption that movement can only be up one point or down one point per period for four periods, the expected value of the 100 call is .75. Another term for expected value is *theoretical value*. Statistically, if the four-period event occurred many times, then the average final outcome would be .75. Of course, no single final outcome would equal .75 because, under the assumptions about the stock price movement, the only possible option price outcomes are whole numbers 0, 2, and 4.

The theoretical value (or expected value) was calculated from the distribution that resulted from the assumption about movement. In the real world, the future is unknown, and the daily price change is not limited to a finite up or down amount. Furthermore, the size of possible movement is not constant. Consequently, in the examples presented above, it was not the known distribution that led to the theoretical values of the 100 call; it was the *expected* distribution. If we substitute the word *volatility* for *distribution*, which is common practice in discussions about options, it can be said that the *theoretical values were calculated using the expected volatility*.

Looking back at Figure 3-6, there are some important observations to be made about the theoretical call value of 1.25, which was calculated after the one-point price rise to 101 in the first time period. This comment may startle some readers, but the theoretical value of 1.25 for the 100 call, with the stock at 101 three periods from expiration, was calculated using the same expected volatility as that used to calculate the .75 theoretical value for the 100 call with the stock price at 100 four periods prior to expiration. The one point up or down movement per period is the same—there is just one fewer period—and the stock has a different starting price. But the expected movement, or volatility, is the same. This is a crucial point to understand: the price of .75 for the 100 call in Figure 3-5 and the price of 1.25 for the 100 call in Figure 3-6 are calculated using the same expected volatility.

IMPLIED VOLATILITY

We now look at the same information from a different angle. Suppose it is four periods prior to expiration, the stock is trading at 100, and the 100 call is trading in the market at .75. It can be said that this price *implies* the distribution in Figure 3-5. Again, substituting the word *volatility* for *distribution*, we can say that the price of .75 for the 100 call (when the stock is trading at 100 four periods prior to expiration) implies the volatility in Figure 3-5.

It follows from this that the price of 1.25 for the 100 call with the stock trading at 101 three periods prior to expiration implies the same volatility.

What, then, can be said about the prices of 1.50 in Figure 3-7 and 2.00 in Figure 3-8? Obviously, these prices imply a larger per-period up or down movement. In other words, these prices imply a higher volatility.

Implied Volatility versus Expected Volatility

Expected volatility is one investor's expectation about what will happen; it is that investor's prediction for the future. For example, given a stock price of 100, if an investor expects the four-period distribution in Figure 3-5, then that investor will calculate an expected (or theoretical) value of .75. Also, given a stock price of 100, if an investor expects the four-period distribution in Figure 3-7, then that investor will calculate a theoretical value of 1.50. Similarly, given a stock price of 101, if an investor expects the three-period distribution in either Figure 3-6 or Figure 3-8, that investor will calculate a theoretical value of either 1.25 or 2.00, respectively.

An investor's expectation, or prediction, may come from experience, from historical data, or from some other source. But as was shown in the preceding discussion, one's choice of volatility will affect the resulting theoretical value.

Implied volatility is the volatility derived from looking at the current market price of an option. Using different words to express the same concept: *implied volatility is the volatility number that justifies the current market price of an option*. For example, if the stock is trading at 100 and the 100 call is trading at a price of .75, this price implies a one point per-period volatility as shown in Figure 3-5. Similarly, if the stock is trading at 100 and the 100 call is trading at 1.50, that price implies a two point per-period volatility as show in Figure 3-7. Also, if there are three periods to expiration, the stock is trading at 101, and the 100 call is trading at either 1.25 or 2.00, then the distributions in Figure 3-6 or Figure 3-8, respectively, in three periods would be implied.

Some traders think of implied volatility as the market's expected volatility. Given the market price of an option, the volatility implied by that price must be, it is reasoned, the expectation of the market.

Changes in Implied Volatility

Changes in implied volatility are often difficult for the options newcomer to understand. It is very important to see that an option price does not imply a direction for the underlying price movement. Calls, puts, and the underlying security always have a price relationship with each other called put–call parity, which was explained in Chapter 2. Consequently, option prices imply a distribution—or volatility—not a direction. If an option price changes without the price of the underlying security changing, then market forces are bringing about a change in the implied distribution or implied volatility.

As an example of a change in implied volatility, consider a situation in which the stock price is 100, there are four periods to expiration, and the 100 call is trading in the market at a price of .75. Now imagine that without any passage of time and without any change in the underlying stock price, the 100 call trades up in price to 1.50. How could such an occurrence be explained? Obviously, an increase in demand for the call has driven up the price. But, in terms of implied volatility, the market has changed its expectation about the distribution of prices of the underlying stock in the upcoming four periods. Specifically, when the price of the

100 call was .75, the market was expecting the distribution illustrated in Figure 3-5. With the 100 call trading at 1.50, however, the market has changed its expectation to the distribution illustrated in Figure 3-7. In other words, the implied volatility changed: the market's expectation for the upcoming four-period distribution of prices of the underlying stock changed from that pictured in Figure 3-5 to that pictured in Figure 3-7.

In the real world, of course, both time and price of the underlying change. To illustrate how implied volatility can fluctuate given changes in time and price of the underlying, consider a situation in which there are four periods to expiration, the stock is trading at 100, and the 100 call is trading in the market at .75. This option price implies the distribution—or volatility—in Figure 3-5 as discussed above. Now imagine that in the course of one period the stock price rises from 100 to 101 and the 100 call price rises in market trading from .75 to 2.00. What has happened?

If the market's expectation for volatility for the next three periods had remained constant, then one would expect the 100 call to have risen in price from .75 to 1.25. In this case, the 1.25 market price would imply the distribution in Figure 3-6. The price of 2.00, however, given three periods to expiration, implies the distribution illustrated in Figure 3-8. The change in price of the 100 call from .75 to 2.00 happened because the market's expectation for volatility in the remaining three periods increased.

To understand what happens when the market's expectation for volatility decreases, consider a situation in which there are four periods to expiration, the stock is trading at 100, and the 100 call is trading at 1.50. Given these circumstances, the market's expectation for volatility is two points per period and is illustrated in Figure 3-7. If the price of the stock trades *up* to 101 and the 100 call trades *down* to 1.25, then the market's expectation for volatility for the next three periods has changed. The market now expects the distribution in Figure 3-6. The market's expectation for volatility in the remaining three periods has decreased from two points per period to one point per period.

Real-World Experience

Can such things happen? This is a reasonable question for an options newcomer to ask. Although such price behavior seems counterintuitive, in fact, changes in volatility are quite common.

The conclusion is obvious: changes in implied volatility can affect the results of trading options. Therefore, traders of options should take time to learn more about this important factor.

THE MEANING OF "20 PERCENT" VOLATILITY

A natural question at this point is, What exactly does the volatility percentage number mean? Essentially, this percentage figure is a statistical measure of the width of the expected distribution of the underlying security. The higher the volatility percentage, the wider the expected distribution. In statistical parlance, the volatility percentage is the standard deviation of the bell-shaped curve that theoretically illustrates the possible price outcomes. For the layperson, this means that a security with 20 percent volatility is expected, two-thirds of the time, to be within a range 20 percent higher or lower than the current price in one year. A security with 30 percent volatility is expected, two-thirds of the time, to be within a range 30 percent higher or lower in one year. Obviously, the security with 30 percent volatility has a greater chance of experiencing a larger price change. Consequently, options on that security will have higher values, assuming other factors are equal.

Rather than learning the mathematics of volatility, option traders should be aware of this concept and incorporate it into their decision-making process. Because trading decisions are largely subjective, learning to incorporate expectations about volatility is just another subjective component.

VOLATILITY AND AVERAGE PRICE MOVEMENTS

Probability tables exist so that if a standard deviation is known (or assumed), then the likelihood or probability of an event occurring can be determined. Although these probabilities can be calculated exactly, the following approximations are sufficient for traders of options.

- Approximately two out of three outcomes will occur within one standard deviation of the mean.
- Approximately 19 out of 20 outcomes will occur within two standard deviations of the mean.
- Approximately 369 out of 370 outcomes will occur within three standard deviations of the mean.

This concept is related to price movement of stock prices in the following manner: *the volatility percentage for a given stock price represents one standard deviation of price movement for a one-year period*. For example, if the stock price is 50 and the volatility is 10 percent, then a price change of one standard deviation for a one-year time period is 5 ($10\% \times 50$). This means that, over three one-year periods, it is probable that in two of three years the stock will be trading in a range between 45 (50 – 5) and 55 (50 + 5) and outside that range in one year, assuming a stock price of 50 at the beginning of each year. Similarly, over 20 one-year periods, it is expected the stock will be trading in a range between 40 (50 – 2 × 5) and 60 (50 + 2 × 5) in 19 of these years and outside that range in one year, once again assuming a 50 price at the beginning of each year. Also, over 370 one-year periods, it is expected the stock will be trading in a range between 35 (50 – 3 × $5) and 65 (50 + 3 × 5) in 369 of these years and outside that range in one year.

Because a one-year time frame is generally not useful for options traders, the following formula can be used to calculate a price change of one standard deviation over a period of time: *The annual volatility percentage divided by the square root of the time (expressed in years) times the stock price.*

Example 1. If a stock is trading at 50 with a volatility of 25 percent, then the *monthly* standard deviation of price change is:

$$(25\% \div \sqrt{12}) \times 50 = 3.62 = 3\tfrac{5}{8}$$

This means that over three one-month periods, it is expected that a stock trading at 25 percent volatility and starting at 50 will trade between 46⅜ and 53⅝ two out of three times and outside that range on the other occasion. It also means that over 20 one-month periods, it is expected that this stock will trade in a range between 42¾ and 57¼ 19 times out of 20 and outside that range once. And, it means there are 369 chances out of 370 that this stock will trade in a range between 37⅛ and 62⅞.

Example 2. If a stock is trading at 50 with a volatility of 25 percent, then the *daily* standard deviation of price change is:

$$(25\% \div \sqrt{252}) \times 50 = 0.78 = \tfrac{3}{4}$$

This means that over three one-day periods, it is expected that a stock trading at 25 percent volatility and starting at 50 will trade in a range between 49¼ and 50¾ two out of three times and outside that range on the other occasion. During 20 one-day periods, it is expected that this stock will trade in a range between 48½ and 51½ 19 times out of 20 and outside that range once. And, there are 369 chances out of 370 that this stock will trade in a range between 47¾ and 52¼.

In the daily calculation, the square root of 252 is used instead of the square root of 365, because there are approximately 252 trading days in a year. Therefore, one trading day is $\frac{1}{252}$ of a trading year.

As a final example, for a 90-day option, which has a life of one-fourth of a year, the volatility percentage is divided by the square root of 4, or 2. Consequently, for the sample $50 stock with a volatility of 25 percent, a move of one standard deviation for 90 days is:

$$(25\% \div \sqrt{4}) \times 50 = 6.25 = 6\tfrac{1}{4}$$

This means that over three one-quarter periods, it is expected that a stock trading at 25 percent volatility and starting at 50 will trade between 43.75 and 56.25 two of the three times and outside that range once. It also means that over 20 one-quarter periods, it is expected that this stock will trade in a range between 37.50 and 62.50 19 times out of 20 and outside that range once. Finally, it means there are 369 chances out of 370 that this stock will trade in a range between 31.25 and 68.75.

VOLATILITY AND IMPLIED PRICE RANGES

Table 3-2 summarizes the estimated standard deviation of price movement indicated by volatility percentages at 30, 60, and 90 days. As an example, consider row four, 30 percent volatility. Given 60 days to expiration, if an option is trading at 30 percent volatility, the implication is that the standard deviation of price movement is 12.26 percent of the stock's current price. If the stock in question is trading at 70, 12.26 percent is 8.58; and the market expects that two-thirds of the time this stock will trade in a range between 78.58 and 61.42 from now until option expiration (in 60 days).

Using This Concept of Price Ranges

This concept of implied price ranges sounds much more difficult to use than it is. First, investors must remember that the option price does not imply a direction; it only implies a possible range of movement—up or down. Second, investors must have a realistic expectation about option price behavior given a price change in the underlying stock. Tables 3-3 and 3-4 show how a $50 call can be expected to behave given changes in stock price and time to expiration under two different volatility assumptions. Table 3-3 shows that if options are trading at a volatility level of 25

TABLE 3-2

Statistical Price Distribution of Volatility

Implied Volatility	Price Range Expectations %		
	30 Days	**60 Days**	**90 Days**
15%	4.35%	6.13%	7.50%
20	5.75	8.17	10.00
25	7.25	10.21	12.50
30	8.65	12.26	15.00
35	10.15	14.30	17.50
40	11.55	16.34	20.00
45	13.00	18.38	22.50
50	14.45	20.43	25.00
55	15.90	22.47	27.50
60	17.35	24.51	30.00

Two-thirds of the time, the underlying will be no more than the indicated percentage up or down from the starting price.

TABLE 3-3

$50 Call—25 Percent Volatility

Stock Price	8 Weeks	6 Weeks	4 Weeks	2 Weeks	Expiration
$51	2⅝	2⅜	2	1⅝	1
$50	2	1¾	1½	1	0
$49	1½	1¼	1	⅝	0
$48	1⅛	⅞	⅝	¼	0
$47	¾	⅝	⅜	⅛	0
$46	½	⅜	3⁄16	1⁄16	0

percent, and if the stock price rises from 46 at eight weeks prior to expiration to 50 at four weeks prior to expiration, then the expectation would be for the $50 call to rise from ½ to 1½. If however, the $50 call were to rise to 2¼, as shown in Table 3-4, then the market's expectation for volatility would have increased from 25 percent to 40 percent. If instead, the $50 call rose to a price less than 1½, then the market's expectation for volatil-

TABLE 3-4

$50 Call—40 Percent Volatility

Stock Price	8 Weeks	6 Weeks	4 Weeks	2 Weeks	Expiration
$51	3¾	3⅜	2⅞	2⅛	1
$50	3¼	2¾	2¼	1⅝	0
$49	2¾	2¼	1¾	1⅛	0
$48	2¼	1⅞	1⅜	¾	0
$47	1⅞	1½	1	½	0
$46	1½	1⅛	¾	¼	0

ity would have decreased. When buying options, an investor would obviously prefer to have volatility increase as the stock moved in the desired direction. Likewise, when selling options, an investor would prefer to have volatility decrease as the stock moved in the desired direction. However, the most important factor is whether or not the stock moves in the predicted direction in the predicted time frame.

OVERVALUED AND UNDERVALUED

Overvalued and *undervalued* are terms that are used and misunderstood as often as volatility. A common misconception is that buying undervalued options and selling overvalued options is good. In fact, *neither is necessarily good*! For example, an investor who purchases an undervalued call can still lose money if the underlying stock declines in price. There are many important factors in option trading decisions, and, frequently, too much emphasis is placed on the misunderstood concepts of overvalued and undervalued.

In order to understand what the terms overvalued and undervalued mean, we must first understand what is meant by *value*. Usually, the terms *theoretical value* or *fair value* are used when talking about value. Recognizing how theoretical values are derived will provide some insight into the meaning of *overvalued* and *undervalued*.

The Basis for Theoretical Value

As discussed above, an option's value depends on the distribution of possible prices of the underlying security in the time remaining to expiration. In other words, the theoretical value of an option depends on future volatility, the price fluctuations of the underlying security that occur between now and expiration. But future volatility is unknown, because one cannot know for certain what the future holds. Consequently, because future volatility is unknown, calculations of theoretical values are only estimates. That's right! So-called theoretical values are only estimates of value, because future volatility is unknown. What, then, is meant when someone says, "The theoretical value of an option is . . ."?

As discussed in Chapter 2, the theoretical value of an option depends on several inputs: the price of the underlying stock, the time to expiration, the strike price, interest rates, and a volatility estimate. Of this list, all are known except the volatility estimate. And, it is important to note, it is only a volatility estimate. Consequently, because one of the most important inputs into theoretical value is only an estimate, that makes the resulting theoretical value only an estimate. This means that when someone states, "The theoretical value of an option is . . . ," only an estimate of theoretical value is being presented based on an estimate of volatility. Taken in this context, option traders might be as wary of option theoretical values as they are of market price forecasts.

Something can only be over or under relative to something else. In the case of an option being overvalued or undervalued, two prices are being compared: the current market price of an option and the price generated by someone's theoretical value calculation. What is the difference between the two? Because all inputs that go into an option's value are known except the volatility estimate, the difference must be related to the volatility estimate. In fact, this is the case. As discussed above, the market price of an option implies a volatility, the so-called *implied volatility*. If that implied volatility figure had been used to calculate the option's theoretical value, then the result of the calculation would be exactly the same as the market price of the option. Consequently, the theoretical value of the option would equal the market price of the option. But, obviously, a different number for the volatility estimate was used because a different price was calculated.

If the volatility estimate used to calculate theoretical value is higher than the volatility implied by the market price, then the calculated option value will be higher than the market price. This leads to the conclusion that the market price of the option is under theoretical value. Therefore, it is reasoned, the option is undervalued.

If the volatility estimate used to calculate theoretical value is lower than the volatility implied by the market price, then the calculated option value will be lower than the market price. This leads to the conclusion that the market price of the option is over theoretical value. Therefore, it is reasoned, the option is overvalued.

Of course, all this assumes one believes the volatility estimate used by the generator of theoretical values. If one does not believe those estimates, then one probably should not place much faith in the theoretical values.

Perhaps the biggest misconception about the strategy of buying undervalued options and selling overvalued options is that, somehow, when an option returns to fair value, a profit will result. Nothing could be further from the truth! Although the volatility estimate is a component of an option's price, a much bigger component is the price of the underlying security.

Refer to Table 3-3 and consider a situation in which it is eight weeks prior to expiration, the stock is 51, and a trader buys the $50 call for 2⅝, which is equal to a volatility level of 25 percent. If the stock declines to 48 over the next six weeks and the call is sold at a volatility level of 40 percent, has the trader made or lost money?

Table 3-4 shows $50 call values assuming 40 percent volatility. At two weeks prior to expiration, with the stock at 48, the $50 call is valued at ¾ (40 percent volatility). Selling at this price results in a 1⅞ loss. While this is less than a 2⅜ loss that would have resulted if the call were sold at the 25 percent volatility level of ¼ indicated in Table 3-3, the result is still a loss. Even though the trader bought 25 percent volatility and sold 40 percent volatility, the result was a loss; because the stock price decline had a greater effect on the call price than did the increase in volatility. This simple example makes a very important point: in the vast majority of cases, a trader's price and time forecast for the underlying is far more important than the forecast for implied volatility.

FOCUS ON IMPLIED VOLATILITY

Although a forecast for changes in implied volatility plays second fiddle to forecasting the change in price of the underlying, knowing about implied volatility may help traders improve results. Rather than focusing on overvalued or undervalued options, traders should be aware of implied volatility levels and how much they can change. Unfortunately, there are no hard and fast rules about what volatility is low and what is high. Knowledge of implied volatility changes should become part of the subjective decision-making process for option traders. There is nothing wrong, per se, with buying a high volatility or selling a low volatility, because price changes in the underlying stock are a far more important factor in option price changes. However, changes in implied volatility may be either the icing on the cake—or the vinegar in the soup: when the trade is completed, the result may be a little bigger or a little smaller because of a change in implied volatility.

SUMMARY

Volatility is a frequently used and frequently misunderstood word. The discussion in this chapter covered several important concepts about volatility. Most investors intuitively understand that volatility relates to price fluctuations in the underlying security, and this is correct. Mathematically, volatility is the annualized standard deviation of daily stock returns. Consequently, option prices imply a range of stock price movement—up or down—not a direction of movement. Higher option prices imply higher stock price volatility, which means a wider possible range of movement.

The stock price volatility implied by an option's price is referred to as the market expectation of volatility or implied volatility. Volatility is the only unknown input in the option pricing formula. Therefore, theoretical value calculations use expected volatility, a trader's estimate of future volatility.

The terms *overvalued* and *undervalued* are properly seen as a comparison of expected volatility and implied volatility. Expected volatility is the percentage used to calculate theoretical values, and implied volatility is the percentage that justifies the current market price of an option. When expected volatility is higher than implied volatility, an option will appear undervalued. When expected volatility is lower than implied volatility, an option will appear overvalued. When expected volatility is equal to implied volatility, an option will appear fairly valued.

Very often investors place too much emphasis on trying to buy undervalued options and sell overvalued options. While volatility is an important component of an option's value, the biggest component of an option's value is the price of the underlying stock. Consequently, most investors should concentrate on market forecasting and stock selection, and place less emphasis on forecasting volatility.

CHAPTER 4

OPTIONS STRATEGIES: ANALYSIS AND SELECTION

Elliot Katz

Investors who add options to their list of investment products to achieve financial goals gain a unique advantage. They increase the number of ways an investor can manage financial assets by giving him or her the power to create positions that precisely reflect his or her expectations of the underlying security and, at the same time, balance risk-reward tolerance.

This means that options increase control over financial assets by providing alternatives that were unavailable in the past. Prior to the establishment of listed options markets, there was no effective way to hedge without disrupting the allocation of assets in the portfolio. In the past, there were only two choices—Buy 'em when you like 'em, sell 'em when you don't. Today, investors discover that they can, for example, sell calls or buy puts to establish a hedge; they reserve selling for the time they no longer desire any involvement with the underlying asset. These investors also discover that options give them more control over the speculative side of their investing; they can select from strategies that run the gamut from limited to unlimited risk and from small to large capital requirements.

Many options investors spend a lot of time acquiring technical knowledge of pricing behavior. This is a worthy goal, but above all, remember that options are derivative instruments. Options cannot exist without an underlying asset. As such, there are only two important factors that direct the investor to the proper strategy. The first and foremost is the investor's opinion of the underlying security. There are strategies to take advantage of a bullish, bearish, neutral, or uncertain opinion. Any decision made without a specific opinion of the potential movement of the underlying security is unsubstantiated and has a high probability of failure. The second factor concerns the profits the investor desires if his opinion is correct and his willingness to accept any losses that accompany his strategy if his opinion is wrong.

From speculation to hedging, options can be used to construct scenarios that maximize payoff for outcomes the investor considers most likely, while controlling exposure to losses from those outcomes he considers less likely. There are ways to express the degree of one's opinion, each with its own particular risk-reward profile. One can buy calls outright to get a leveraged bullish position, or use strategies that require a smaller cash outlay but have less profit potential. This requires understanding the trade-offs that go with every decision and the motivation one brings to strategy selection.

Options are often thought of as merely speculative instruments. This is not true. An option strategy by itself is neither speculative nor conservative. The investor using options must also understand that it is the way a strategy is selected, managed, and capitalized that determines its "personality." The goal of this and following chapters is to help the investor gain the understanding to use options to their fullest.

This chapter describes essential options strategies in words and illustrations. The pictures show how each strategy evolves from the day it is established until the options' expiration date. The discussion that accompanies each diagram explains how the strategy works, why an investor might select the strategy, and how to evaluate a particular risk-reward profile. The reader is encouraged to refer to this chapter periodically for review, especially if he or she has formulated an opinion on the underlying security and is ready to enter the options market but remains unclear about the risk-reward profiles of the strategies being considered.

DESCRIPTION OF DIAGRAMS

Each strategy diagram is constructed using the following conventions:

1. The horizontal axis represents the price of the underlying security advancing from lower prices at the left to higher prices toward the right.
2. The vertical axis represents profit-loss; profit is in the area above the horizontal axis, loss below.

3. To emphasize the importance of the effects of time decay, arrows point in the direction of decreasing time until expiration. The straight-segment line represents the profit-loss of the strategy at options' expiration, the dashed line the profit-loss when the position was initiated.
4. Strike prices of options used in the strategy are indicated by roman numerals; lower strikes by lower numbers and vice versa.
5. A break-even point is always located at any point where the position plot crosses the horizontal axis.

CALL BUYING

Without question, call buying is the most popular option strategy. It is a way to put a bullish opinion into action. If the price of the underlying security advances enough, the long-call buyer can have profits many times the initial investment of the option premium. The potential to reap such large profits is what attracts investors. It is an attraction that often results in frustrating and disappointing experiences. A better understanding of the ways call buying should be used will go a long way toward lessening the possibility of disappointment from the start.

Call buyers must know from the outset what they want from their position. If they don't, they can get results very different from what's expected. For instance, if they do not want a speculative position, yet unknowingly use the reasoning of the speculative buyer, they expose themselves to much greater risks than acceptable. If the underlying security moves in their favor, their profits are much greater than they expected, and this result usually leads to the investors' thinking they chose wisely. They are then apt to continue buying calls in a speculative manner only to be extremely dissatisfied when their forecast for the behavior of the underlying is wrong. The problem is that they did not start with an honest appraisal of their motives.

This section is divided into segments that discuss the motives, expectations, and concerns that accompany speculative and nonspeculative call-buying decisions. You will soon discover why it is important to know the difference.

Expected Behavior

A call buyer is subject to profits and losses outlined in the summary and Figure 4-1, all at expiration. Throughout this section, a $50 strike price

FIGURE 4-1

Long Call

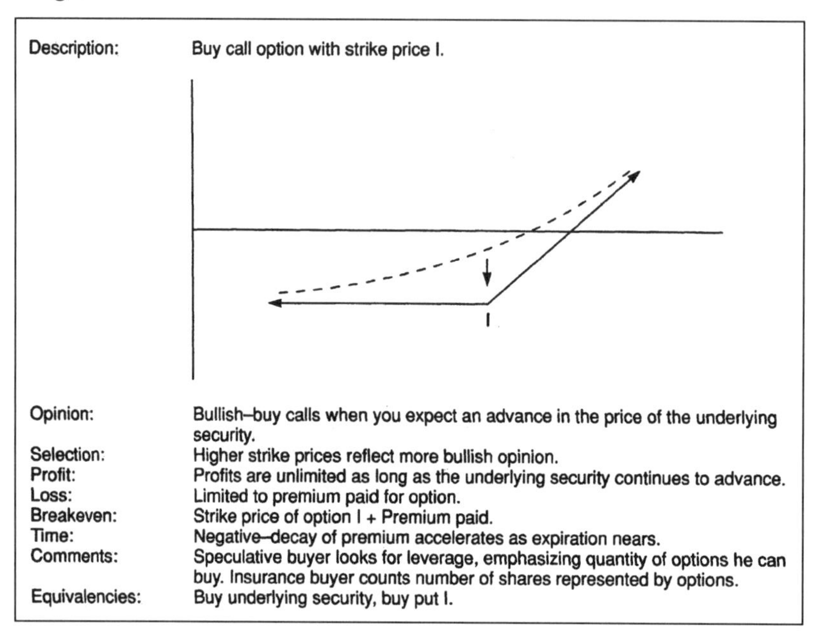

Description:	Buy call option with strike price I.
Opinion:	Bullish–buy calls when you expect an advance in the price of the underlying security.
Selection:	Higher strike prices reflect more bullish opinion.
Profit:	Profits are unlimited as long as the underlying security continues to advance.
Loss:	Limited to premium paid for option.
Breakeven:	Strike price of option I + Premium paid.
Time:	Negative–decay of premium accelerates as expiration nears.
Comments:	Speculative buyer looks for leverage, emphasizing quantity of options he can buy. Insurance buyer counts number of shares represented by options.
Equivalencies:	Buy underlying security, buy put I.

call option on XYZ stock is used as a model. As an example, start with XYZ at $50 per share and the 90-day call at $3. Given this option price, you can calculate the break-even points and maximum profit and loss at expiration as follows:

Breakeven: $53 = Strike price ($50) + Premium ($3).

Maximum loss: $3 ($300 per option) = The premium paid if XYZ is below $50 at expiration.

Maximum profit: Unlimited given by (Price of XYZ – $50 – $3).

At-expiration analysis, however, does not provide insight into what happens to a call option before the expiration date. Take a look at the expected effects of time and price movement on the $50 call option in Table 4-1.

The Break-Even Point

You can see that, regardless of when the option is purchased, the underlying security must be above its current price for a long call to break even. This is, after all, a bullish strategy, and one expects the price to rise. However, the break-even price (strike price plus premium) at expiration is simply the highest price the underlying must reach for the position to break even. Before expiration, the break-even point is lower. For example, if XYZ is $48 and you buy the $50 call for $1 15/16 with 90 days left before expiration, the option can still break even at $50 per share after 45 days. In contrast, the breakeven at expiration is approximately $52 per share.

The Speculative Call Buyer

The speculative call buyer uses options for their leverage. He is attracted to the percentage gains he can achieve on his investment from a rise in the price of the underlying. He also understands that if the underlying does not make the move he needs, he can lose most, if not all, of the investment. His goal is to select calls that balance his desire for profits with the possibility of the option's expiring worthless.

To achieve this balance, the speculative call buyer must address the issues of intrinsic value, *i.e.*, out-of-the-money or in-the-money, time decay, and delta.

In-the-Money versus Out-of-the-Money

If XYZ is $48 per share, the 90-day, $50 call is out-of-the-money. The option has a premium of $1 15/16 and an expiration break-even price of about

TABLE 4-1

Expected Price Variation of $50 Strike Call

Price of XYZ	Days to Expiration 90	75	60	45	30	15
$56	7 1/2	7 1/4	7	6 5/8	6 3/8	6 1/8
54	5 3/4	5 1/2	5 1/4	4 7/8	4 1/2	4 1/4
52	4 1/4	4	3 5/8	3 1/4	2 15/16	2 7/16
50	3	2 11/16	2 3/8	2	1 9/16	1 1/16
48	1 15/16	1 5/8	1 3/8	1 1/16	3/4	5/16
46	1 1/8	15/16	11/16	1/2	1/4	1/16

$52, 4 points above the current price. If XYZ is $52, the call is in-the-money. The option has a premium of $4¼ and a break-even price of $54¼, only 2¼ points above the current price.

Because the out-of-the-money call needs a larger advance in XYZ to reach the break-even point, buying it is more bullish than the purchase of in-the-money calls. A more bullish position is riskier. For accepting more risk, the speculative buyer of out-of-the-money calls should have better results if the stock behaves as expected. This is exactly what happens. First, the cost of the out-of-the-money option is lower, $1 15/16 versus $4¼ for the other kind. For equivalent dollar amounts, he could buy more out-of-the-money options. This is a key to understanding the thinking of the speculative call buyer. The desire for big profits leads to an emphasis on the quantity purchased. Second, an out-of-the money call performs relatively better if XYZ moves higher in price shortly after the option is purchased. After 15 days, the out-of-the money call can double in price, to $4, if XYZ rises from 48 to 52. For the in-the-money option to double, *i.e.*, trade from 4¼ to 8½ in 15 days, XYZ would have to trade up to 57½. The emphasis on quantity validates the percentage comparisons the speculative option buyer uses.

This does not mean that the out-of-the-money call is better. As time passes, the break-even point approaches the price at expiration. The prospective out-of-the-money call buyer should be aware that a decline in XYZ makes it difficult for the call to break even, even if the stock rallies. Track the call price through the following path:

1. When XYZ is 48, buy the 90-day call for 1 15/16.
2. After 30 days, XYZ is 46; the call is trading at 11/16.
3. By the end of the next 30 days, XYZ must rally to 50¾, or 10 percent, for the call to break even.

Out-of-the-money calls are on their own if the underlying makes this kind of a move; they have no intrinsic value and can lose 100 percent of their premium. In-the-money calls start with the support of their intrinsic value. The next segment addresses the trade-offs between time decay and intrinsic value.

Time Decay and Intrinsic Value

For the speculative call buyer, the advantages of buying at-the-money or out-of-the-money calls are balanced by the premium decay they experience. These types of call options lose 100 percent of their premium if the price of the underlying is not above the option's strike price at expiration. On the other hand, for in-the-money options to lose all of their premium, the underlying must have declined to the strike price at expiration. The difference is the intrinsic value carried by the in-the-money options and the impact its presence has on the decay of the total premium.

Go back to the table and concentrate on the price changes of the $50 call option if the price of XYZ were to remain at $46, 4 points out-of-the-money, for the entire 90 days until expiration. The option has no intrinsic value, and this call is trading at only 1⅛. The premium drops to 11/16 ($68.75) after 30 days. This is a loss of about 40 percent of the initial premium. Over the next 30 days, the decay is even greater. The option decays to ¼, a further decline of 65 percent.

In-the-money call options also experience premium decay over time, but less than 100 percent of the total option premium. This is because these options have intrinsic value. Only time premium can decay; intrinsic value cannot. Going back to the price table, with XYZ at 54, the $50 call is 5¾. If the stock is $54 at expiration, the call will be worth its intrinsic value of $4. This limits the loss to $1¾, or 30 percent of the original premium.

Given equivalent dollar amounts, a call option position becomes less speculative the farther the call is in-the-money. The $5¾ premium of the in-the-money one is almost twice that of the at-the-money. The dollar profit of the in-the-money option from an advance in XYZ is closer to the profit from owning the stock, but when compared to the number of at-the-money or out-of-the-money calls that can be purchased, this gain is made on fewer options. The smaller quantity and the smaller likelihood of the option's expiring worthless makes a position in in-the-money calls less speculative.

Delta and the Speculative Call Buyer

The delta of an option is defined as a local measure of the rate of change in an option's theoretical value for a change in the price of the underlying security. Delta underlies the speculative call buyer's emphasis on quantity. Table 4-2 shows the deltas of the $50 call for various prices of XYZ.

As a call goes deeper in-the-money, it becomes more sensitive to price changes in XYZ. At $54 per share, the $50 call can be expected to change about 80 cents for every point change in XYZ. At $50, the sensitivity is 59 cents. If the speculative buyer has $600, he can buy two at-the-money calls or one in-the-money call. The at-the-money position has a delta of 1.20 (2 × .60). The in-the-money call has a delta of .80.

TABLE 4-2

Deltas

Price of XYZ	Price of 90-Day Call	Option Delta
$54	$5¾	.80
52	4¼	.70
50	3	.59
48	1^{15}⁄16	.46
46	1⅛	.33
44	⅝	.21

Furthermore, the delta of the at-the-money calls can go from .59 to 1.00. The delta of the in-the-money can also go to 1.00, but it will do so starting from .80. This exemplifies the leverage the speculative buyer gets when buying more options for the same amount of money.

Conclusions about Speculative Call Buying

Speculative buyers have to balance many aspects of call behavior to create the right position. Most important of all is for investors to recognize that they are using calls speculatively. If they are concerned about how time decay, intrinsic value, and delta affect the performance of their investment, there is a good chance that they are speculative buyers. If they compare the number of options of different strike prices they can buy with a fixed number of dollars, they are definitely speculative buyers.

The Insurance Value of Calls

However, the investor who stands ready to buy 1,000 shares of XYZ at $50—but instead spends $3,000 for 10 at-the-money calls—is not speculating. He is making an investment decision to use the call options to protect a possible $50,000 commitment.

Why can call options be described as a kind of insurance? The upcoming section on put buying presents evidence that a long put plus long stock is equivalent to a call option. The protection provided by the put option is easy to comprehend: Puts increase in value as the underlying security falls farther below the strike price, protecting against part of the loss in the long position in the underlying. The insurance analogy, however, is sometimes more difficult to grasp when applied to calls. The following excerpt by Herbert Filer may bring the point home:

> It happened in November 1957, after the market had had a severe break. A man with a southern drawl and wearing a big ten-gallon hat, walked into our office and wanted to speak to the "boss." "You know," he said, "I bought a lot of your calls and I tore them up-lost my money." I thought maybe he was going to pull a gun on me . . . he said, "Don't worry—how lucky it was that I bought calls instead of stock. If I had bought the stocks way up there I would have gone broke!" [1]

Indeed, calls resemble insurance, too. The buyer wants to protect the cash he would have used to buy the underlying security and preserve the opportunity to profit from an advance. Nor is insurance free. A premium must be paid.

The "insurance investor" has a view of call options that is different from the speculative buyer's. To the insurance buyer, the premium of a call option is the cost of wanting insurance on his capital. He understands that the seller of a call option says, in effect, to the buyer, "If, for the next three months you don't want to lose money from a direct investment in this stock should it go down, and still participate if it goes up, you are going to have to compensate me for assuming this risk instead of you." Part of that compensation is a risk premium measured by the expected volatility of the stock in exchange for the ability to let the call expire worthless. The rest is an interest payment to the seller.

In-the-Money or Out-of-the-Money?

The investor who uses call options as insurance sees out-of-the-money and in-the-money options as different types of insurance policies. In contrast, the speculator sees out-of-the-money calls as an opportunity to gain more leverage by buying increasing quantities of less expensive options. The insurance buyer of an out-of-the-money call sees it as a cheap policy with a large deductible. The deductible is the amount the underlying must move for the option to be worth anything at expiration. It is the amount of the increase in the price of the underlying the investor is willing to give up. By contrast, the in-the-money call is like an expensive policy with no deductible. It is already worth its intrinsic value and lets the investor participate now. A wise insurance buyer always balances the deductible with the premium he pays.

The Real Difference

The biggest difference between the speculative and the insurance buyer is that the insurance buyer does not count the number of options he can

---

[1] Herbert Filer, *Understanding Put and Call Options* (New York: Crown Publishers, 1959).

buy. Because he is protecting a capital commitment to a predetermined number of shares, he counts the round lots of stock the options represent at expiration. He buys the same quantity, no matter which strike price he selects. The investor using calls this way has no need for leverage.

Which Approach?

Neither approach to call buying is more correct than the other. There is room in the option market for speculative and insurance call buyers. The most important thing is to understand which one you are at any given time. If you do that, you will make better decisions and will have better experiences with this strategy.

PUT BUYING

A long put position by itself expresses bearishness. The underlying security must go down in price for this position to be profitable. The investor who buys puts in expectation of a decline in the underlying security is attracted by the potential for a very large percentage gain on the premium he pays. This is the speculative approach to buying puts. However, a put option can be used conservatively. Puts can be very effective as term insurance policies to protect the value of an investment in the underlying security. Speculative and insurance buyers view put options differently. We'll explore different ways to use put buying, motives behind their use, and expected results in this section.

General Behavior

At expiration, the put buyer is subject to profits and losses as outlined in the summary and Figure 4-2. We'll use the $50 strike price put option on XYZ stock as an example, and start with XYZ at $50 per share and the 90-day put at $2. Given this option price, you can calculate the breakeven points, and maximum profit and loss at expiration as follows:

Breakeven: $48 = Strike price ($50) – Premium ($2).
Maximum loss: $2 ($200 per option) = The premium paid for the option.
Maximum profit: $48 = Strike price ($50) – Premium ($2).
Occurs if the underlying goes to $0.

To get a better feel for how put options behave over time, refer to Table 4-3, summarizing the expected effects of price and time on the premium of the $50 put.

FIGURE 4-2

Long Put

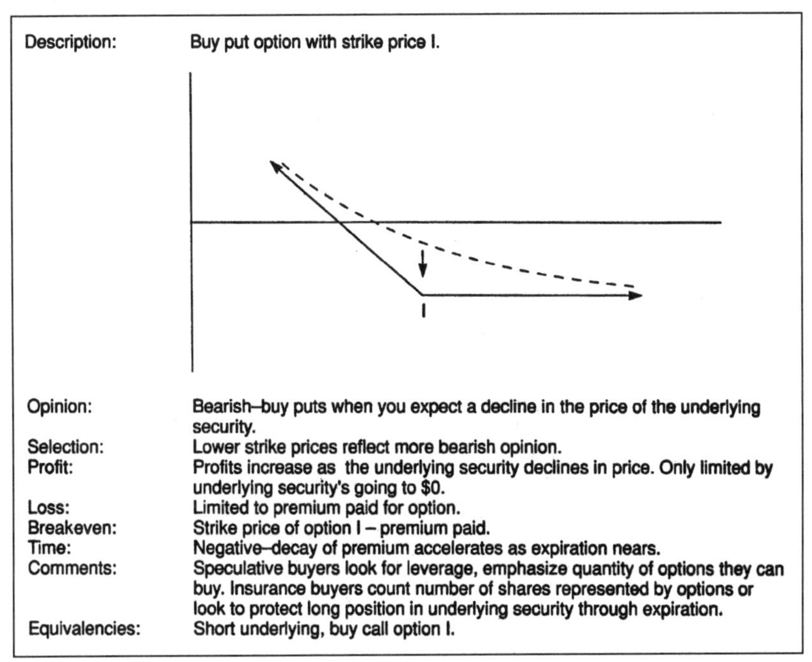

TABLE 4-3

Expected Price Variation of $50 Strike Price Put Option

Price of XYZ	Days to Expiration					
	90	75	60	45	30	15
$54	13/16	9/16	1/2	3/8	1/4	1/16
52	1 1/4	1 1/8	1	13/16	9/16	5/16
50	2	1 7/8	1 11/16	1 1/2	1 1/4	15/16
48	2 15/16	2 13/16	2 11/16	2 9/16	2 3/8	2 3/16
46	4 1/4	4 1/8	4	4	4	4
44	6	6	6	6	6	6

Many investors believe that put options behave exactly like calls, but in the opposite direction. This is an incorrect assumption. There are differences. Lack of awareness of these differences can lead to unrealistic expectations or mismanagement of put option positions.

The most obvious difference is that at-the-money calls and puts do not have the same price. The effect of dividends notwithstanding, at-the-money calls will be priced higher than at-the-money puts. The put–call parity equation shows why this is true. In the example given, XYZ does not pay a dividend, the risk-free rate is 8 percent, and the expected annual volatility is 25 percent. Based on these conditions, the 90-day at-the-money call is $3 and the put is $2.

Time Decay

The astute investor can draw some interesting conclusions regarding the behavior of put options from the pricing table. First of all, puts decay slower than calls do. In 45 days, if XYZ remains at $50 per share, the $50 put loses ½ point of time premium and is worth $1½. The $50 call, which starts with a premium of $3, loses 1 point. The put declines 25 percent; the call declines 33 percent. Over the next 30 days, the call loses an additional 15⁄16 or 47 percent. The put only loses 9⁄16 or 38 percent. This should give some solace to the out-of-the-money put buyer, but don't get too comfortable. The erosion of the premium eventually attacks the put as well. There is no escape!

In another observation, as puts go farther in-the-money, they lose their time premium faster than calls do. At $44, the $50 put, with 75 days left, has a value of $6. If XYZ were $56, the $50 call, also 6 points in-the-money, has a value of $7¼. The call retains 1¼ points of time premium.

Out-of-the-Money versus In-the-Money Puts

To the speculative buyer, the bearishness of a long put position is directly related to the amount a put is out-of-the-money. The underlying has to decline farther for an out-of-the-money put to break even at expiration than for an in-the-money put to break even. The price table shows that if you buy the $50 put when XYZ is $52, the break-even is $48¾. If you buy the same put when XYZ is $48, the break-even is $47. XYZ must decline 3¼ for the out-of-the-money, but only 1 for the in-the-money to break even.

There is always balance in the options market. The advantage the speculator gets for buying out-of-the-money puts is the potential for large percentage gains. For example, XYZ is $46 per share at expiration; the $50 put is priced at $4. The investor who went long on this put 90 days prior when XYZ was $54, paid $\frac{13}{16}$ for the option. The increase of $3\frac{3}{16}$ in the price of the put is 392 percent of the original premium. By contrast, the investor who waited 15 days and went long on the same put when XYZ had already dropped to $50, paid $1\frac{7}{8}$. The put now gains $2\frac{1}{8}$ from its purchase price, or 113 percent of the starting premium. A speculative buyer can make this percentage comparison because his or her tendency is to maximize the number of options purchased with a given amount of premium dollars to spend. Given the relationship of the premiums in the above example, the investor can purchase 23 out-of-the-money puts for every 10 at-the-money puts.

The speculative buyer making this comparison to maximize the leverage of his position must be aware that the chance of losing the entire premium increases with the distance the puts are out-of-the-money. In-the-money puts provide less leverage; they are more expensive, so the speculative buyer cannot buy so many. However, they have a greater chance of providing some profit and a smaller chance of expiring worthless. This is the omnipresent trade-off between in-the-money and out-of-the-money options with which the speculative buyer must contend.

A Strategic Application

We observed that put options go to parity more quickly than calls. This makes in-the-money put options an attractive strategic alternative to the short sale of stock. If XYZ were $46 with 90 days before expiration, the $50 put would not have much time premium remaining. Buying the put at $4\frac{1}{4}$ is similar to being short the stock—the option is very sensitive to price changes in XYZ. This position has advantages over a short sale. First, no stock needs to be borrowed and no margin balance is created. Second, should XYZ rally, the risk of the put is limited to the premium paid. A short sale of XYZ is exposed to unlimited losses.

Term Insurance

The other major use for put options is like insurance. An investor who owns XYZ stock is concerned that there can be a drop in the stock's price over the next 90 days. This conclusion may be the result of technical or fundamental analysis on the stock or on the market. Whatever the reasoning, this investor needs to protect his investment over the time period in question, but he does not want to sell the stock and give up further profits if his forecast is wrong. This investor can purchase a term insurance policy on XYZ for the next 90 days by buying the $50 put option for $2 in the same quantity as the number of round lots he owns.

The $50 put allows the holder to sell the stock at $50 per share anytime between the day the option is bought and expiration. The out-of-the-money $45 put, which allows him to sell the stock 5 points lower, is less valuable because the investor is leaving his stock open to more of a decline. The $45 put is $0.50. The in-the-money put, which allows him to sell the stock at $55 is more valuable. The investor is not picking up any part of a decline. And the $55 put is $5.

This does not mean that the $45 put is an inferior choice. If the owner of XYZ wants very low-cost insurance and is willing to accept $5 more of the risk of XYZ himself, the out-of-the-money put might be the better choice. Here is where the real decision is made. If you want more protection, you also have to give up more of the upside. The $50 put is $2. So the investor gives up the first 2 points of any rally in the stock in exchange for being able to sell the stock at $50. The $45 put has only a 50 cent upside opportunity cost.

Difference between Speculation and Insurance

The motivation of the investor using puts like insurance is different from that of the investor using puts to profit from a forecasted decline in the underlying security. The speculative put buyer has to balance the characteristics of put option behavior, time decay, intrinsic value, and delta, to come up with the proper position. The insurance buyer only needs to address the deductibility he is willing to accept, given the costs of the insurance. Furthermore, buying a put as insurance on a stock is certainly not bearish. The insurance buyer would much rather see his stock rally substantially than have to use the policy. If this investor were truly bearish, he would be wise to simply sell the stock.

An Interesting Equivalence

The reader with more options experience may have noticed that the purchase of a put, while holding a long position in the stock, is a strategy with a limited loss and (after subtracting the put premium) unlimited profit. Figure 4-3 is the profit-loss diagram of the result of owning XYZ at $50 and a $50 put bought for $2 at expiration. It looks exactly like the diagram of a call option at expiration. Indeed, a long put, long stock position is the same as a long call option position. When a put is bought on a stock as insurance, it turns that stock into a call for the life of the option.

FIGURE 4-3

Buy Stock, Buy Put

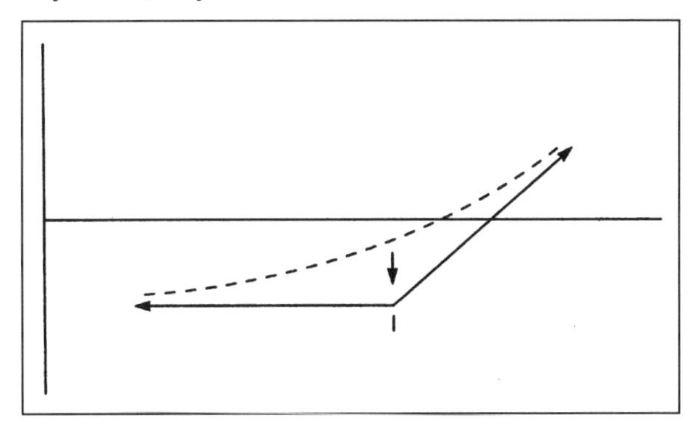

CALL SELLING

Call selling achieves its maximum profit if the price of the underlying security is below the call's strike price at expiration. However, call selling is not necessarily a bearish posture. The sold call only needs to expire worthless for the investor to realize his maximum profit, and the underlying may not have to decline to satisfy this criterion. In fact, call selling can be very successful when used as a neutral strategy. Remember that, in the options market, the opposite of a bullish strategy is not a bearish strategy. Instead, it is a *not*-bullish strategy! To illustrate, call selling is a not-bullish strategy. (The method discussed here does not involve ownership of the underlying security. This strategy, named *covered call selling*, is discussed under its own heading.)

Still using the XYZ, 90-day, $50 strike price, at-the-money call option with a $3 premium, the seller is taking a position that has the following characteristics at expiration:

Breakeven: $53 = Strike price ($50) + Premium ($3).
Maximum profit: $3 ($300 per option) =
The premium received for the option.
Maximum loss: Unlimited.

Option selling techniques are limited profit strategies. Regardless of how far below $50 the price of XYZ goes, the $3 is as good as it's going to get. The call cannot do any better for the seller than expire worthless. This is why the payoff diagram of call selling becomes horizontal as the price of XYZ declines below the strike price of the call (Figure 4-4).

FIGURE 4-4

Short Call

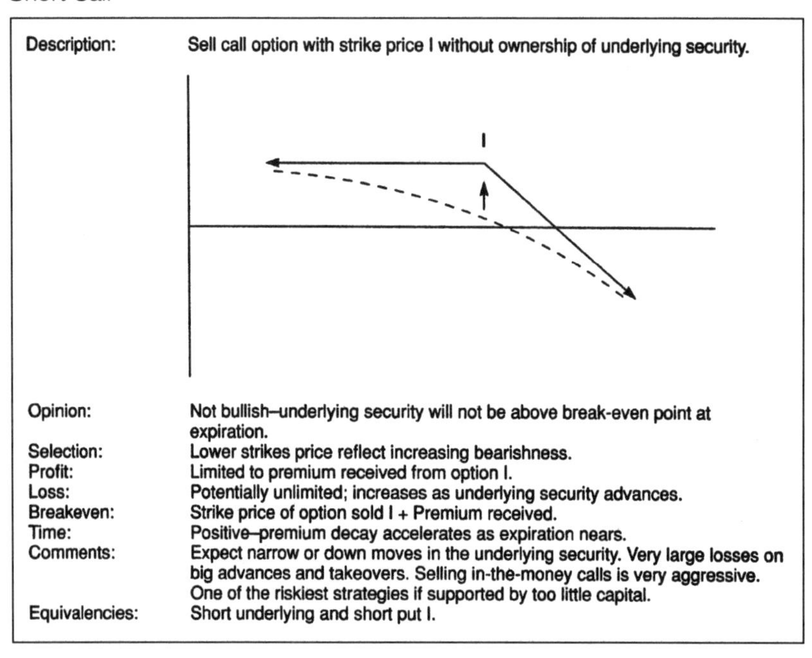

TABLE 4-4

Expiration Prices of Various Call Options

Option Strike Price	Closing Price of XYZ at Expiration				
	45	49	53	57	61
$45	0	4	8	12	16
50	0	0	3	7	11
55	0	0	0	2	6

Furthermore, a short call position is exposed to large losses if the underlying advances far beyond the strike price of the option. Examine Table 4-4 of expiration values of various call options over a range of closing prices for XYZ at expiration. One can see that the call option seller cannot be bullish!

Motivation of the Call Seller

It seems that call selling is a strategy with very high risk for a small potential return. How, then, can the call seller justify using this technique? The answer is twofold. First, he is agreeing with the option market's estimate of the potential movement of the underlying. (The call buyer is the one looking for a greater move.) You have seen that expected volatility, or price movement, has a large influence on an option's premium. The call seller believes that the potential price movement of the underlying will not be greater than that implied by the premium. If he thought that it could be greater, he would do something else!

Second, the seller is acutely aware of the negative effect of passing time on option premiums. Unlike the call buyer, the call seller eagerly awaits the last weeks prior to expiration when premium decay accelerates. In Table 4-5, here's what happens to an at-the-money call.

In-the-Money, At-the-Money, and Out-of-the-Money

The call seller must consider the relationship of the price of the underlying to the strike price of the option sold. The at-the-money call, because the strike price is close to the current price of the underlying, makes its maximum profit if the underlying stays where it is, *i.e.*, at about the strike price, and can make some profit even if the underlying advances slightly. This is because the break-even point is above the current price. The sale of an at-the-money call happens to be a neutral strategy that is also profitable if the underlying drops in price.

Out-of-the-money calls have the greatest probability of expiring worthless. Their premiums are lower, resulting in less potential profit, but the seller of out-of-the-money calls gets a cushion against a rise in the price of the underlying. It can be very enticing to sell cheap, out-of-

TABLE 4-5

Expected Price of $50 Strike Call

	Days to Expiration					
Price of XYZ	**90**	**75**	**60**	**45**	**30**	**15**
$50	3	2 11/16	2 3/8	2	1 9/16	1 1/16

the-money calls continually and watch them expire. The trouble is that they do not always expire worthless.

Consider the following. XYZ has traded at around 52½ per share for the last 120 days. During this time, an investor sold the $55 calls twice, *i.e.*, over two expirations, for a total of $275. Both times, the options expired worthless and the $275 was kept as profit.

Recently, the investor has sold another $55 call with 45 days before expiration for 1¼ ($125). The stock is at 53½. Prior to the next morning's opening, favorable news has come out on the stock. It could be a new buy recommendation or, even worse for the naked short call, a takeover bid. The stock finally opens at 57¼ and the $55 calls are 4⅛ ($412.50). Buying the short option back at 4⅛ creates a net loss for the investor after four months.

In-the-money calls have higher premiums than at- or out-of-the-money options. The sale of such an option is more bearish than the sale of either of the other two types because the strike price of an in-the-money call is lower than the current price of the underlying. The underlying must decline in price for a short, in-the-money call to reach its maximum profit. The break-even point is usually not much higher than the current price of the underlying, and it is lower than the break-even point of the at-the-money call.

Sale of an in-the-money call is an aggressive strategy that can lose quite a bit of money very quickly if the underlying begins to move higher. In-the-money calls have less time premium than at-the-money calls. If the $50 call is trading at $3, the $45 call is about $6⅜. This in-the-money call does have a high dollar premium, but it has only $1⅜ of time premium and a breakeven of $51⅜. There is not much room for error.

Another difficulty with in-the-money calls is that they become less sensitive to declines in the price of the underlying. This is because the delta of a call option decreases if the price of the underlying goes down; the in-the-money call starts to look more and more like an at-the-money call. The option continues to drop in price but loses less on each subsequent drop. Table 4-6 shows the prices and deltas of the $45 call with 75 days left before expiration. If the investor is truly bearish, there is a better strategy to express that opinion.

It should be of more than passing interest that this discussion of the differences between the three types of call options has shown that the aspects of call behavior that hurt buyers will help sellers (and vice versa).

TABLE 4-6

How Delta Changes

XYZ Price	Option Price	Option Delta
$49	$5¼	.83
48	4½	.78
47	3¾	.72
46	3	.65
45	2⁷⁄₁₆	.58

How Much Premium Is Enough?

It would seem that it is always best for the seller of naked calls to collect as much premium as possible. This could not be farther from the truth. Do not be enticed by large premiums. Higher call option premiums (the effect of interest rates notwithstanding), as a percent of the price of the underlying security, come from an expectation of higher volatility. This, in turn, implies an expectation of greater price movement. There are many investors who consistently sell calls for high premiums. They also consistently buy back in-the-money calls for *even greater* premiums. At no time should an investor put premium level above break-even analysis. The call seller must always ask this question, "Do I think the underlying will be below the break-even price of the call option I am going to sell at expiration?" If the answer is an unequivocal, "Yes," then, and only then should a call be sold.

Capital Commitment and Risk

It is important to note that naked call selling is a capital-intensive strategy. Without a position in the underlying, short calls are considered *naked*. Most firms require that a high minimum amount of equity be present in the account before allowing naked call selling to begin. There is also a margin requirement for each short call in the account, which could increase as the underlying increases.

The amount of capital used to support call selling determines the risk of the strategy. Investors who get into trouble with call selling usually do so because they short more options than they are willing to be short stock. If there is a big rise in the price of the underlying, the investor, who initially sold as many calls as his capital allowed, might be unable to support the now in-the-money calls.

PUT SELLING

A short put position is profitable if the underlying security closes above the break-even point (*i.e.*, strike price minus premium) at option expiration. The maximum profit is limited to the premium received and is achieved when the price of the underlying is above the strike price at expiration (horizontal line in payoff diagram, Figure 4-5).

Assuming the $50 strike price put with 90 days to expiration is $2, the short put has these characteristics at expiration:

Breakeven: $48 = Strike price ($50) – Premium ($2)
Maximum profit: $2 ($200 per option) at prices above $50
Maximum loss: $48 = Strike price ($50) – Premium ($2).
Occurs if the underlying goes to $0.

FIGURE 4-5

Short Put

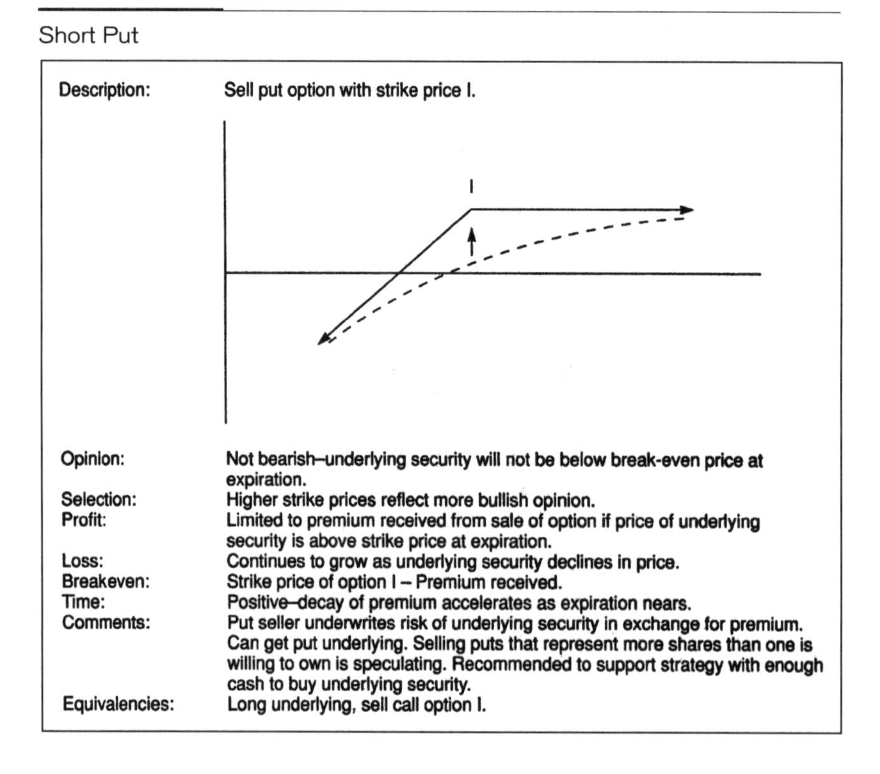

Put Selling and Insurance

Put selling can be compared to the activity of insurance underwriting. Insurance companies receive premiums from those they insure and, in return, accept certain risks. The insurance could be on a person's home, car, or life. Businesses insure inventories, plants, and equipment. Should the policyholder suffer a loss, he can make a claim and collect from the insurance company. If he never makes a claim, the insurance company keeps the premium. This is the essence of put selling. In exchange for the put option premium, the seller accepts the downside risk of the underlying security.

Insurance companies spend a lot of time and money to determine how much risk is in a policy. Their goal is to charge a premium that accurately reflects the probability that the policy will be paid off. This is no different from the options market. When an investor sells a put option, he has entered into a position that requires him to pay off the buyer of the put by accepting the stock at the strike price. The premium he receives is the option market's assessment of the potential price movement of the underlying for the time remaining to expiration of the option.

Motivation to Sell Puts

One goal of an investor who chooses to sell puts might be to act as an underwriter and collect option premiums. He makes money if the underlying is above the break-even point at expiration; however, he must be ready to accept substantial losses if the underlying declines far below the strike price. If XYZ is below $50 at expiration, the short put is worth the difference between the final price of XYZ and the $50 strike price, as shown in Table 4-7. Except for the $2 premium received, the put seller has assumed the risk of the stock's price falling below the strike price.

Another goal could be to acquire stock. A short put position can result in assignment, requiring the purchase of the underlying security at the strike price. Let's say a short XYZ $50 strike put option results in the purchase of 100 shares of XYZ stock at $50 per share. The $2 premium received from selling the put lowers the purchase price of the XYZ stock to $48. Assignment of a short put always results in the purchase of stock at a net price that is the break-even point. Should XYZ move beyond $50 at expiration, the put expires worthless and the $2 premium is kept as a consolation profit.

Regardless of his motives for selling puts, it is imperative that the put seller not be bearish. No investor wants to buy a stock he doesn't like.

TABLE 4-7

Expiration Profit/Loss of Short Put

XYZ at Expiration	Value of $50 Put	Profit/ (Loss)
$50	$ 0	$ 200
47	300	(100)
44	600	(400)
41	900	(700)
35	1,500	(1,300)

The Lost Opportunity

There is a second break-even price on the upside for the put seller. From our example, the upside break-even point is $52. A position in XYZ or the short put has a $2 profit if XYZ is $52 per share at expiration. Above $52, a long position in XYZ stock continues to earn profits. The put profit of the short put is limited to the $2 premium.

There is no financial loss above the upside break-even price; but there is an opportunity loss. The upside break-even price is the figure beyond which a long position in the underlying does better than the short put. The investor who wants to benefit from an increase in the price of the underlying above the upside break-even price may not find put selling the proper strategy. As for any option selling strategy, the expiration price range defined by the financial and opportunity break-even prices should match or encompass the investor's expectations.

Speculative or Conservative?

The investor who stands ready to purchase the underlying security is not speculating. This investor has the capital necessary to establish a long position in the underlying security. The assignment making him long the stock may not be his goal, but it is not a cataclysm either.

Put selling becomes speculative when the investor sells puts that represent more shares than he is willing to own. He can do this because there is only an initial margin requirement for each short put. (For margin purposes, a short put is considered uncovered regardless of the amount of capital supporting the activity.) This investor has the possibility of a bigger loss than he is prepared to accept. That is speculation.

COVERED CALL SELLING

A *covered call* combines a short call with a long position in the underlying security. The two covered call strategies—*covered writing* and *overwriting*—are explained here. Covered call selling, whether via buy-writes or overwrites, allows an investor to profit from a correct forecast of the limit of price performance of the underlying.

In exchange for receiving the option premium, the covered call seller has limited the profit he can realize from a gain in the price of the security on which he has sold the call. The call buyer can take possession of the underlying at the strike price via exercise, so a covered call seller makes his maximum profit if the underlying security is above the call's strike price at expiration. The profit is the option premium received plus the gain from the advance, if any, of the price of the underlying up to the strike price. If the sold call is in-the-money, the loss from selling the underlying below its purchase price must be subtracted. The option premium also provides a partial hedge against a decline in the price of the underlying (Figure 4-6).

Covered Writing

When an investor buys stock and sells call options simultaneously, the strategy is called *covered writing* or *buy-writing*. Say that an investor buys 500 shares of XYZ stock at $40 per share and at the same time sells five of the 90-day $40 calls for $2¼ each. The total premium, excluding commissions, is $1,125. The cost of the stock, also excluding commissions, is $20,000. At expiration, his profit-loss can be seen in Table 4-8.

Above $40, at expiration, the profit of the covered write is limited to the premium received for the options. Below $40, the position loses $1,125 less than the stock investment. Whenever the call is assigned, the covered writer sells XYZ at $40 and realizes the maximum profit of $1,125.

Profits or Protection?

Like any other option strategy, the covered write can be used to express varying degrees of one's opinion. At-the-money buy-writes are neutral. They realize most, if not all, of their profit from the option premium.

Out-of-the-money buy-writes are bullish. Say, for example, that the stock was purchased at $37½ and the option sold for about $1⅛. The option premium is lower because the call is out-of-the-money. However, the position has greater potential profit than the at-the-money buy-write because there is profit to be made on the advance of XYZ to the strike price. The maximum profit on this buy-write, for 500 shares, is $1,812.50.

FIGURE 4-6

Covered Call

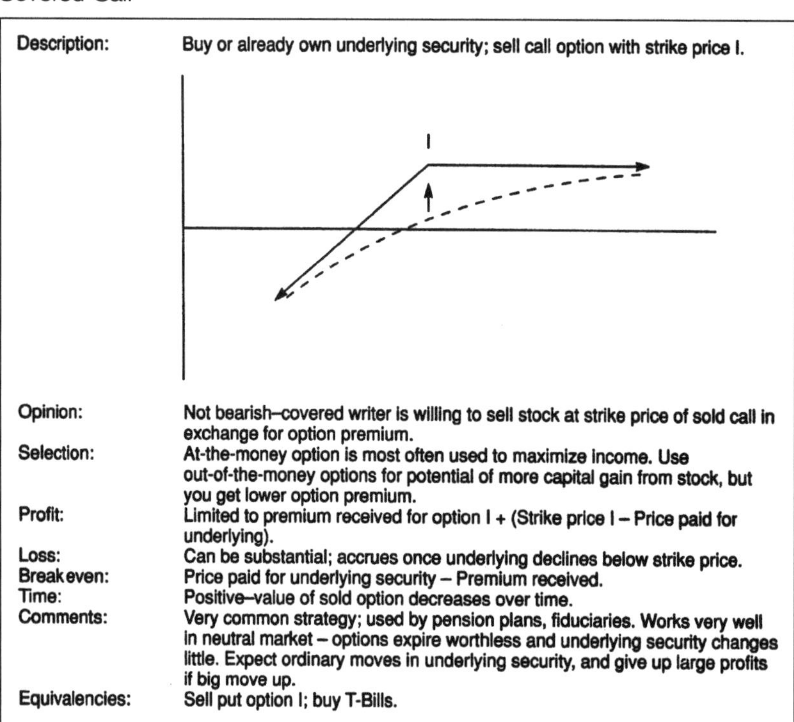

TABLE 4-8

Profit/Loss

Price of XYZ	Value of $40 Call	Profit/(Loss) on Calls	Profit/(Loss) on Stock	Total Profit/(Loss)
$30	$ 0	$1,125	$(5,000)	$(3,875)
35	0	1,125	(2,500)	(1,375)
39	0	1,125	(500)	625
40	0	1,125	0	1,125
41	1	625	500	1,125
45	5	(1,375)	2,500	1,125
50	10	(3,875)	5,000	1,125

The option premium provides $562.50; the stock, $1,250. In the first example, the stock provided no additional profit.

Another difference in the two positions is that the out-of-the-money provides less protection against a decline ($1⅛ per share versus the at-the-money's $2¼ per share). This is always the trade-off in selecting a covered write in a stock. The potential for larger profits is balanced by the smaller premium.

How Much Premium?

High premiums alone do not make a good covered write. Option premium levels and risk, as defined by volatility, are directly related. If there are two $25 stocks, HVL (high volatility) and LVL (low volatility), their respective 120-day call premiums might be $2⅛ and $1½, representing 8½ and 6 percent of the stock price. The covered write on HVL is not necessarily the superior one. Its possible return is higher, but it comes with the additional risk of owning a stock that could have a wider range of price movement than LVL.

Your main criterion should be your opinion of the underlying. If you do not want to own the underlying if it is not called away, you should not purchase it for the sole purpose of collecting high option premiums.

Overwriting

The other type of covered call selling is *overwriting*. In this strategy, the investor already owns the underlying security and now sells call options. He or she may have bought the security yesterday, six months ago, or six years ago.

The investor now feels that the probability of a move in the price of the underlying is small. Without options, this investor can take only one of two possible actions—sell it or hold it. Only selling allows him to extract any money from his holding.

Overwriting provides a third alternative. This investor can select among expirations to balance the premium received, the selling price *if called*, and the timeframe over which he expects the underlying to underperform. There can be 50-, 80-, and 140-day call options available with strike prices 2 points and 7 points out-of-the-money. For a stock at $53 per share, the following call options and prices might exist:

Option Strike Price	50-Day Call	80-Day Call	140-Day Call
$55	1⅜6	1¾	2¾
60	¼	9⁄16	1 7⁄16

Whichever option he chooses, the overwriter must be aware that, in exchange for the premium received, he has *sold* the right of ownership of his stock at the strike price to *someone else*. He has capped the appreciation of the stock at the strike price. The premium is not "free money." It comes with an opportunity cost that prevents additional profits above the strike price plus premium and a financial cost of providing only as much protection as the premium. The overwriter concludes that the financial and opportunity break-even prices given by the call he sells limit the range of prices he expects.

Covered or Naked?

Addition of a long position in the underlying security makes the covered call seller very different from the naked call seller. That is, the investor who sells naked calls has a position with limited profit if the underlying is below the strike price and unlimited risk if it rallies substantially. He is neutral with a bearish bias. The covered call writer has limited profit if the underlying is above the strike price and risk if the underlying declines substantially. He is neutral with a bullish bias.

The difference is the price range over which the investor wants to take his risk. It is very important that an investor with a neutral opinion also consider on which side of the market he is willing to take risk if he is wrong.

Covered Writing and Put Selling

Compare the expiration profit/loss diagrams of covered call selling and put selling. They have the exact same shape. This can come as a surprise to many investors who feel that any option selling strategy is extremely risky.

If covered call writing is the equivalent strategy, it should have the same characteristics as put selling:

1. The covered call seller is willing to forgo any additional profits from the stock on movements beyond the strike price. So is the put seller.
2. The put seller underwrites the risk of the stock through expiration. In covered call selling, in exchange for the premium of the call option, the investor maintains the risk of the underlying.
3. The break-even point of both strategies is lower than the current price of the underlying.

BULL SPREADS

A *bull call spread* is created by purchasing one call option and simultaneously selling another call option with a higher strike price, as in the following example (XYZ at 80¾).

Buy 1 XYZ 90-day $80 call	$5¼
Sell 1 XYZ 90-day $85 call	− 2⅞
Net cost of bull call spread	$2⅜

A bull call spread is a combination of two options—one long, one short. The result is a position that has limited risk and profit. The limited risk is the net premium paid for the spread. If XYZ is below $80 at expiration, both call options will expire worthless and the investor will lose the $2⅜ he paid for the spread. The profit from the spread is limited because the short $85 call has value if XYZ is above $85 at expiration. This value erases any additional profits gained from the long $80 call. In Table 4-9, you can see what the $80 to $85 bull call spread looks like at expiration. This example demonstrates that a spread can only be worth as much as the difference between the strike prices of the options. For prices of XYZ between $80 and $85, the $80 call advances without any offsetting effect from the $85 call. Above $85, this effect limits the spread's value. Table 4-9 shows the expiration points of maximum profit, and breakeven. The maximum profit is $2⅝ ($262.50) per spread—the maximum value of the spread, $500, minus

TABLE 4-9

Bull Spread

Price of XYZ	Value of Long $80 Call	Value of Short $85 Call	Value of Spread
$75	$ 0	0	$ 0
80	0	0	0
81	100	0	100
82	200	0	200
83	300	0	300
84	400	0	400
85	500	0	500
90	1,000	($500)	500
95	1,500	(1,000)	500

FIGURE 4-7

Bull Spread

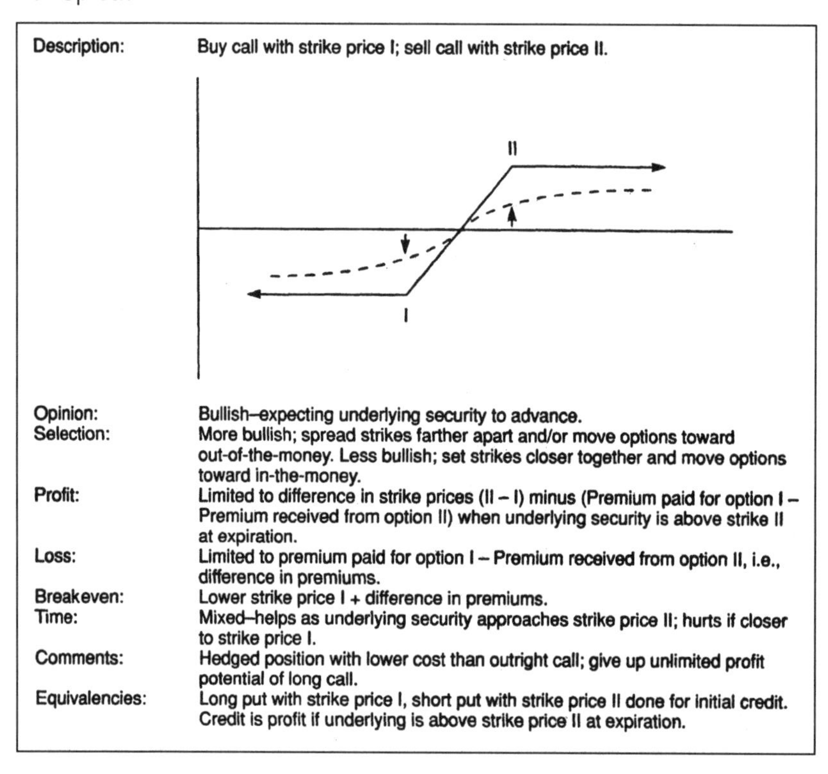

Description:	Buy call with strike price I; sell call with strike price II.
Opinion:	Bullish–expecting underlying security to advance.
Selection:	More bullish; spread strikes farther apart and/or move options toward out-of-the-money. Less bullish; set strikes closer together and move options toward in-the-money.
Profit:	Limited to difference in strike prices (II – I) minus (Premium paid for option I – Premium received from option II) when underlying security is above strike II at expiration.
Loss:	Limited to premium paid for option I – Premium received from option II, i.e., difference in premiums.
Breakeven:	Lower strike price I + difference in premiums.
Time:	Mixed–helps as underlying security approaches strike price II; hurts if closer to strike price I.
Comments:	Hedged position with lower cost than outright call; give up unlimited profit potential of long call.
Equivalencies:	Long put with strike price I, short put with strike price II done for initial credit. Credit is profit if underlying is above strike price II at expiration.

the initial cost of $237.50. Maximum profit is realized at every price of XYZ above $85 at expiration. The break-even point is $82⅜. At $82⅜, the long $80 call is worth $2⅜, and the short $85 is worth nothing. The break-even point of a bullish call spread is the strike price of the long call plus the net premium paid for the spread. The bull call spread strategy is drawn in Figure 4-7, showing the limited loss, limited profit characteristics.

Motivation of the Bull Spreader

The bull spread is used by the bullish investor who may not be entirely comfortable with a long call position. He may not be bullish enough to warrant the purchase of a call option straight out. He may have a target price (thus being willing to forgo additional profits from any advance beyond that target) or he may be uncertain as to the size of the advance, hesitating to "pay" to profit from every price of XYZ above the break-even point of a long call. Given these opinions, a hedged strategy is ap-

TABLE 4-10

Expected Value of $80–$85 Call Spread

Price of XYZ	Days to Expiration 75	60	45	30	15
$78	1 11/16	1 9/16	1 7/16	1 3/16	13/16
82	2 1/2	2 1/2	2 1/2	2 7/16	2 3/8
86	3 5/16	3 3/8	3 1/2	3 5/8	3 15/16
90	3 15/16	4 1/16	4 1/4	4 7/16	4 3/4

propriate. The bull call spread is the particular hedged option position that satisfies this investor.

Many investors with bullish opinions delay buying calls when premiums seem a little too rich for them. The decision to wait can result in a lost opportunity. The bull spread lets these investors establish a bullish position now. Later on, naked long calls can be purchased. There's nothing wrong with starting with a hedged position.

The Trade-off of Spreading

The major trade-off of spreading is the elimination of the unlimited profit potential that goes with a long option position. In the example, the cost of the spread versus the long $80 call option is lowered because of the premium received from the short $85 call option. Therefore, the call spread has less risk, on a dollar-for-dollar basis, than the long call. For this lower cost, the buyer of the spread gives up additional profits from every price of XYZ above $85 per share at expiration.

There is another important trade-off besides the limitation of profits. It is the time trade-off. Table 4-10, summarizes the performance of the $80 to $85 spread over different prices and time. If XYZ advanced to $90 per share in 15 days (75 days to expiration), the spread is worth about 3 7/8 and the investor has to wait 75 days for the remaining 1 1/8. The $80 call, on the other hand, is trading at 11 7/8. The spreader trades in immediate performance for a lower cost of entry.

Types of Bull Call Spreads

Some bull call spreads are more bullish than others. Bullishness is determined by the strike price of the short call. This determines how much the underlying has to move for the spread to achieve its maximum value. For example, the $85 to $90 call spread is more bullish than the $80 to $85 call spread. The more bullish spread is more likely to expire worthless, but it can be more profitable. The $85 to $90 call spread costs only $1 7/16 and has a maximum profit of $3 9/16.

The name *bull spread* can be somewhat inaccurate. Returning to XYZ at $80 3/4, the $75 can option is $8 1/2. The bus spread of a long $75 call and a short $80 call will cost $3 1/4 ($8 1/2 – $5 1/4). This 5-point spread has a maximum profit of $1 3/4 (5 – 3 1/4) if XYZ is at least $80 at expiration. The stock is currently $80 3/4. This is a bull spread that is profitable, even if XYZ does not move. The spread works because the time premium in the short option is greater than that in the long option. The time premium, however, makes the spread a position that does not reach its maximum profit if the underlying advances quickly. The full advantage of selling more time premium than is bought is not gained until the expiration date is very close.

But the *spread* between the options' strike prices can be larger than 5 points (for stocks with fractional strike prices, spreads less than 5 points can be possible). A $75 to $85 call spread costs $5 5/8 ($8 1/2 – $2 7/8). Breakeven is $80 5/8. The distance between the strikes is 10 points and, therefore, the maximum profit is $4 3/8 if XYZ is above $85 at expiration ($10 – $5 5/8).

The break-even point of this spread is very close to the current price of the stock—$80 3/4. At expiration, the spread has the same results as the stock between $75 and $85. Below $75, the spread is worthless and the loss is $562.50; the owner of XYZ stock continues to suffer losses. Above $85, the spread is worth $10 and the profit is limited to $437.50. The owner of XYZ stock continues to profit. This spread might be suitable for an investor who wants to profit from a rally in the stock to at least $85 over the next 90 days, but who does not want to be subject to additional losses if XYZ drops below $75. Figure 4-8 compares the spread to owning the stock or the $75 call.

The Bullish Put Spread

In the put market, the bull call spread has an equivalent—sell the higher strike put and buy the lower strike put. This creates a net credit because the premium of the sold put is higher than that of the purchased put. The maximum profit of this position is the credit. If the price of the underlying is above the strike price of the short put at expiration, both options expire worthless. On the risk side, if the price of the underlying is below the lower strike price at expiration, the seller of this spread can lose the difference in the strike prices minus that initial credit.

FIGURE 4-8

Comparison of Spread, Long Call, and Long Stock

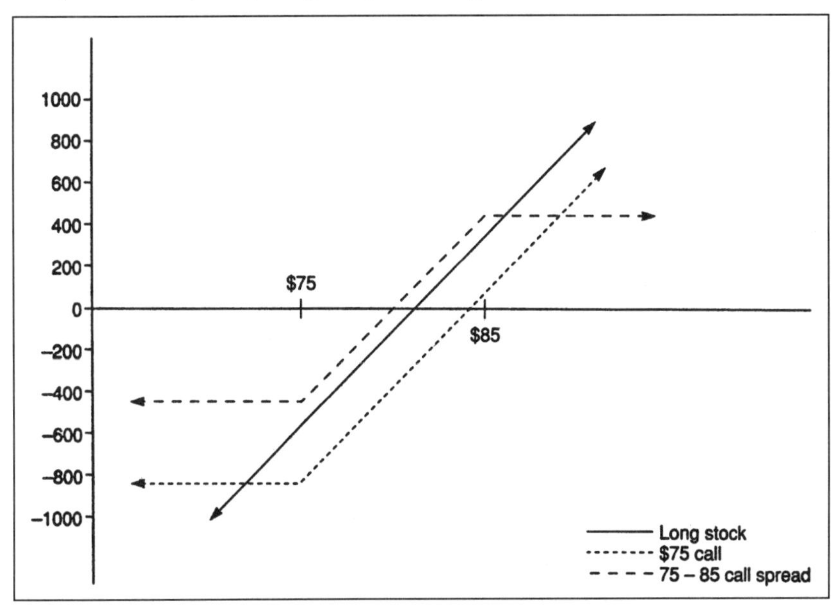

Selling the $85 put and buying the $80 put is a *bullish put spread.* With XYZ still at 80¾, the prices of 90-day puts would be $5⅜ and $2⅞, respectively. The "sale" of the put spread creates an inflow of $2½ ($250) per spread. If XYZ is above $85 at expiration, both options are worthless and the investor retains the $250 as profit. Below $80, the $85 put is always worth 5 points more than the $80 put. The 5 point maximum value of the spread results in the maximum loss of $2½. Recall that the call spread cost $2⅜ and had a maximum profit of $2⅝.

The two positions are equivalent. The reason for the smaller profit in the put spread is explained in the discussion of the box spread arbitrage in Chapter 8.

The bullish *short put* spread can have a distinct advantage over the bullish *long call* spread. Because the long call spread results in an initial debit, it must be paid for in full when it is established. The short put spread results in an initial credit. Under current margin rules, it is only necessary for the risk of the spread, *i.e.*, the credit received minus its maximum value, to be available in the account. There need not be an initial cash outlay if the margin account has excess equity.

Two Special Risks of Early Assignment

Dividends concern the call spreader. If the options of a call spread are far enough in-the-money the day before the ex-dividend date of the stock, the short call can be assigned. There is no stock risk because the long option can be exercised the next day. However, the stock that is delivered against the assignment is delivered cum-dividend, *i.e.*, with the dividend. The stock that is purchased via exercise is purchased ex-dividend. The risk to the investor in this situation is the dividend. This is a sudden additional cost that lowers his profit in the position.

The second special situation that a spreader should be aware of involves spreads of American index options (see Chapter 2). If the short option in an index spread of this type is assigned early, the cash settlement mechanism of index options creates a debit in the account for the amount that the short option is in-the-money at the end of that business day, adjusted by the multiplier of the index. When the long option is exercised, the amount credited to the account will be determined by the settlement value of the index on the DAY it is exercised. There is a full one day's risk if the long option is not sold at some time during the next trading day.

BEAR SPREADS

The *bear spread* is commonly established with puts. It is a hedged strategy consisting of a long put and a short put with a lower strike price. The cost of the spread is the difference between the premium paid for the long option and the premium received for the short option:

Buy 1 XYZ 90-day $80 put	$2½
Sell 1 XYZ 90-day $75 put	– 1
Net cost of bear put spread	$1½

Like the bull call spread, the bear put spread is a position that has limited risk and limited profit (Figure 4-9). Risk is limited to the net premium paid for the spread. If XYZ is above $80 at expiration, both put options expire worthless, and the investor loses the $1½ he paid for the spread. The profit from the spread is also limited because the short $75 put has value if XYZ is below $75 at expiration. This value erases any additional profits gained from the long $80 put. The $80 to $75 bear put spread in Table 4-11 looks like this at expiration.

The value of the spread at expiration, and hence, its profit, is limited at prices of XYZ below $75. The value of the $80 put is unaffected by the $75 put until the latter is in-the-money. The spread is worthless for prices above $80. Table 4-11 also shows that the break-even point for a

FIGURE 4-9

Bear Spread

Description: Buy put with strike price II; sell put with strike price I.

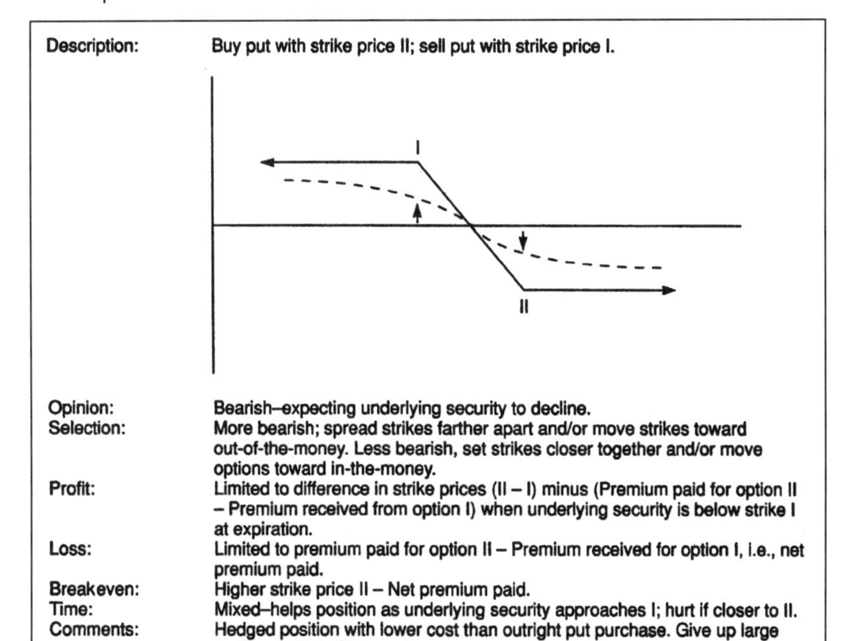

Opinion: Bearish–expecting underlying security to decline.

Selection: More bearish; spread strikes farther apart and/or move strikes toward out-of-the-money. Less bearish, set strikes closer together and/or move options toward in-the-money.

Profit: Limited to difference in strike prices (II – I) minus (Premium paid for option II – Premium received from option I) when underlying security is below strike I at expiration.

Loss: Limited to premium paid for option II – Premium received for option I, i.e., net premium paid.

Breakeven: Higher strike price II – Net premium paid.

Time: Mixed–helps position as underlying security approaches I; hurt if closer to II.

Comments: Hedged position with lower cost than outright put purchase. Give up large profit potential of long put.

Equivalencies: Long call with strike price II; short call with strike price I done for initial credit, which makes a profit if underlying security is below strike price I at expiration.

TABLE 4-11

Bear Spread

Price of XYZ	Value of Long $80 Put	Value of Short $75 Put	Value of Spread	Profit/(Loss) in Spread
$85	$ 0	$ 0	$ 0	($150)
80	0	0	0	(150)
79	100	0	100	(50)
78	200	0	200	50
77	300	0	300	150
76	400	0	400	250
75	500	0	500	350
70	1,000	(500)	500	350
65	1,500	(1,000)	500	350

bear put spread is the strike price of the long put minus the net premium paid to establish the spread. In this example, the break-even price, at expiration, is $78½.

Motivation of the Bear Spreader

The rationale for using the bear spread is similar to that for the bull spread. Investors feel the cost of entry into a long put position is too high, but they want to establish a position now. Investors who may have a price target on the underlying are willing to give up the substantial profits that accompany a long option position. The difference is that these investors select a bear spread when they expect the underlying to decline.

Trade-Offs of Bear Spreading

Besides limited profits, there is a performance trade-off because of time. Before expiration, the time premium remaining in the options prevents the spread from reaching its full value if the underlying declines soon after the spread is purchased. When choosing a spread over a naked long position, the spread needs to be held. Table 4-12 shows the value of the $80 to $75 put spread for various prices of XYZ in 15 day intervals.

If XYZ were to decline to $74 with 60 days remaining to expiration, the $80 put would be worth $6⅛ and the $75 put, $2 15/16. The spread would be worth $3 3/16. It takes the entire 60 days for the spread to gain the remaining $1 13/16. The spreader trades in immediate performance for a lower cost of entry into a position!

TABLE 4-12

Bear Spread

Price of XYZ	Days to Option Expiration				
	75	60	45	30	15
$82	1⅜	1 5/16	1 3/16	1	11/16
80	1¾	1¾	1 11/16	1 9/16	1 5/16
78	2 3/16	2 3/16	2 3/16	2 3/16	2 3/16
76	2⅝	2 11/16	2¾	2⅞	3⅛
74	3 1/16	3 3/16	3 5/16	3⅝	4 1/16

Types of Bear Put Spreads

A bear spread is not limited to 5 points. The $85 to $75 put spread costs about $4½. The break-even price for this position at expiration is $80½, *i.e.*, strike price $85 minus $4½. If XYZ is below $75 at expiration, the spread is worth 10 points. It is worth comparing this spread to the $80 to $75 spread in which both options are out-of-the-money. The profit of each spread is truncated when the stock reaches $75 at expiration. The out-of-the-money spread has a maximum value of $5 and a maximum profit of $3½. The $85 to $75 spread has a maximum value of $10 and a maximum profit of $5½.

Spreads can be used speculatively. For example, if the investor places similar dollar amounts into each spread, the $80 to $75 spread in which both options are out-of-the-money is a more speculative position and should have a greater potential return. This out-of-the-money spread can have a 233 percent profit ($3½ on $1½). The spread is worthless if XYZ is above $80 at expiration. The 10-point spread can have a 122 percent profit ($5½ on $4½), breaks even at $80½, and only expires worthless if XYZ is above $85 at expiration.

Taking the comparison one step farther, the $85 to $80 put spread costs $2⅝. The break-even point of this spread is $82⅜, and the maximum profit is 90 percent ($2⅜ on $2⅝). However, the stock is only $¾ above the price at which the spread achieves its maximum value of 5. As a position that is less bearish than either the $80 to $75 or the $85 to $75, this spread should have less potential than either, and it does. The three positions are compared in Figure 4-10.

The bearishness of a put spread is determined by the strike price of the short option. The farther out-of-the-money that put is, the farther the underlying must drop to bring the spread to its maximum value. The most aggressive spreads have both options out-of-the-money and a large difference in their strike prices. The least aggressive spreads have both options in-the-money and a small difference in their strike prices.

The Bearish Call Spread

Calls can be used to establish *bear spreads* just as puts can be used to create bull spreads. Let's go back to the $75 to $80 bull call spread from the previous section. The $80 call was $5¼ and the $75 call was $8½. The bear call spread is the result of selling the $75 call and buying the $80 call for a credit of $3¼. Should XYZ be below $75 at expiration, both options expire worthless. If XYZ is above $80, the spread is worth $5, resulting in

FIGURE 4-10

Comparison of Three Put Spreads

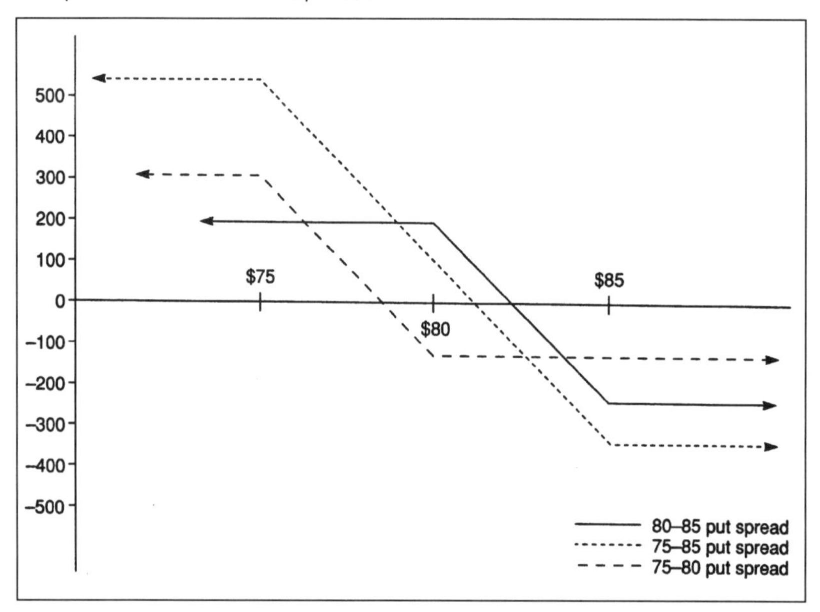

a loss of $1¾. The put spread had a maximum profit of $3½ for an investment of $1½. Both positions are equivalent. As a credit spread, the risk of the position has to be available and can usually be satisfied with excess margin.

Summary of Spreading

The general rules of spreading are simple:

1. A spread can only be worth as much as the difference of the strike prices of the options that define it.
2. If a spread is purchased, the maximum profit is the maximum value of the spread minus the net amount paid. The maximum loss is the amount paid.
3. If a spread is sold, the maximum profit is the net amount received. The maximum loss is the maximum value of the spread minus that amount.
4. Spreads must be done in a margin account. If a spread is sold, the difference between the credit received and the maximum value of the spread must be available.

LONG STRADDLE

A *long straddle* consists of a long call and a long put with the same strike price and expiration date. By now you know that long calls are profitable if the underlying goes up, and long puts are profitable if the underlying goes down. A long straddle is profitable if the underlying goes up or down substantially.

Motivation

The long straddle is best suited to an environment in which the investor feels a large move is possible, but is unsure as to its direction. Consider a long straddle in a situation where an upcoming earnings report on a stock is expected, or the report of an economic statistic which can move the market is due. A technician can buy a straddle on a stock that has reached a very significant support or resistance level and is expecting the stock to either punch through that level or bounce right off it. In each case, the underlying could respond with a large move in either direction.

At Expiration

XYZ is at $80. The 90-day $80 call is $4¾; the 90-day put is $3¼. The $80 straddle costs $8 ($800 per straddle). Table 4-13 shows what the profit/loss looks like at expiration. The graph of this table gives the characteristic *V* shape profit-loss profile of the long straddle. It appears to straddle the strike price (see Figure 4-11).

The break-even points of a straddle at expiration are the strike price plus or minus the total premium paid for the call and the put. In this example, the break-even points are $72 ($80 strike price minus $8 cost) and $88 ($80 strike price plus $8 cost).

Risks of the Long Straddle

The risk of this strategy is quite large. Should the underlying not budge, both premiums are lost. However, the maximum loss only occurs at one price at expiration—the strike price of the options. Any move away from that single price immediately begins to lessen the loss because one of the options will have value.

Because of the opportunity to make money from a move in either direction, this strategy can seem too good to be true. In a way, it is. A long call option has a break-even point of the strike price plus the premium; a

TABLE 4-13

Peculiar Straddle Shape Evident

Price of XYZ	Long $80 Call	Long $80 Put	Total Value of Straddle	Profit/(Loss) on Straddle
$65	$ 0	$1,500	$1,500	$700
70	0	1,000	1,000	200
75	0	500	500	(300)
80	0	0	0	(800)
85	500	0	500	(300)
90	1,000	0	1,000	200
95	1,500	0	1,500	700

FIGURE 4-11

Long Straddle

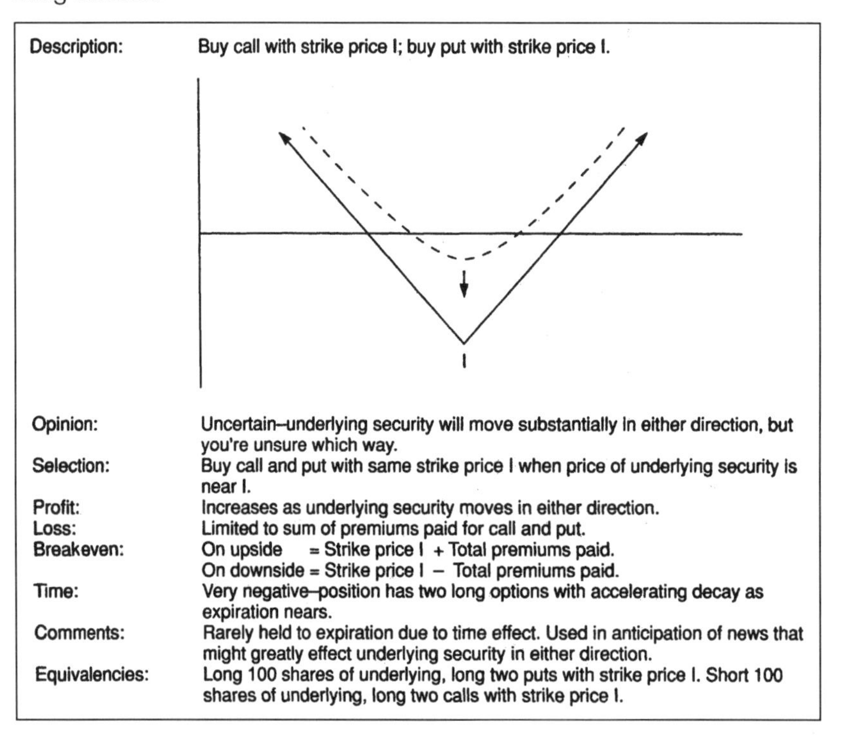

Description: Buy call with strike price I; buy put with strike price I.

Opinion: Uncertain–underlying security will move substantially in either direction, but you're unsure which way.
Selection: Buy call and put with same strike price I when price of underlying security is near I.
Profit: Increases as underlying security moves in either direction.
Loss: Limited to sum of premiums paid for call and put.
Breakeven: On upside = Strike price I + Total premiums paid.
On downside = Strike price I − Total premiums paid.
Time: Very negative–position has two long options with accelerating decay as expiration nears.
Comments: Rarely held to expiration due to time effect. Used in anticipation of news that might greatly effect underlying security in either direction.
Equivalencies: Long 100 shares of underlying, long two puts with strike price I. Short 100 shares of underlying, long two calls with strike price I.

long put, the strike price minus the premium. The breakeven points for the long straddle are the strike price plus and minus the sum of both premiums. A long straddle is the same as buying a call for the price of a call and a put, and buying a put for the price of a put and a call. At expiration, the underlying must have moved enough in either direction to compensate for the total premium paid. In the example, XYZ must be more than 10 percent away from its current price for the position to be profitable. The long straddle buyer must honestly assess the chances of the underlying to make a move of the magnitude required by the break-even prices.

Before Expiration

If time is the enemy of long option strategies, it is literally twice as bad for the long straddle.

The long straddle has two clocks running against it, one for the call and one for the put. It should not be a surprise that the long straddle is very sensitive to time decay. Table 4-14 shows the expected value over time of the 90-day, $80 straddle, purchased for $8 as the price of XYZ changes. Because of a long straddle's extreme sensitivity to time, the investor long a straddle can decide to take profits if the underlying makes a large move soon after the straddle is purchased. If XYZ runs up to $90 in 15 days, the straddle is worth $12¼. If it remains at $90, the position

TABLE 4-14

Expected Value of $80 Straddle

Price of XYZ	*Days to Option Expiration*				
	75	60	45	30	15
$95	16⅝	16¼	15⅞	15½	15¼
90	12¼	11¾	11¼	10¾	10¼
85	9	8¼	7½	6⅝	5½
80	7¼	6½	5½	4½	3¼
75	7½	7	6⅜	5¾	5¼
70	10⅝	10½	10¼	10	10
65	15⅛	15⅛	15	15	15

begins to lose time premium. The investor who waits for a continuation of the move finds that he has given back some of that initial profit.

Another reason to consider taking profits is that the position no longer reflects the investor's initial intentions. The straddle now consists of an in-the-money call and an out-of-the-money put. The investor would much rather see XYZ continue to rally. Of course, another 5-point rise in XYZ increases the value of the straddle to twice its purchase price, but the position is also susceptible to a pull-back of part of the move.

Buying a straddle close to expiration does not solve the time decay problem. As expiration approaches, there is less time for the underlying to make a move. If the expected volatility of the XYZ has not changed, the $3¼ cost for the 15-day straddle accurately reflects this.

Before rushing to buy this straddle close to expiration ask yourself, "Do I think that XYZ will be above $83¼ or below $76¾ in 15 days?"

Which Stocks?

One might conclude that the long straddle is best for stocks that show a tendency to make larger moves. However, it may not be wise to base one's choice solely on that basis. As you have seen numerous times, the expected movement in a stock affects option premiums directly via a higher volatility number in the pricing model.

The XYZ $80 straddle in the example had an $8 total premium. If XYZ were a more volatile stock, the premiums of the call and put would be higher. This effect increases the upside break-even point and lowers the downside break-even point by the amount of the additional premium paid. At expiration, XYZ will need to have moved farther for the straddle to be profitable.

LONG STRANGLE

The *long strangle* is very similar to the long straddle. Like the straddle, it is a strategy that reflects one's forecast of a large move in the underlying in either direction. The difference between a strangle and a straddle is that the strike prices of the call and put that make up the strangle are not the same.

At Expiration

If the underlying is between strikes, there are two straddles and two strangles from which to choose. Let's say that XYZ is $57 and the 90-day options available are:

Option	Option Price	Option	Option Price
$55 call	$4⅝	$60 call	$2⅛
$55 put	$1½	$60 put	$3⅞

Two strangles can be created from these four options. One consists of the long $60 call and the long $55 put, called the *out-of-the-money strangle*. The other contains the long $55 call and the long $60 put, called the *in-the-money strangle*. The profit-loss diagram of a strangle shows that this strategy's maximum loss occurs over the range of prices between strike prices. In Table 4-15, you can see how the out-of-the money strangle looks at expiration. A strangle has break-even points at expiration of the strike price of the call plus the total premium paid and the strike price of the put option minus the total premium paid. For the strangle in this example, the break-even points are $51⅜ and $63⅝. This is a range of plus and minus 10 percent of XYZ's current price of $57 (Figure 4-12).

In-the-Money or Out-of-the-Money?

The out-of-the-money strangle of the $60 call and $55 put costs $3⅝. By contrast, the in-the-money strangle of the $55 call and $60 put costs $8½. Quite a difference in cost, but the positions are the same! For one, their break-even points are the same. The out-of-the-money strangle has break-even points of $63⅝ and $51⅜. The in-the-money strangle has break-even points of $63½ (the $55 call strike plus the $8½ total premium), and $51½ (the $60 put strike minus the $8½ premium). If the po-

TABLE 4-15

Strangle

Price of XYZ	Value of $55 Put	Value of $60 Call	Value of Strangle	Profit/Loss of Strangle
$45	$1,000	$ 0	$1,000	$637.50
50	500	0	500	137.50
55	0	0	0	(362.50)
60	0	0	0	(362.50)
65	0	500	500	137.50
70	0	1,000	1,000	637.50

FIGURE 4-12

Long Strangle

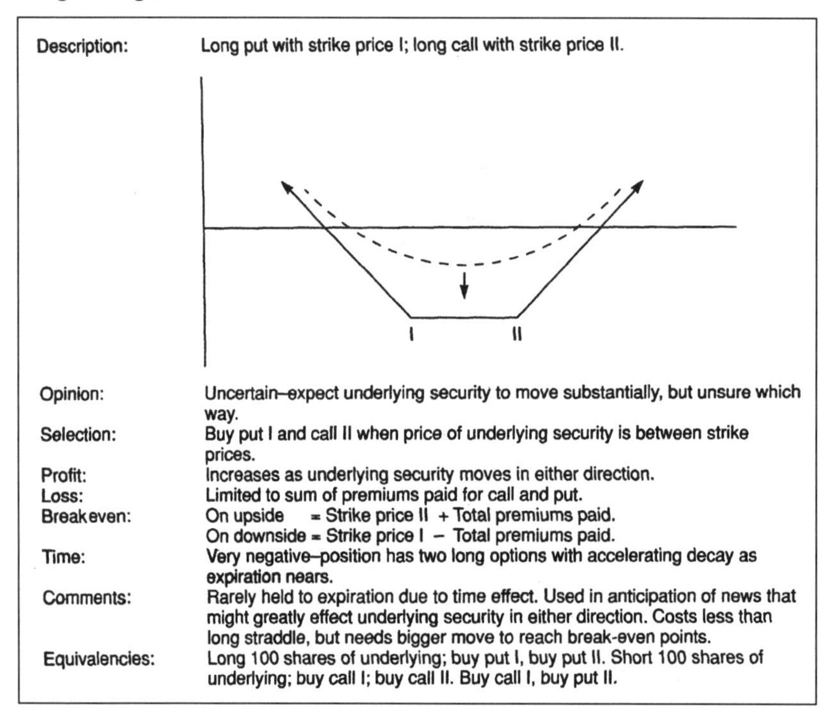

Description:	Long put with strike price I; long call with strike price II.
Opinion:	Uncertain–expect underlying security to move substantially, but unsure which way.
Selection:	Buy put I and call II when price of underlying security is between strike prices.
Profit:	Increases as underlying security moves in either direction.
Loss:	Limited to sum of premiums paid for call and put.
Breakeven:	On upside = Strike price II + Total premiums paid. On downside = Strike price I − Total premiums paid.
Time:	Very negative–position has two long options with accelerating decay as expiration nears.
Comments:	Rarely held to expiration due to time effect. Used in anticipation of news that might greatly effect underlying security in either direction. Costs less than long straddle, but needs bigger move to reach break-even points.
Equivalencies:	Long 100 shares of underlying; buy put I, buy put II. Short 100 shares of underlying; buy call I; buy call II. Buy call I, buy put II.

sitions are the same, they must also have the same risk. The maximum loss of the out-of-the-money strangle is $3⅝; the maximum loss of its counterpart is $3½.

Why is the risk of the in-the-money strangle only $3½ if it costs $8½? If XYZ is somewhere between $55 and $60, at expiration, the sum of the values of the put and call are $5. For instance, XYZ is $58 at expiration. The $55 call would be $3, and the $60 put is $2. At $56, the call is $1, and the put is $4. Because the sum of the call and put premium will always be $5, the $8½ premium paid for the strangle can only decrease by $3½. This is the risk of the in-the-money strangle.

Strangle or Straddle?

A strangle of an out-of-the-money call and put costs less than the straddle of the at-the-money call and put. As with any long option position, loss is limited to the premium paid for the options. Do not consider buying a strangle in favor of a straddle just because the initial investment is lower. The strangle may cost less, but it is vulnerable to losses over a wider range of prices than the straddle. The straddle only suffers complete loss of the premium paid if the underlying lands at the strike price. Absolute dollar risk should not be the only deciding factor.

Now compare the strangle with the two straddles available. The two straddles are the $55 call and put, and the $60 call and put. The $55 straddle costs $6⅛ and has break-even points of $61⅛ and $48⅞ at expiration. The $60 straddle costs $6 and has break-even points of 66 and 54 at expiration. The strangle costs $3⅝ and has break-even points of $63⅝ and $51⅜.

Which is the best of the three? When a stock is between strike prices, such as $57 in our example, the strangle creates a position which balances the profit potential from a movement of the underlying in either direction. The break-even points of the strangle are 6⅝ points above and 5⅝ points below the current price. Neither of the straddles does this. The break-even points of the $55 straddle are 4⅛ points above and 8⅛ points below XYZ's current price; the $60 straddle, 9 points above and 3 points below. All three positions are presented in Figure 4-13.

What if XYZ were $55? The 90-day options might be priced as follows:

Option	Price	Option	Price	Option	Price
$50 call	$6¾	$55 call	$3⅜	$60 call	$1⅜
$50 put	$ ⅝	$55 put	$2¼	$60 put	$5⅛

Now the $55 straddle is $5⅝ with break-even points of $60⅝ and $49⅜. The $50 to $60 strangle is $2 with breakevens of $62 and $48.

Which is better? Only investors can make that decision. Do they put more emphasis on break-even analysis, absolute dollar risk, or leverage? In terms of leverage, investors can establish three $50 to $60 strangles for almost the same dollar risk as one straddle. The strangles will be a position with six long options (three calls and puts) versus the straddle's two (one call and put). If XYZ makes a big move, the three strangles are much more profitable than the one straddle. However, there is one very important trade-off: the entire premium paid for the strangles is lost if XYZ is between the strike prices of the options. The straddle only suffers total loss of the premium paid at the singular point of the strike price. This is a very aggressive and speculative way to com-

FIGURE 4-13

Comparison of Straddles and Strangle

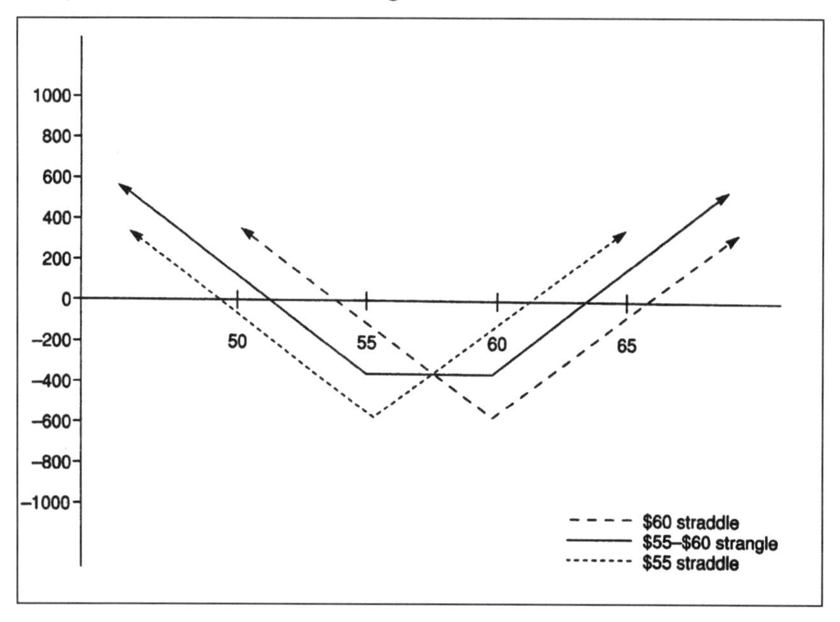

pare strangles to straddles. Figure 4-14 shows why this type of comparison can be so dangerous.

SHORT BUTTERFLY

The *butterfly,* so named because of the "wings" in its profit-loss diagram, is a four-option position. That is, the common short butterfly is created from the purchase of two calls of the same strike price and the sale of one each of the closest in-the-money and out-of-the-money call options.

Assume XYZ is at $55. Use these 90-day option prices from the long strangle discussion.

Option	Price	Option	Price	Option	Price
$50 call	$6¾	$55 call	$3⅜	$60 call	$1⅜
$50 put	$ ⅝	$55 put	$2¼	$60 put	$5⅛

The short butterfly is:

Short 1 $50 call	$6¾	
Long 2 $55 calls	(6¾)	(2 × 3⅜)
Short 1 $60 call	1⅜	
Net credit	$1⅜	

At Expiration

The $1⅜ credit, the maximum profit of this strategy, is achieved if XYZ is above $60 or below $50 at expiration. Below $50, all of the options expire worthless, and the investor keeps the entire net premium received. Above $60, the total value of the two long $55 calls are the same as the total value for the short $50 and $60 calls. For example, at $65, the long $55s are $10 each, for a total of $20. The short $50 is $15 and the short $60 is $5, for a total of $20. The break-even prices for the butterfly are the strike price of the short in-the-money call plus the credit received, and the short out-of-the-money call minus the credit received (Figure 4-15).

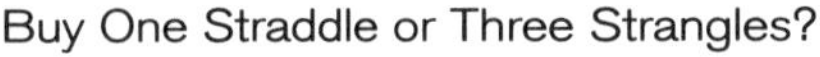

FIGURE 4-14

Buy One Straddle or Three Strangles?

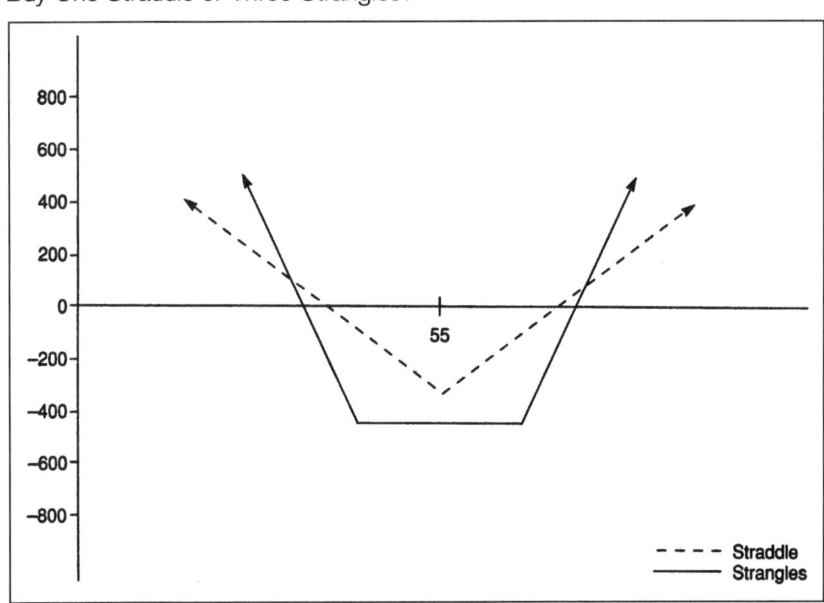

FIGURE 4-15

Short Butterfly

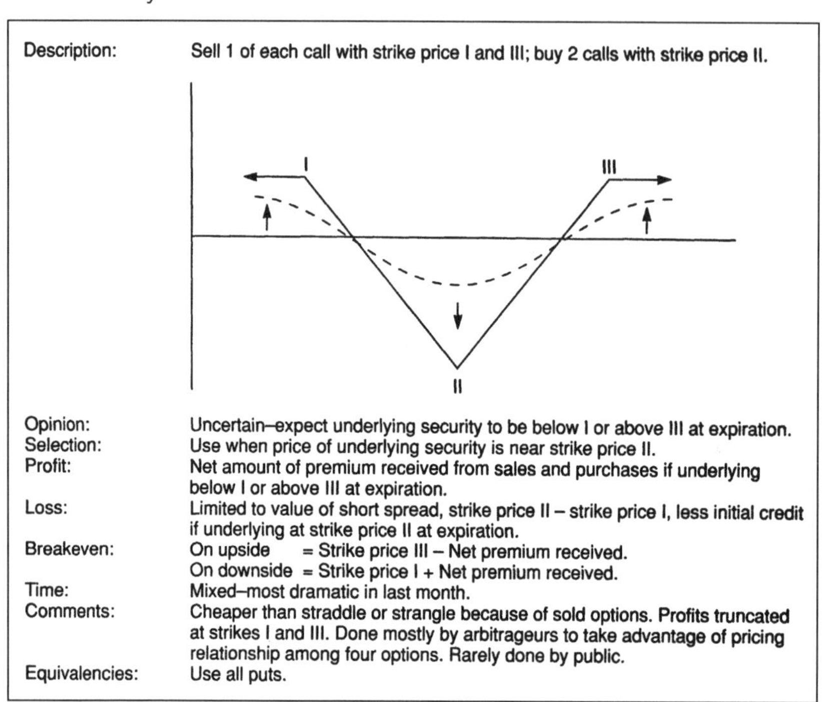

The maximum loss occurs at the strike price of the long options, the middle strike. At this price, the investor has a position that is short two options, one worth 5 points and another that is worthless. Unfortunately, the two that were purchased are also worthless. The maximum loss, then, is the value of the in-the-money short option minus the credit collected. In this example it is $3⅝ ($5 – $1⅜).

1 Butterfly = 2 Spreads

A short butterfly is a strategy that combines two spreads: a bullish long and a bearish short. This description confirms the limited profit, limited loss of this technique. Spreads were discussed earlier in the chapter. They were described as having both limited profit and limited loss. As a combination of spreads, the butterfly retains these characteristics.

If you break down the example into its spread components, the $1⅜ credit also results. The $50 to $55 call spread is sold for a $3⅜ point credit. The $55 to $60 call spread is bought for a $2 point debit. Above $60, each spread is worth its full value, in this example 5 points. Because the strategy is composed of one short spread and one long spread, the 5-point value of the short spread is counteracted by the 5-point asset in the long spread leaving the $1⅜ credit as profit.

When to Use

There is no good reason for the public investor to use this strategy. The short butterfly is mostly used by professional traders to exploit mispricings among options. It is rarely done in the public domain.

Here's why. The short butterfly loses money if the underlying stays about where it is—assuming the long options are at-the-money. It loses less as the underlying moves away from the middle strike. The short butterfly looks like the long straddle at prices around the middle strike, and, based on this comparison, it appears that the short butterfly should be used when the investor expects the underlying to be somewhere other than the middle strike price at expiration. The $55 straddle costs $5⅝. The short butterfly has a maximum loss of $3⅝ at $55. The butterfly looks better at these prices.

On the other hand, the failure of the short butterfly is its inability to provide additional profit should the underlying make a substantial move. The long straddle has the potential for very large profits if the underlying moves well beyond a break-even point. The maximum profit of the butterfly, $1⅜, is reached very close to the break-even points. The short butterfly hedges away too much of the profit that can be made from a move in the underlying away from the middle strike.

SHORT CONDOR

The *short condor* is a variation of the short butterfly. It also requires four options and is the combination of a short and a long spread. The difference is that the strike prices of the long options are separated. The name condor comes from the larger "wingspan" of the profit-loss diagram, which distinguishes it from the butterfly (Figure 4-16).

An example of a short condor position would be:

Short 1 XYZ $85 call: $9⅞
Long 1 XYZ $90 call: 6½
Long 1 XYZ $95 call: 4
Short 1 XYZ $100 call: 2¼

FIGURE 4-16

Short Condor

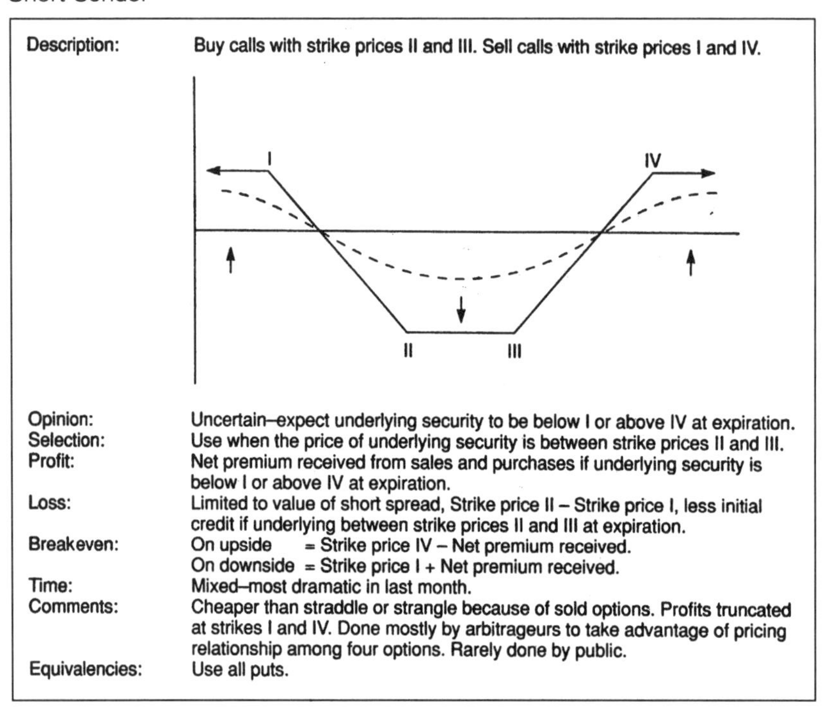

Description:	Buy calls with strike prices II and III. Sell calls with strike prices I and IV.
Opinion:	Uncertain–expect underlying security to be below I or above IV at expiration.
Selection:	Use when the price of underlying security is between strike prices II and III.
Profit:	Net premium received from sales and purchases if underlying security is below I or above IV at expiration.
Loss:	Limited to value of short spread, Strike price II – Strike price I, less initial credit if underlying between strike prices II and III at expiration.
Breakeven:	On upside = Strike price IV – Net premium received. On downside = Strike price I + Net premium received.
Time:	Mixed–most dramatic in last month.
Comments:	Cheaper than straddle or strangle because of sold options. Profits truncated at strikes I and IV. Done mostly by arbitrageurs to take advantage of pricing relationship among four options. Rarely done by public.
Equivalencies:	Use all puts.

These are 90-day option prices when XYZ is $92. The net credit of these four transactions is $1⅝. As with the short butterfly, it is also the maximum profit if XYZ is above $100 or below $85 at expiration. The short call spread in this example is $85 to $90; the long spread is $95 to $100. The break-even prices for the condor are the strike price of the short in-the-money call plus the credit received, $86⅝, and the short out-of-the-money call minus the credit received, $98⅜.

The short condor suffers its maximum loss over the range of prices between the two long call strike prices. For the short butterfly, it is a single price. Here the risk is $3⅜—the 5 points the short $85–$90 call spread will be worth between $90 and $95 minus the maximum $1⅝ potential profit. Above $95 at expiration, the $95 call gains in value. This gain is unchecked by the short $100 call until XYZ goes above $100.

The short condor, with a distance between the long strikes, can be compared to the strangle. The $90 put is $2⅞, so the $90 put, $95 call strangle costs $6⅞. Break-even points are $83⅛ and $101⅞. The $6⅞ is lost between $90 and $95. The conclusion reached is that the premium saved by establishing a short condor may not be worth the limited profit potential of the position. Indeed, one could say that the short condor is a strangle that has had its wings clipped!

SHORT STRADDLE

The *short straddle* earns very large profits in a neutral market. No other option strategy is so profitable. Of course, using this technique means very unique and substantial risks if the market does not remain neutral.

A short straddle consists of a short call and a short put of the same strike price. XYZ is at $80. The 90-day $80 call is $4¾; the 90-day put is $3¼. The $80 straddle can be sold for $8 ($800 per straddle). In Table 4-16, there is a profit-loss picture. The only difference between the short and the long straddles is that the profits have been changed to losses and vice versa.

The graph of the short straddle is the inverted *V* shape profit-loss profile of the short straddle appearing to straddle the strike price (Figure 4-17).

Maximum profit is the total premium received; however, there is only one price where this maximum profit is realized—the strike price of the sold options. If the underlying is anywhere other than the strike price, but still within the break-even points, there is a profit, but it is smaller. Outside the break-even points, the potential loss in this strategy is unlimited. The straddle seller has given the potential for unlimited profits *to the buyer*.

TABLE 4-16

Short Straddle

Price of XYZ	Short $80 Call	Short $80 Put	Total Value of Straddle	Profit/(Loss) on Straddle
$65	0	$1,500	$1,500	$(700)
70	0	1,000	1,000	(200)
75	0	500	500	300
80	0	0	0	800
85	500	0	500	300
90	1,000	0	1,000	(200)
95	1,500	0	1,500	(700)

FIGURE 4-17

Short Straddle

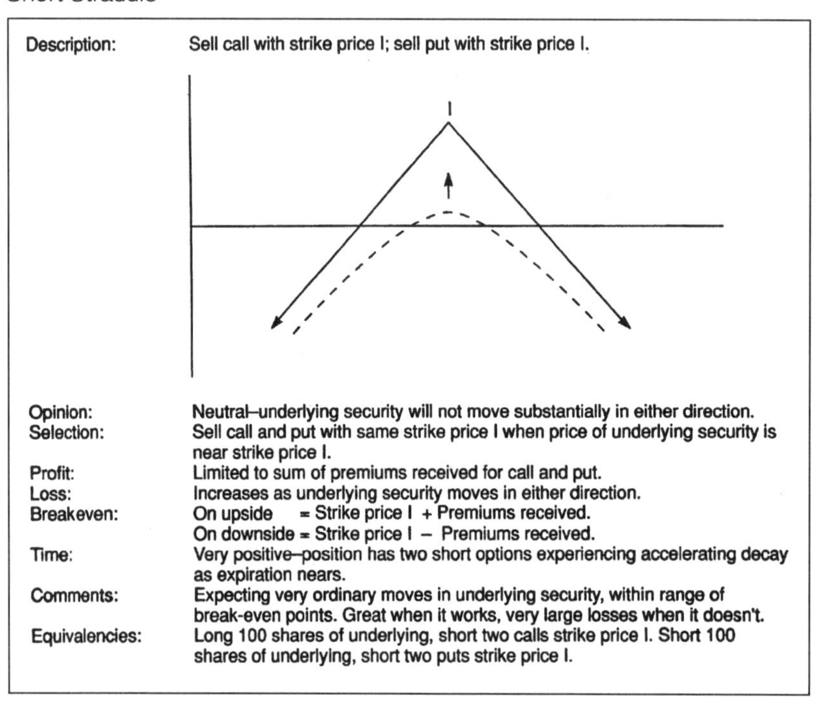

The break-even prices at expiration are the strike price of the options sold plus and minus the total premium received.

Motivation

Before options, there was no way for an investor to profit directly from correctly forecasting a narrowly trading market. The only way to improve investment performance was to allocate assets properly. However, asset allocation provides only relative opportunity profit by maximizing investment in sectors that are outperforming and minimizing investment in those underperforming. Without options, there is no way to make money. The short straddle and the three strategies that follow are designed to be profitable when very little happens.

The seller of a straddle expects (needs) the underlying to move in a narrow range around the strike price. In this example, the break-even points are \$72 (\$80 strike price minus \$8 credit) and \$88 (\$80 strike price plus $8 credit), the same as for the long straddle, some 10 percent away from the current price.

The investor should select this straddle position only if he expects XYZ to trade within plus-or-minus 10 percent of $80 over the next 90 days. Many investors continually forget that selection of the proper option strategy must address the expected price movement of the underlying over time-and the consequences of unexpected outcomes. The short straddle is a strategy that could have serious financial results if the underlying moves substantially away from the strike price, *e.g.*, as a result of a takeover bid or a dividend cut.

Time Decay

Say that the straddle sold at $8. If XYZ were to remain at $80, the straddle would drop to $6½ in 30 days. After another 30 days, the straddle decays to $4⅝. The overall decay is 42 percent. Time is the best friend of all short option strategies!

Risks and Capital Management

The short straddle assumes all the risk of long stock at prices below the downside break-even point at expiration and all the risk of a short position in the stock for prices above the upside break-even point. In effect, the position gets "longer" as prices go down, and "shorter" as prices go up. In more technical terms, as the underlying moves lower, the delta of the put approaches –1; as the underlying moves higher, the call delta approaches +1.

The possibility of extensive losses cannot be overemphasized. Short straddles and strangles are very attractive techniques. They beckon the investor with the lure of "free money." After all, these strategies require only a margin deposit, which can very often be satisfied with the loan value of securities in the investor's account. This loan value can be used over and over again, producing very large returns. This reasoning often results in investors using too much of that loan value in naked option strategies.

Consequences of this type of capital management can be disastrous. It only takes the sale of one ill-timed, short position for the investor to feel the full force of his error. The lack of additional margin to support the increasing requirement can force him to close some or all of the position early. As described earlier, the loss from buying back a short option could eliminate much, if not all, of not only the profits of previous successful positions, but the investor's capital as well.

Capital Required

Because this strategy has naked options, it needs to be supported by an initial margin deposit that is based on the current requirements. This can be higher than the exchange minimums, and your brokerage firm can impose additional criteria. However, since a short straddle can only be unprofitable on one side of the trade (*i.e.*, the underlying is either going up or going down), the requirement will only be the greater of the put or call margin plus the premium of the other side.

SHORT STRANGLE

The goal of the short strangle is to profit from movement in the underlying security over a range of prices. These price limits are defined by the strike prices of the options sold—one call and one put per strangle.

Unlike the short straddle, the strike prices of the call and put are different. For example, XYZ is $57 and the 90-day options are:

Option	Option Price	Option	Option Price
$55 call	$4⅝	$60 call	$2⅛
$55 put	$1½	$60 put	$3⅞

The out-of-the-money strangle, made up of the $55 put and the $60 call, brings in $3⅝ in total premium. Break-even points at expiration are $63⅝ and $51⅜. These prices are 10 percent above and 9.8 percent below the current price. When the underlying is between strike prices, the strangle more accurately reflects the investor's neutral opinion (Figure 4-18).

Other comparisons between the short straddle and short strangle are:

1. The short strangle usually brings in less premium because it is most often done with out-of-the-money options.
2. The maximum profit is achieved if the underlying is between the strike prices at expiration. For the straddle, it is at a singular price.
3. As with the short straddle, the short strangle is exposed to unlimited losses if the underlying makes a substantial move to the upside or the downside. The position gets longer as the underlying goes lower and gets shorter as it goes higher.

FIGURE 4-18

Short Strangle

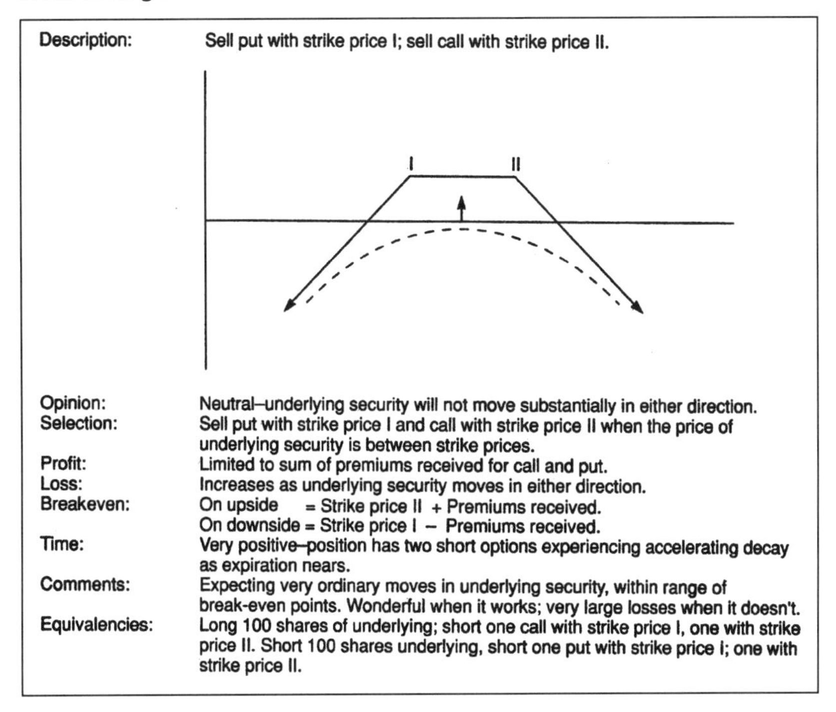

Description: Sell put with strike price I; sell call with strike price II.

Opinion: Neutral–underlying security will not move substantially in either direction.
Selection: Sell put with strike price I and call with strike price II when the price of underlying security is between strike prices.
Profit: Limited to sum of premiums received for call and put.
Loss: Increases as underlying security moves in either direction.
Breakeven: On upside = Strike price II + Premiums received.
On downside = Strike price I − Premiums received.
Time: Very positive–position has two short options experiencing accelerating decay as expiration nears.
Comments: Expecting very ordinary moves in underlying security, within range of break-even points. Wonderful when it works; very large losses when it doesn't.
Equivalencies: Long 100 shares of underlying; short one call with strike price I, one with strike price II. Short 100 shares underlying, short one put with strike price I; one with strike price II.

A sensible alternative is to start with some protection in place when the position is established. This is examined in the next two sections.

LONG BUTTERFLY

A *long butterfly* is the sale of two at-the-money calls and the purchase of an out-of-the-money and an in-the-money call. Recall that the short butterfly was described as the combination of two spreads that limited the profit and loss of the position. The same is true for the long butterfly. The difference is that the spreads are reversed. Instead of the middle strike being the long calls, they are the short calls.

XYZ is at $85; the 60-day option prices are:

Option	Price	Option	Price	Option	Price
$80 call	$7¼	$85 call	$4	$90 call	$2
$80 put	$1⅛	$85 put	$2¾	$90 call	$5¾

The long butterfly is:

Long 1	$80 call	$7¼	
Short 2	$85 calls	$8	(2 × $4)
Long 1	$90 call	2	
Net debit		$1¼	

The net debit is the maximum loss of this position if XYZ is below $80 or above $90 at expiration. Below $80, all options expire worthless. Above $90, the long $80 to $85 spread is worth 5 points; so is the short $85 to $90 spread. The values cancel each other out, resulting in the 1¼ loss of the initial investment.

Profit, maximized at $85,is 3¾. At this price, the $80 call is worth 5 points, and all the other options expire worthless.

Break-even points for the long butterfly are the lowest strike price plus the initial debit and the highest strike price less the initial debit. In this example, these prices are $88¾ and $81¼ respectively (Figure 4-19).

FIGURE 4-19

Long Butterfly

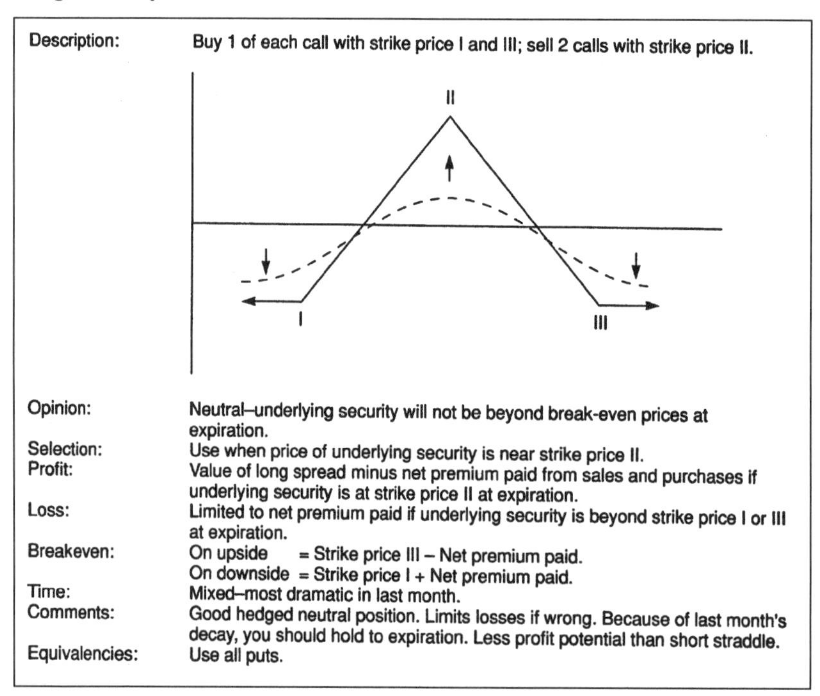

Motivation

The long butterfly is used by the investor who wants to profit from a forecast of a narrow trading range for the underlying but is unwilling to accept the risk of unlimited losses that comes with the sale of either a straddle or strangle.

The long butterfly, then, can be an attractive alternative to the short straddle. First, though, one might ask, "How can a butterfly spread of call options be compared to a straddle composed of a call and a put?" The answer is something that the investor using options should always remember when analyzing and deciding among strategies: "Place the emphasis on the risk-reward implications of the strategies!" Options can be combined in many different ways to achieve similar results. Put-call parity shows why this is true.

The $85 straddle can be sold for $6¾ ($675), the sum of the call and put premiums. This is 3 points more premium than the maximum profit of the butterfly. Instead of selling one straddle for $6¾, the investor can buy two butterfly spreads for a total investment of $2½. The $1¼ outlay per butterfly is the maximum loss. The maximum profit is $750 ($3¾ per butterfly), $75 more than the single straddle.

Break-even points for the straddle are $91¾ and $78¼. For the butterflies, they are $88¾ and $81¼. The performance difference is evident in Figure 4-20. One straddle is being compared to two butterflies. Therefore, the profit from the butterflies declines twice as fast. In the diagram, the slopes of the lines from the point of maximum profit demonstrate this effect. The long butterfly can be an effective neutral strategy especially for those who would find the excessive losses possible with short straddles hard to accept. The trade-off is the increased speed at which the butterfly achieves its maximum loss.

The Directional Butterfly

The long butterfly has a unique and interesting application for the investor with a directional bias. A long butterfly that can be placed entirely out-of-the-money can be very profitable if the underlying moves "into" the profit range of the position. XYZ is $105 and 90-day options are:

$110 call: $3½

$115 call: $17⁄8

$120 call: $1

FIGURE 4-20

Comparison of Two Long Butterfly Spreads to Short Straddle

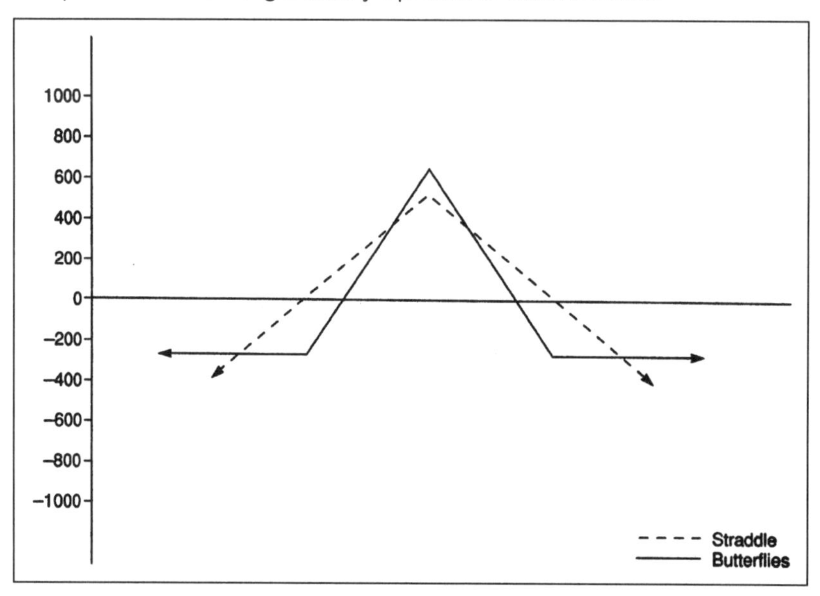

The butterfly of a long \$110, long \$120, and two short \$115 calls costs \$¾. That is an investment of \$75, excluding commissions. If XYZ closes at \$115 on expiration Friday, the butterfly is worth its maximum value of 5 points, a 4-to-1 profit. The break-even points are \$110¾ and \$119¼.

The caveat is that the investor selecting this type of position cannot be too bullish. Profits will decline between \$115 and \$119¼, and the \$¾ is lost if XYZ is above \$120 at expiration. This might be acceptable to the investor with a specific target price and the time frame in mind.

This long butterfly has a very low cost of entry—lower than any of the individual calls or spreads. It is the epitome of the power of options:

> Options allow the investor to maximize his participation in outcomes he considers most likely by creating positions that minimize or eliminate participation or create losses for outcomes he considers less likely.

LONG CONDOR

We need not elaborate on the *long condor*. It has the same risk-reward characteristics as the long butterfly except for the separation of the strike prices of the sold options (Figure 4-21).

Because of the long distance between the two long options, perhaps 15 points or more, the long condor is more effective when used with higher-priced securities. For instance, the long condor on a $52 stock could be the long $45 call, short the $50 and $55 calls, and long the $60 call. This 60-day condor would cost $3¼ and have only $1¾ points potential profit. The 60-day long condor for an $82 stock—long $75 call, short $80 and $85 calls, long $90 call—would cost $2⅜ and have 2⅝ points potential profit. Individual 60-day option prices could be:

XYZ at $52		Position	XYZ at $82	
$45 call	$7¾	Long	$75 call	$8¾
50 call	3⅝	Short	80 call	5
55 call	1 3/16	Short	85 call	2½
60 call	5/16	Long	90 call	1⅛
Cost:	$3¼		Cost:	$2⅜

FIGURE 4-21

Long Condor

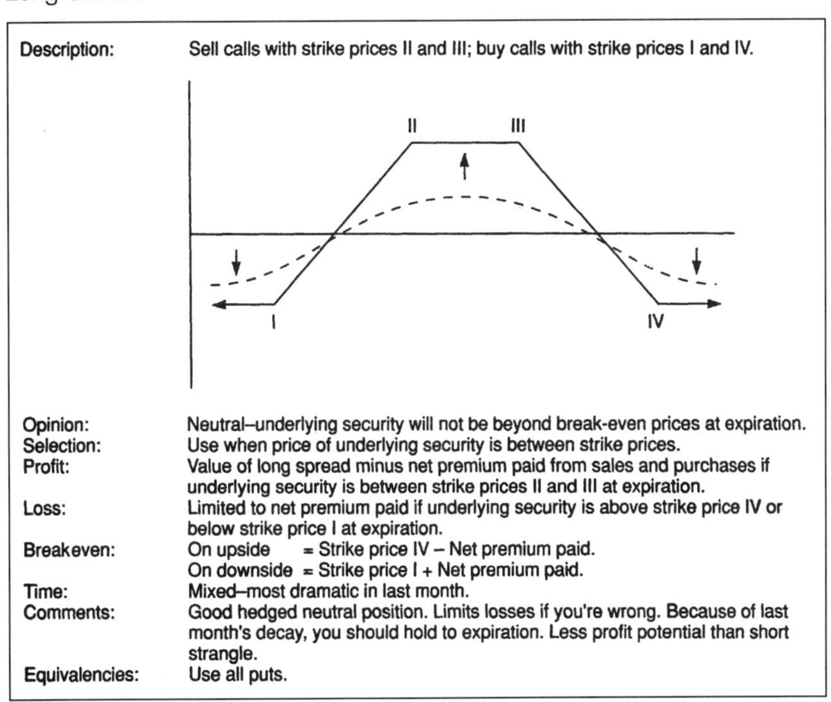

The difference in profitability comes not from the absolute distance between the long options, but from the relative distance. The $45 call is 7 points or 13.5 percent below the stock price. The $75 call is also 7 points below the stock price, but is only 8.5 percent below the current price. Again, the option market has proved that prices accurately reflect options' relation to the price of the underlying. The seemingly better profit-loss of the long condor on the $82 stock comes at the expense of bringing the points of maximum loss closer to the current price.

PART TWO

INVESTING AND TRADING STRATEGIES

CHAPTER 5

INVESTING AND TRADING STRATEGIES FOR THE INDIVIDUAL INVESTOR

J. Marc Allaire

There may be no clear cut demarcation between an "investor" and a "trader." The same individual will more than likely act as an investor at one point and as a trader at another. Nevertheless, we will adopt the following working definitions: an *investing strategy* is one where buying, holding, and/or selling the underlying equity is the primary focus; in other words options are being used in a subsidiary fashion to establish or manage a stock position. In contrast, a *trading strategy* is one where the primary vehicle is the options and there is no intent to buy, hold, or sell the underlying stock.

Covered writing, in which calls are sold on a share-for-share basis against purchased or owned stock, is the quintessential investing-oriented option strategy. Buying calls in order to be positioned for an expected earnings announcement, however, falls under the trading strategy banner.

INVESTING STRATEGIES

Investing Strategy 1: Using Put Options to Buy Stock

The most basic, plain vanilla investing strategy is "buy stock." At times, the rationale for purchasing a particular stock is unequivocal and an investor feels compelled to simply buy that stock. On other occasions, a stock may appear attractive, but not at the current price. How often have we heard, or uttered, the phrase, "I really like this stock, but I'm not willing to pay the going price. I'll buy it if it comes down 3 or 4 points"?

In such circumstances writing put options to acquire the underlying stock may be the appropriate strategy. Consider a situation in which, on January 28, Feather Steel Manufacturing (FSM) is trading at 42. We are hesitant to purchase FSM at this level but we would not be adverse to adding the stock to our portfolio in the $39 range. The March 40 puts (which have 52 days to expiration) are bid at 1¼. We decide to write the March 40 puts at 1¼. Why?

Let's quickly review what selling an equity put entails. The writer of an equity put option assumes the obligation to purchase the underlying stock at the option's exercise price, until the option's expiration date. In our example, this means that we could be *assigned, i.e.* required to purchase, FSM at $40, and we could be assigned at any time until the third Friday in March. If we are fully prepared to buy this stock, then assuming this obligation should not be a problem.

When we write a put option there are only two possible outcomes: either the puts will expire worthless if they are out-of-the-money at expiration, or they will be assigned and we will be required to purchase the underlying stock. Because equity options are American-style options, the risk of early assignment cannot be ignored.

To continue with our example, if the puts are assigned, we must then purchase FSM at $40, a price which will be greater than the current market price. The adjusted-cost basis on this would be $38.75, *i.e.* the option's strike price of $40 less the premium received of 1¼, and this price may be higher or lower than the current market price. Is this an acceptable price? It was on January 28 when we said we would be writing the March 40 Puts to purchase FSM in the $39 range. Of course, by March FSM could be trading substantially below 39; but this is the risk that every investor who purchases stock runs: the price may decline.

The second possible outcome is that at the March expiration FSM is trading above $40 and the puts expire worthless. In this case we would then keep the option premium of 1¼ and, at that point, we may want to ask ourselves if we should then write the May 40 puts. Writing further-dated puts is only appropriate if we are still looking to acquire FSM.

The risk of the puts expiring worthless is an opportunity risk. If FSM has risen substantially above $40 by the March option expiration, then we will have missed the opportunity of purchasing the stock at the price of $40, an attractive price if the current price is substantially higher. This is also a risk that stock investors know: when waiting for a stock to pull back in order to buy it, there is always the possibility that the stock will rally before we purchase it. But, under this scenario, the put writer is not left completely empty-handed: the premium received represents our "return on investment."

REALITY CHECK—WRITING PUTS TO ACQUIRE STOCK

If you are thinking of selling put options to add a stock to your portfolio, you may want to take the following reality check. A "no" to any of the following questions may indicate the strategy is inappropriate for you or for the stock/option in question.

- Are you financially and psychologically able to purchase this stock? Do you have the cash and/or buying power? Would owning this stock prevent you from sleeping nights?
- Is the adjusted cost of the stock (option strike price minus premium received) an acceptable entry point for this stock?
- Do you realize that if the stock starts running up after you sold the put, your maximum profit potential will be limited to the put option's premium?
- Do you understand that you are assuming the full risk of stock ownership, *i.e.*, you could be the proud owner of a stock that has just declined substantially on some unexpected news?

Recap

Writing equity puts is a strategy suitable for investors looking to purchase stocks at prices lower than the current market. There are, however, no guarantees that the stock will be purchased. In the event that it is purchased, the investor must then bear the full risk of ownership.

Investing Strategy 2: Using Call Options to Sell Stock

Just as at times investors are looking for a pull-back in the price of a stock to buy it, at other times they are waiting for a rally in order to dispose of it. Does the following have a familiar ring to it: "I'm not quite ready to sell at this price, but if the stock goes up a couple of points . . ."

Writing *covered call options* is a strategy that may be of help in the scenario described above. On June 4 Artificial Golf Courses (BOGY) is at $77. In this example, we assume that this stock was purchased some time ago at a lower price and is now approaching our target price of $80. Of course, one possible solution is to wait for BOGY to reach 80 and then sell it. But we could also write the August 80 calls (with 79 days to expiration), currently bid at 4⅝.

An investor who writes a call option is assuming the obligation to sell the underlying stock at the option's exercise price at or prior to expiration, or $80 in our example. And, when an investor writes a covered call, there are only two possible outcomes at the option's expiration: either the stock will be above the exercise price, in which case the option will be assigned and the stock will be sold at the strike price ($80 in this example), or the stock will be below the strike price, the option will expire worthless, and the investor will still hold the underlying stock.

Let's return to our example. If, on the third Friday in August, BOGY is trading above 80, then one can expect the calls to be assigned and, consequently, the stock will be sold at 80. We would also keep the option premium of $4⅝, so our total proceeds from selling our stock would be $84⅝.

BOGY could be at 80¼ in August, or it could be at 100¼. In both cases the calls will be assigned. In the second case though, many investors would start second guessing themselves. "Why did I sell the calls!?! I am now required to sell at $80 a stock worth more than $100! If I just hadn't sold those calls . . . !!"

But what if we hadn't sold those calls?

It is impossible to say for sure, but we may have sold the stock when it reached $80. Wasn't that our target price? And how many of us have seen a stock rise even further after we sold it? When selling calls to dispose of a long stock position, the first question an investor must ask is: "Am I satisfied selling this stock at this price?" If the answer is no, writing calls may not be the appropriate strategy.

The second possible outcome is that at the August expiration BOGY has not reached $80. At this point the call options would expire worthless and our obligation to sell would be terminated. The call option premium of 4⅝ is ours to keep and we would then look towards a follow-up strategy. Are we still looking to sell BOGY around 80? Assuming BOGY is still trading around 77, then selling the November 80 calls may make sense to us. Of course the possibility exists that, by August expiration, BOGY will have declined to, say, 68. In this case it may not make economic sense to sell the November 80 calls, because their premium may be minimal.

Because many investors who are new to covered writing feel locked in to their position, it makes sense to ask, "Does covered writing eliminate an investor's flexibility to sell a stock they own whenever they want, in order to take advantage of any unexpected development?" or, "Does writing a two-month covered call mean there is nothing that can be done over the next 60 days?" The answer is an unequivocal no!

Although it is true that writing a covered call does create an obligation to sell the underlying stock until the expiration date, this obligation can be terminated simply by closing the position, *i.e.,* buying a call to close the open short call position. Why would we do this? There are at least two possible reasons. First, we could have different expectations for the stock (maybe as a result of unexpected news). In this case, we might be satisfied selling the stock at the current price. Second, we may revise upward our target price, perhaps as a result of some positive development. In either case it is possible to close our obligation by repurchasing the calls we sold. We might do this in order to sell the stock immediately or to continue holding the stock with the hope of selling it at a higher target price.

How much will it cost to repurchase the calls? The answer to this question depends on how much time has elapsed since we sold the option and the current price of the stock. If the stock's price is unchanged or down from the day when we wrote the calls, we will probably be able to repurchase the options at a profit. But if the stock has already rallied because of some new development, the options could be trading at a much higher premium than our original sale price, thus forcing us to cover an option position at a substantial loss.

REALITY CHECK—WRITING CALLS TO SELL A STOCK

Thinking of writing covered calls to sell a stock you own? Make sure you agree with all the points in our reality check.

- You will be satisfied with the effective selling price (strike price of the option plus premium received) even though the stock may have risen to a higher price by the time you are assigned.
- You understand that the stock could reach your target price, then decline before the expiration date and you would not necessarily have sold your stock.
- Selling your stock before the option's expiration date will require you to either purchase the call you have sold (maybe at a loss), or else remain uncovered on the calls sold, an extremely risky strategy.

Recap

Writing covered call options against a long stock position is a strategy suitable for investors looking to sell a stock they own at a price that is effectively higher than the current market level when the option is sold. But the seller of covered calls must keep in mind that there is no guarantee that the stock will be sold, and that they maintain the downside risk of owning the stock.

Investing Strategy 3: The Buy-Write

In the previous strategy an investor who had previously purchased a stock wrote covered call options when that stock neared the target selling price. The *buy-write* also combines a long stock/short call position, but both positions are usually *established simultaneously*, usually more with an eye to expected returns than with an eye to a stock reaching a target price.

To establish a buy-write, an investor buys shares of a specific stock, and *at the same time* writes call options on this same stock; hence the name of the strategy. Most brokerage firms, in fact, will accept two-part orders, known as "buy-write" orders, in which stock is purchased and calls are sold at an established net price or not at all. A buy-write order is similar to a *limit-price buy order* for stock, which assures the customer that the stock is purchased at or below the stated price or not at all.

Investors who employ the buy-writer strategy frequently approach this strategy from a return perspective; they want to know what kind of returns they will earn if the stock remains at the price at which it was purchased (the *static return*), and the return they would obtain if the stock rises above the option's exercise price and the calls are then assigned (the *if-called return*). Examples of how these returns are calculated are presented in the Recap section below.

Before examining the return calculations in detail, however, it is important to review the reasons that would motivate an investor to establish a buy-write on a particular stock:

- The stock: The most important element of the buy-write strategy is the stock-purchase decision. No buy-write should be considered based solely on returns. The starting point must be the underlying stock, and the crucial question that each investor must answer is, "Am I comfortable holding this stock as part of my portfolio?" There is no justification to do a buy-write on Advanced Bio-Neuro-Gizmo if owning that stock will keep you awake at night.

- Outlook for the stock: most buy-writes are established using relatively short term options; a two- to four-month range is common. The forecast for the underlying stock over this same period should be neutral to bullish. If the stock's price remains relatively flat (neutral outlook), then the buy-write will return approximately the static rate of return. If the stock price rises above the strike price of the call (bullish outlook), then the buy-write will return the if-called return. A buy-write is totally inappropriate for a stock whose price is expected to drop over the near term.
- The returns. A buy-write should offer rates of return that an investor feels is adequate compensation for the amount of risk taken. Note that every buy-write involves a compromise between the two rates of return: sell a further out-of-the-money option and you will increase the if-called return, but, at the same time, reduce the static return; sell an at-the-money option and you will increase the static return, but also reduce the if-called return.

The buy-write remains one of the most-used option strategies by individual and institutional investors. The strategy is conservative, and the following reasons explain its enduring popularity.

- *Limited downside protection.* By selling calls at the same time the stock is purchased, the initial cost of the position is lowered. For example, if a stock is purchased at $48 and the October 50 calls are simultaneously sold for $2, the investor need only invest $46 to establish the position. The premium received from selling the calls can be used to partially pay for the stock. This initial net cost of $46 is the investor's break-even point on the strategy, thus there is downside protection of $2 from the stock's current price of $48. To reiterate a point made earlier, any investor who enters into a buy-write should have a neutral to bullish forecast for the underlying stock. If this forecast turns out to be incorrect and the stock declines, it is nice to know that the buy-write offers at least partial protection against a fall in the price of the underlying stock.
- *Reduced volatility.* To understand this concept, assume that two investors own the same ten stocks; the first owns the stocks

outright. The second investor writes covered call options on all ten stocks. If the volatility of these two portfolios is compared over time (*i.e.*, the fluctuations of the portfolios' total value), it would be seen that, over time, the portfolio with covered calls has smaller changes in value, *i.e.*, it is less volatile. Reducing portfolio volatility, of course, is not achieved for free. This can only be accomplished at the cost of accepting lower expected returns. Of the two portfolios described above, we should expect the portfolio of outright stock to outperform the portfolio with stocks and covered calls, because the covered-call portfolio has lower risk.

- *Outperform in flat markets*. Even if lower returns are expected over the long term, one can still expect this strategy to outperform in a market moving sideways. In such markets, stocks generate few, if any, gains. But under the same circumstances, the writer of covered calls will earn the buy-write's static return. Covered writing's return profile (*i.e.*, the fact that it outperforms equity ownership in flat markets, but underperforms in rising markets) makes it a valuable diversification tool. If part of an investor's equity investment is dedicated to covered writing, the pure long stock positions and the covered writes tend to complement one another in varying market conditions.

REALITY CHECK—THE BUY-WRITE

Buy-writes look appealing? Make sure you agree with the following points:

- Will owning the underlying stock keep you awake at night? If so, you may be taking on more risk than you can tolerate.
- Is your forecast for the stock neutral or moderately bullish?
- Do you understand that the most you can earn on this strategy is the if-called return? If the stock explodes to the upside, you are committed to selling it at the option's exercise price.
- Do you think the static return compensates you fairly for the risk of owning the underlying stock?

Recap

The buy-write is a conservative strategy ideal for stable to moderately rising stocks. Attention should be focused first on stock selection. Then, on the static and if-called returns.

The Buy-Write: Calculating Potential Returns

For this example, assume the following:

$$\text{Stock Price} = \$43$$
$$\text{May 45 call (91 days to expiration)} = 1\tfrac{3}{4}$$
$$\text{Quarterly dividend} = 0.20$$

Static return (assumes the stock price is unchanged at expiration in 91 days):

$$\text{Return for Period} = \frac{\text{Option Premium} + \text{Dividends Received}}{\text{Initial Investment}}$$

$$= \frac{1\tfrac{3}{4} + 0.20}{43 - 1\tfrac{3}{4}} = 0.0473 \text{ or } 4.73\%$$

This return for 91 days can be converted to an annual rate of return as follows:

$$\text{Annualized Rate of Return} =$$
$$\text{Return for Period} \times \frac{\text{Days per Year}}{\text{Days to Expiration}} = .0473 \times \frac{365}{91} = .190 = 19.0\%$$

If called return (assumes the stock price is above the strike price at expiration, the call is assigned and the stock is sold):

$$\text{Return for period} =$$
$$\frac{\text{Option Premium} + (\text{Exercise Price} - \text{Purchase Price}) + \text{Dividends Received}}{\text{Initial Investment}}$$

$$= \frac{1\tfrac{3}{4} + 2 + 0.20}{43 - 1\tfrac{3}{4}} = 0.0958 \text{ or } 9.58\%$$

This return is converted to an annual rate of return as follows:

$$\text{Annualized Rate of Return} =$$
$$\text{Return for Period} \times \frac{\text{Days per Year}}{\text{Days to Expiration}} = .0958 \times \frac{365}{91} = .384 = 38.4\%$$

A note of caution: Be wary of annualized returns that are calculated using very short term options. A buy-write with 2-week options may produce seemingly great potential returns, but the calculation that annualizes returns assumes that the same returns can be repeated continuously for a full year. That would be 26 times over the course of a year if the returns from a 2-week buy-write are annualized. This is impossible, because there are only 12 expiration dates per year! Another caution about risk: if the potential returns from a particular buy-write are very high, this is probably the market's way of telling you that the risk level of this trade is also very high!

Investing Strategy 4: Buying and Selling Stock

Have you ever been in the following situation: you currently hold some shares of a stock you like and: 1) if the stock rallied, you wouldn't mind selling the shares you own and taking your profits; or 2) if the stock pulled back, you wouldn't mind purchasing additional shares to add to your position.

Of course, in this situation, you could wait for the stock to move up or down and then sell your shares on a rally or buy additional shares on a dip. Alternatively, you could use one of two variations of an option strategy: *selling covered straddles* or *selling covered combinations.*

To understand the mechanics of these strategies, recall the investor's situation and objectives. The investor owns some shares which he is willing to sell at a higher price than the current market price. Also, the investor has cash and is willing to purchase some additional shares at a lower price. If an investor holds a stock and is willing to sell at a higher price, the indicated option strategy is to write covered calls. If the investor is looking to purchase stock at a lower price, the indicated option strategy is to write puts. So, for the investor looking to sell higher or purchase lower, both calls *and* puts should be sold. This is known as selling a *covered straddle* (if the exercise price of the call and the put are the same) or a *covered combination*[1] (if the exercise prices are different). An example illustrates this:

An investor holds 300 shares of Medium Rare Metals (GEM), currently trading at 55¼. This investor is willing to sell the stock 10 percent higher at about 60 or to purchase another 300 shares 10 percent lower at

---

[1] Sometimes known as a "combo," or as a *strangle*.

about 50. Instead of waiting for the stock to rally to 60 or pull back to 50, this investor can enter into the following option positions[2]:

Sells 3 Dec 55 calls @ 3
Sells 3 Dec 55 puts @ 2½

Assuming that these sold options have about 2½ months, or 75 days, to expiration, and further assuming that neither option has been assigned early, which is a risk of selling American-style options, let's fast forward to the December expiration and look at the possible outcomes.

The stock will trade above 55, or it will trade below 55. Or, there is the possibility that the stock will close at exactly 55. In this latter case, one, both, or none of the scenarios as described below will occur. It is impossible to tell, because the option writer's fate is in the hands of buyers who may or may not decide to exercise their options.

Should the stock price be above 55 at expiration, the calls, in all likelihood, will be assigned and the investor will be required to sell the stock at the option's exercise price of 55. But when a call option is sold, the premium is collected and kept, in this example $3. Remember, also, that the put was sold at 2½ and these will expire worthless if the stock closes above 55. So another 2½ is kept. The stock is then effectively sold at:

Effective Selling Price of Stock

= Strike Price of Call + Call Premium + Put Premium

= 55 + 3 + 2½

= 60½

The price of 60½ is slightly higher than the initial objective of 60, so one of our objectives has been met. The important point is that the stock price does not need to rally to 60½ to be sold at 60½. The stock price simply needs to close above 55 at expiration. Attaining one's objective has been made easier through the use of options!

The second possible outcome is that the stock price will be below 55 at expiration. In this case, the written calls expire worthless, but the written puts are be in-the-money and will more than likely be assigned. Consequently, the investor will be required to purchase an additional

[2] These two orders can be entered as one "package" at most brokerage firms.

300 shares at 55. But once again, the premium from the options sold is kept and the effective purchase price is calculated as follows:

$$\begin{aligned} &\text{Effective Purchase Price of Stock} \\ &= \text{Strike Price of Put} - \text{Call Premium} - \text{Put Premium} \\ &= 55 - 3 - 2\tfrac{1}{2} \\ &= 49\tfrac{1}{2} \end{aligned}$$

Once again, one of the investor's objectives is attained and stock is effectively purchased for under $50, not including transaction costs. There is no need for the stock to pull back all the way to 50 for the investor to buy stock at 50. The stock price simply needs to close below 55 at option expiration.

Of course when option expiration rolls around the stock could be substantially above 60, or substantially below 50. At that point, hindsight may lead the investor to think that "do nothing" would have been a better strategy. That is the beauty of hindsight. But this also leads us back to our objectives. Were they met? Yes! And what role did the options play? They simply helped us meet these objectives.

A variation on the covered straddle strategy is the *covered combination*. In this strategy the written calls and puts have different strike prices. The selling or buying of stock at expiration will be similar to that of the covered straddle if the stock price rises above the call's strike price, or if it falls below the put's strike price. But, in the case of a covered combination, there is a third possibility: the stock price could close somewhere between the two strike prices at expiration. If this happens, then both the call and put will expire worthless. Consider the following example.

Long	500 Hog Farms (PIG)	@ 32¼
Sold	5 PIG Feb 35 Call	@ 1
Sold	5 PIG Feb 30 Put	@ ¾

If, at the February expiration, the stock closes above 30 but below 35, then both options expire worthless. The investor keeps the 500 share position in the stock and keeps the 1¾ per share option premiums. But then the investor is at another decision point. Should another combination with a further out expiration be sold? Should the stock be sold and the position closed completely? Or should the investor do nothing and simply maintain a long 500 share position in the stock? There is no "right" answer to this question; each investor must make a personal decision based on a personal forecast for the stock and a personal willingness to assume the consequent risk.

Because there is the possibility that both options in a covered combination will expire worthless, some investors prefer covered combinations over covered straddles. But, again, there is no "better" strategy in an absolute sense.

REALITY CHECK—COVERED STRADDLES AND COVERED COMBINATIONS

When writing a covered straddle or combo an investor enters into two obligations: to sell the stock currently held, and to purchase additional shares. These obligations should be well understood. Since the investor is writing both a put and a call the reality check list for put writing and for selling covered calls should be reviewed in addition to the points below.

- Do you understand that you could be forced to sell the stock you currently hold? Are you satisfied with the effective selling price you will realize if this happens?
- Do you understand that you could be forced to purchase shares in addition to those you currently hold? Are you both psychologically and financially prepared to purchase these shares?
- Do you realize that if the stock you hold falls in price your position, risk and losses will be doubled since you will have to buy more shares?

Recap

Selling covered straddles or covered combinations should be used by investors who currently hold a stock and are looking either to sell these shares at a higher price or to add to their current position by purchasing additional shares at a lower price.

Investing Strategy 5: Making Good on a Stock Gone Bad

Purchasing stock is one of the most basic investment strategies, and it sometimes leads to one of the most basic investment problems: the purchased stock goes down instead of up.

This sometimes creates a situation where an investor's goal is simply "to get out even." The original objective of making money on the investment now has changed to one of not losing money. What alternatives does an investor in this unfortunate situation have?

One alternative for an investor with a stock that has declined in price is to use the old "hold and wait" strategy. No action is required to implement this strategy. The investor simply hangs on and hopes that the stock rallies back to the initial purchase price, at which point the shares can be sold.

Another alternative is commonly known as "doubling up" or *lower-cost averaging*. This strategy requires that the investor purchase an equal number of shares to the original position at the current, lower price. The benefit of this strategy is that the stock needs to rise only half way back to the initial purchase price for the entire position to achieve breakeven. But, doubling up has some big negatives. First, an additional investment is required. Second, risk is increased, because for every continued one dollar drop in the stock's price, the investor will now lose two dollars.

A third alternative involves options, and, once again, we find that options can be a very valuable tool in helping an investor attain an objective. Consider the following example.

Assume an investor holds 400 shares of Fly-By-Night Airlines (JET) now trading at $40. The problem is, the stock was purchased for $50 a few months ago. Now, the investor would be happy to get back to breakeven, recoup the original investment, and move on to the next investment. In other words: "Please, please I don't want the cheese anymore! Just let me out of the trap!"

The investor in this example is willing to listen to ideas, but there are two things the investor is reluctant to do. First, the investor does not want to commit additional funds to the position which, of course, would be required with an average down strategy. Second, the investor does not want to take on additional downside risk. In other words, if the stock were to continue going down, the risk cannot exceed that of owning 400 shares, which is the current position.

Is there hope? Consider the following option transaction:

Buy 4 Oct 40 calls @ 3½
Sell 8 Oct 45 calls @ 1¾

This is known as the *stock repair strategy*. Before looking at what this strategy does, it is important to address the investor's concerns. First, the investor does not want to commit additional funds to the position, and the two-part option strategy does not require any (excluding commissions). Buying the 4 Oct 40 calls at 3½ each will cost $1,400, but selling the 8 October 45 calls at 1¾ each will generate $1,400 fully paying for the purchased calls.[3] No additional funds are required.

The second concern is increased downside risk. What happens to the option position if the price of the stock continues to decline? The option position contains only call options; if the stock is below the lower strike of 40 at expiration, then all the call options expire worthless. Because the position was initiated at no cost, the net result at expiration (assuming the stock is below 40 in this example) would be breaking even on the options: no initial cost, no ending value.

So what does the strategy do? The following table illustrates the strategy components and the overall position at various prices for the underlying stock at option expiration:

	Gains (losses) on			
Stk Price at Exp.	Long 400 Shrs ($50 × 400 = $20,000)	Long 4 40 Calls ($350 × 4 = $1,400 DR)	Short 8 45 Calls ($175 × 8 = $1,400 CR)	Total Profit/Loss
39	($4,400)	($1,400)	+$1,400	($4,400)
40	($4,000)	($1,400)	+$1,400	($4,000)
41	($3,600)	($1,000)	+$1,400	($3,200)
42	($3,200)	($ 600)	+$1,400	($2,400)
43	($2,800)	($ 200)	+$1,400	($1,600)
44	($2,400)	+$ 200	+$1,400	($ 800)
45	($2,000)	+$ 600	+$1,400	-0-
46	($1,600)	+$1,000	+$ 600	-0-

So what has this option strategy accomplished? It has lowered our break-even point without increasing risk. If the investor holds on and prays (alternative one), the stock must rise to the initial purchase price of $50 in order for the investor to break even. If the investor doubles up (alternative two), then the stock only has to rise to $45, or half way back to the initial purchase; but then there is, a required increased investment and increased risk. The option strategy (alternative three), however, offers the advantages of both without the disadvantages: the stock needs

[3] Commissions will have to be paid, but keep reading and you may find this a reasonable cost to enter into this strategy. Very often this position can be entered into for "even" money, a small debit and at times a small credit. Given what the strategy offers, investors may be willing to pay ¼ or ½ to enter into this strategy.

to rally only to $45 for the investor to break even, and there is neither additional commitment of funds nor additional risk if the stock price declines further.

Does the strategy sound too good to be true? Obviously, there must be a disadvantage, so what is it? The bottom line in the table above shows what happens if the stock rallies past the upper strike price of $45. It doesn't get any better than that! Above a stock price of $45 at expiration, the long 40 Calls are exercised, and the short 45 calls are assigned. The net result is that the entire position is liquidated and the investor is left with cash equal to the original investment. The negative of this is that, without a long position in the stock, the investor cannot benefit from a price rise. And that is the disadvantage: the best this strategy can do is if the stock price rises to the strike price of the short calls. But is this so bad? Can you recall what the original goal was?

Forget about the original goal and consider human nature. What do you believe an investor will think if JET stock starts approaching the original purchase price of $50? The reasoning will be along the lines of: "Why did I do this? The stock is about to start making me money, and I'm about to get forced out at breakeven!"

Could we change our mind? Of course, but at what cost? Consider what happens if the stock rallies past the higher strike and back to our original cost of $50. We could then purchase the 8 short 45 calls at $5 for a total cost of $4,000. At the same time, we would sell the 4 40 calls at $10 for gross proceeds of $4,000. So should the stock rally back to the original purchase price, we could change our mind, unwind the option position, and the cost would be limited to transaction fees.

The possibility of changing one's mind illustrates another aspect of options' flexibility.

REALITY CHECK—STOCK REPAIR STRATEGY

Bought a stock? Watched it drift down? Think the stock repair strategy might help? Make sure you understand the following points.

- Do you understand that the stock repair strategy will be of no help if your stock keeps going down? It won't hurt, but it won't help!

REALITY CHECK—STOCK REPAIR STRATEGY (CONTINUED)

- Are you happy "just breaking even"?" Although you may be able to change your mind later and unwind the position, you should realize that the best case scenario is only getting your money back.
- No promises: in order for you to break even, the stock will have to rally to the higher strike price. If it doesn't, you may get some of your money back, but there are no guarantees that you will break even.

Recap

The stock repair strategy is ideal for investors holding an equity position where they have an unrealized loss and where they would be satisfied to simply break even on the stock. The option strategy helps by reducing the investor's break-even point for little or no out-of-pocket cost. In return, the investor gives up any upside potential beyond the strike price of the short calls.

Investing Strategy 6: Purchasing Stock on Margin

Purchasing stock on margin is a common bullish strategy used by aggressive investors: a stock is bought using borrowed funds to pay for part of the purchase price. If the stock rises, the borrowed funds, frequently referred to as *"leverage,"* increases the returns on the invested capital. Of course, leverage works both ways; if the stock declines, the losses on the invested capital will also be multiplied. In either case the investor also incurs the interest expense payable on the borrowed funds.

An alternative to buying shares on margin is the purchase of in-the-money *LEAPS call options*. LEAPS are, essentially, long-term options, and their unique features are described in the table that follows:

In the following example assume that Medium Tech Conductor (BIT) is trading at $64 and the 18-month 50 calls are trading at $20. An aggressive investor decides to purchase these calls, as an alternative to buying the stock on margin. As we all know buying calls is a well-known strategy that offers leverage. So two questions need to be answered: why LEAPS, and why in-the-money?

WHAT MAKES A LEAPS OPTION?

What are equity LEAPS? The short answer to this question is that they are exactly like short term equity options, except that their expiration date will be anywhere from 9 to 31 months away. But investors may want to note the following minor differences:

- Limited strike prices. LEAPS strike prices are generally set at wider intervals than those for short-term options. Strike intervals of $10 or $15 are typical.
- Different ticker symbols. LEAPS trade under different tickers than the regular short-term options. At any given time there are two LEAPS expirations available (*e.g.* January 2002 and January 2003)) , each with its own unique root ticker symbol.
- Pricing issues. When pricing short-term options there is one unknown variable: the stock's volatility. For a two-year LEAPS, the stock's volatility is still unknown, but so are the levels of the risk-free interest rate over the life of the option, and the dividends that the stock will pay over the next two years. Because of this increased uncertainty, some option professionals consider LEAPS more difficult to price than shorter-term options.

One of the advantages a stock owner has over the buyer of a call option, is that there is no expiration date on the stock. Except for the opportunity cost of lost interest on the funds invested in a stock, there is no cost to the stock owner of waiting a little longer. Call buyers, however, especially those who purchase the shortest-term calls, are, figuratively speaking, constantly hearing the ticking of the clock as expiration nears. And how many option buyers have seen a stock rise in price nicely, after their call options expired?

So LEAPS are purchased to "buy more time." There is no need for the stock to move up right away, although LEAPS still have a limited life. But why in-the-money?

There is a twofold answer to this question. First, in-the-money LEAPS have relatively high deltas. This means that, if the underlying stock moves up as forecast, then the LEAPS will capture most of this increase in price. Second, the amount of time premium purchased can be minimized by using in-the-money options. Refer back to the BIT example. BIT is trading at $64 and we purchased the 50 LEAPS calls for $20. In this example, the time premium of the 50 calls is $6 [$20 – (64 – 50)]. Remember, the time value portion of an option's price is the amount greater than the option's intrinsic value.

But the investor who purchases a stock on margin also has a cost above the "intrinsic" value of the stock. That cost is the interest that must be paid on the borrowed funds. Assume that stock was bought on 50 percent margin, or $32 of borrowed funds in the example above. Further assume the stock is held for 18 months and the cost of borrowing is 8 percent. Given these assumptions, the interest expense would be $3.84 per share ($32 borrowed × 8% interest per year × 1½ years).

Comparing the time value of the LEAPS to the borrowing cost associated with purchasing the stock on margin is an inexact science at best. But it lets one estimate approximately how much of the LEAPS premium is a function of the stock's volatility and not the cost of carry.

Of course the investor who buys the stock on margin could sell it before 18 months and reduce their interest expense. But the holder of the LEAPS could also close the position, *i.e.*, sell before expiration and recapture part of the time premium paid.

A side benefit to purchasing the LEAPS call is that there is no possibility of a margin call, which is always a possibility when a stock is purchased on margin. Like all option buyers, the purchaser of LEAPS must pay for these fully up-front. But, if the stock declines, no additional funds will be required. The owner of a LEAPS option will never be forced out of a position by a margin call.

REALITY CHECK—BUYING IN-THE-MONEY LEAPS

Does buying in-the-money LEAPS calls sound appealing? Check out these questions.

- Would you buy the underlying stock? If not, you should probably pass on buying the LEAPS calls, because they will have a risk/reward profile similar to a long stock position over a fairly wide price range.
- Do you understand that the final value of your LEAPS calls could be zero? LEAPS are options, and if the stock is below the exercise price at expiration, they will expire worthless.
- High delta, does not mean 1:1. As the stock rises, in-the-money LEAPS calls will capture most, but not all, of the stock's appreciation.
- Because you do not own the stock, you will not receive the dividends. And the dividends may be significant for the stock underlying the LEAPS calls you are considering buying.

Recap

For aggressive investors, buying in-the-money LEAPS calls may represent an interesting alternative to purchasing stock on margin. Buying LEAPS calls is a strategy that offers leverage, good participation in the underlying stock price movement because of the relatively high delta, and eliminates the possibility of margin calls. Of course, like all options, LEAPS calls do have an expiration date, and unlike a stock, will not "last forever."

TRADING STRATEGIES

For emphasis, the definition of trader is restated: A trader has no position in and no interest in establishing a position in an option's underlying stock. Traders purchase calls with the intent of selling them at a profit, not to exercise them at some point in the future.

Trading Strategy 1: Buying Calls and Puts

Outright option purchases are probably the most frequently used trading strategies. Bullish? Buy calls. Bearish? Buy puts. These strategies are well known, but so are three attendant problems. First, if the underlying stock does not move, then the options will erode over time. Second, even if the stock moves in the desired direction, if it does not move enough, then purchased options will not realize a profit. And, third, even if the stock moves enough in the desired direction, it must do so within the life span of the option, or the option might expire worthless.

These three problems stem, in part, from the fact that options are wasting assets. Options have limited lives and, therefore, an expiration date. Time decay[4] is the number one enemy of the option buyer. Because of this, our next trading strategy will focus on a technique used to reduce the negative impact of time decay.

Trading Strategy 2: Bull Call Spreads and Bear Put Spreads

The general definition of a *spread* is a multiple-part position in which at least one option is bought and at least one different option is sold. This section will discuss the *bull call spread*, a two-part option strategy in which one call is purchased and a second call is sold. The second call has the same expiration but a higher strike. The following comments about the bull call spread which is a bullish strategy; can be applied to a *bear*

---

[4] For a more detailed discussion of time decay, see Chapter 2.

put spread which is a bearish strategy. A bear put spread is a two-part option strategy in which one put is purchased and a second put is sold. The second put has the same expiration but a lower strike.

As a starting point, consider an investor who is bullish on Fill & Drill Dental Explorations (OUCH) stock. The first strategy that comes to mind is: buy calls. But OUCH is a very volatile stock, so OUCH options are very expensive. In this example we assume 60 days to April expiration and the following prices:

		Implied Volatility
OUCH Stock	$78	54%
April 75 Calls	8	52%
April 80 Calls	5⅝	55%
April 85 Calls	3⅞	58%
April 90 Calls	2½	56%

A trader who is bullish on OUCH stock may be hesitant to purchase any of these call options because of the high implied volatility of these options. This is a situation in which spreading can be a useful tool.

If our trader initiates an April 75–85 bull call spread by buying one April 75 call at 8 and, simultaneously, selling one April 85 call at $3⅞, then the cost of initiating the position is $4⅛ or $412.50 per spread, not including commissions.

First, let's analyze the spread in terms of rights and obligations. By purchasing the April 75 call, the trader has the right to buy OUCH stock at $75. And, by writing the April 85 call, the trader has assumed the obligation to sell OUCH stock at $85. The combination of this right and this obligation determines the maximum value of the spread: $10. The overall risk profile of the position is determined by the initial cost and maximum value; and, in this case, the risk profile is described as limited risk and limited profit potential. The risk is limited to the initial amount paid of 4⅛, or $412.50, not including commissions, and this amount is lost if the stock is below $75 at expiration and the options expire worthless. The profit potential is limited to 5⅞, or $587.50, which is calculated by subtracting the spread's initial cost of 4⅛ from its maximum value of $10.

This *"limited profit potential"* feature of bull call spreads is problematic for some traders. "Why should I limit my gains?" they ask, "Doesn't one purchase options with the objective of realizing huge profits?" These questions deserve answers.

Why would someone purchase a bull call spread? First, consider the initial cost. Purchasing the 75 call outright has a cost, and a risk, of $8. In contrast, the April 75–85 bull call spread in this example has a cost and risk of $4⅛. The lower cost and, therefore, the lower risk are two advantages for the bull call spread.

"But what about purchasing the April 85 calls?" an astute observer might ask. Purchasing the April 85 calls at 3⅞ has nearly the same cost and risk as purchasing the April 75–85 call spread and the profit potential of the purchased April 85 call is not limited. So, the logic goes, if one is willing to risk approximately $4, why not purchase the 85 calls instead of the 75–85 bull call spread?

As the table below shows, even though it has limited profit potential, the 75–85 bull call spread outperforms the purchased 85 call under a number of scenarios. If the stock is unchanged at April expiration in 60 days, the spread will have a value of 3, the amount by which the 75 call is in-the-money, and result in a loss of 1⅛. In contrast, the purchased 85 call, in contrast, will expire worthless and results in a loss of the entire premium paid of 3⅞.

Another example of where the bull call spread in this example also outperforms the purchased 85 call is if the stock rises to $85 at expiration. In this case the spread would achieve its maximum value of $10 and result in its maximum profit potential of 5⅞. The 85 call, however, still expire worthless if the stock price is $85 at expiration.

When does the purchased April 85 call outperform the April 75–85 bull call spread? The answer is: Only when the stock rises above $94¾.

Stock Price at Expiration	Value of 75 Call	Value of 85 Call	Value of 75/85 Spd	Gain/(Loss) 75-85 Spd	Gain/(Loss) Long 85 Call
74	0	0	0	(4⅛)	(3⅞)
75	0	0	0	(4⅛)	(3⅞)
77½	2½	0	2½	(1⅝)	(3⅞)
80	5	0	5	⅞	(3⅞)
82½	7½	–	7½	3⅜	(3⅞)
85	10	–	10	5⅞	(3⅞)
87½	12½	2½	10	5⅞	(1⅜)
90	15	5	10	5⅞	1⅛
92½	17½	7½	10	5⅞	3⅝
94¾	19¾	9¾	10	5⅞	5⅞
95	20	10	10	5⅞	6⅛
96	21	11	10	5⅞	7⅛

The conclusion is that in flat to moderately bullish scenarios, the bull call spread will outperform a purchased out-of-the-money call. In a very bullish scenario, a stock price rising above $94¾ in this example, then the out-of-the-money call will generate a higher profit. What each trader must determine is: How bullish am I?

For many traders, the smaller loss if the stock falls, is unchanged, or rises to $78⅞ and the greater or equal profit if the stock rises to between $78⅞ and $94¾ make it reasonable to give up the greater profit potential offered by the long call above $94¾. But this is a personal decision which every trader must make.

The decision between purchasing bull call spreads and purchasing outright calls is a decision about trade-offs: lower breakeven, better returns over a limited range of the underlying stock, but no possibility of very large profits versus a higher break-even point and the opportunity for unlimited profits.

Another aspect of spreads that need discussion is how that spread values change as time changes and as the underlying stock changes in price. The concept of delta was explained in Chapter 3. *Delta* is an estimate of how much an option's price will change for a change in price of the underlying stock. Since bull call spreads involve two option positions, it is possible to use delta to anticipate how a spread may react to a change in the price of the underlying stock. In other words, a spread has a delta.

The delta of a bull call spread is the net delta of the option positions that create the spread. In the example above, assume that the 75 call had a delta of +0.63 and that the 85 call had a delta of +0.39. The net delta would then be (+0.63) – (+0.39), or +0.24. Remember, the 85 call is sold, consequently its delta is subtracted from the delta of the 75 call that was purchased.

Because the delta of the spread in this example is lower than the delta of either a purchased 75 call or a purchased 85 call, the spread will change in price more slowly than either of the two options. This is why spreads are sometimes described as "slow moving animals." Traders must realize that a bull call spread will, as a rule, reaches its maximum value only if the stock has risen above the higher exercise price and there is very little time left to expiration.

BULL CALL SPREAD VS. LONG CALLS (PROS & CONS)

Pros:

- Lower initial cost, lower risk
- Reduced breakeven

BULL CALL SPREAD VS. LONG CALLS (PROS & CONS) (CONTINUED)

- Range over which the bull spread will outperform is lower than range over which long call will outperform

Cons:

- Limited upside potential (can't hit a home run)
- Total value of spread is relatively slow moving (lower net delta)
- The Maximum value is reached only close to expiration or if all options are deep in-the-money
- Higher transaction costs since more options are traded

Recap

A less aggressive strategy than an outright call purchase, the bull call spread helps minimize the time premium paid for a position and limits the impact of time erosion. The strategy has limited risk, but also limited profit potential.

NOTE FOR THE BEARISH INVESTOR
BEARISH? DO PUTS SEEM TOO EXPENSIVE?

The same reasons that were used to purchase the bull call spread instead of purchasing a call outright can be used to justify the purchase of a bear put spread instead of purchasing a put outright. The bear put spread strategy consists of purchasing one put and selling a second put. The second put has the same expiration but a lower strike price.

Trading Strategy 3: Calendar Spreads

In bull call spreads and bear put spreads, the spreading technique was used to minimize the impact of time decay. In the following example, spreading is used to limit risk while taking advantage of time decay.

Assume that a trader's forecast for That Burger Joint (FAT) is for the stock price to trade in a "narrow range" over the next 45 days. Selling calls and/or puts outright is the typical strategy of choice to take advantage of such a forecast. However, if the forecast is wrong, and the underlying stock trades outside of the forecasted narrow range, then positions involving uncovered options have unlimited risk in the case of short calls and substantial risk in the case of short puts.

Two of the neutral market strategies traders frequently ask about are *butterflies* and *condors*, and profit-loss diagrams of these strategies appear in Chapter 3. These strategies, in addition to having catchy names, will benefit from the passage of time while simultaneously limiting risk should the underlying stock move up or down sharply. But these strategies have a major drawback for the "average" individual investor: the number of options involved and, therefore, high transaction costs. Butterfly spreads are comprised of three option positions and condor spreads are comprised of four. This means that three or four commissions must be paid when entering into the strategy and possibly the same when closing. For many individual investors these spreads appear attractive at first, but end up being impractical due to commission costs.

Calendar spreads, however, offer the advantages of butterfly and condor spreads—profiting from time decay with limited risk—without the disadvantage of three or four commissions. Calendar spreads involve the purchase of one option and the simultaneous sale of another option with the same underlying and same strike price but with a different expiration. Because only two options are involved, it is a more realistic strategy for the "average" individual investor.

How do calendar spreads work? Returning to our example of FAT, assume the stock price is currently $49 and the following call options are listed for trading:

March 50 Calls (28 days): 1½
April 50 Calls (56 days): 2⅜
May 50 Calls (91 days): 3⅛

While selling the March 50 calls uncovered, *i.e.* without owning the underlying stock, maximizes the benefit of time decay, it is also a strategy with unlimited risk. To limit this risk, a calendar spread is created by purchasing another call with a later expiration date. If the May 50 calls are purchased at 3⅛, for example, then a calendar spread is established for a net debit of $1⅝ ($3⅛ – $1½).

Before estimating the profit potential of this calendar spread, consider the risk under two large-movement scenarios, a big down move in the underlying stock and a big up move. If the price of FAT drops substantially, to $30 at March expiration, for example, then the March 50 calls would expire worthless, and the May 50 calls would have little value or, at worst, no value. The result is that the amount initially invested in the position, 1⅝, has decreased to zero for an unrealized loss of

1⅝. The loss would not be realized until the May 50 calls are closed or expire worthless.

If the price of FAT rises substantially, to $70 at March expiration, for example, then the March 50 calls would be worth intrinsic their value of 20 in this example. The May 50 calls, being deep in-the-money, would also be trading at their intrinsic value or, at best, slightly higher than their intrinsic value. The result is, with both calls trading at or near 20, the calendar spread would be trading near zero. Consequently, the amount initially invested in the position, 1⅝, will have decreased to zero for a loss of 1⅝.

In either extreme scenario—a substantial rise or a substantial decline in the price of the underlying stock—the loss on the calendar spread will not exceed the initial cost of the position. Thus, the risk of a calendar spread is limited to the amount invested.

So much for the worst case scenario. What is the best thing that can happen? The answer is: nothing! That is, no movement in the price of the underlying stock until the expiration of the shorter-term option. Assume, for example, that at March expiration, the price of FAT is $49, unchanged from when the calendar spread was established. In this case, the short March 50 call will expire worthless, and the long May 50 call will have a positive value—but how much value?

At the expiration of March options, the May options will still have two months until they expire. Consequently, their value will depend on how the market is pricing FAT options, *i.e.*, a function of expected volatility. Although this is impossible to determine with certainty in advance, it is possible to make an educated estimate making reasonable assumptions. If the implied volatility of FAT options does not change, then it is reasonable to expect that May 50 calls will be trading at or near a price of $2⅜, which is where the April 50 calls were trading one month ago when the March–May calendar spread was established.

Given this assumption about the value of 2⅜ for the May 50 call, then a profit of ⅞ can be estimated. The calendar spread was initially created for a net debit of 1⅝, and, with the May 50 call at 2⅜ and the March 50 call expiring worthless, the calendar spread has a value at March expiration of 2⅜. 2⅜ minus 1⅝ results in a profit of ¾, not including commissions.

Back to the example. Assume everything goes as planned, *i.e.*, at March expiration the price of FAT is unchanged at $49 and the March 50 call expires worthless. What might we do? We have three choices. First, we can unwind the position by selling our long May 50 call. The profit would then be $¾ ($2⅜ − $1⅝), assuming unchanged volatility.

Second, if we are no longer neutral on the stock but have turned bullish, we can hold on to our long May 50 call. We then have $1⅝ at risk and the full upside potential of the May 50 call.

Third, if we are still neutral on FAT, we could establish a second calendar spread by selling an April 50 call at $1½, once again assuming the implied volatility has remained unchanged. Our position would then be long one May 50 call and short one April 50 call.

The choice as to whether to liquidate, hold the long call, or re-establish another time spread, must be guided by our forecast for the price of FAT for the next month.

A question that is frequently asked is, "Should puts or calls be used when establishing a calendar spread? Does it matter?"

If a stock is trading exactly at a strike price and the forecast is "dead neutral," then there is not a significant difference in risk or profit potential between establishing a calendar spread with calls or with puts. However, the choice gives a slight bias to the strategy.

Assume a stock trading at $80 on which we have a neutral forecast and on which we want to establish a calendar spread. If we have a slightly bullish bias, we should use puts. Why? If our bias is correct and the stock drifts up slightly, our short put option will expire worthless and we will not have to trade out of it.

By the same token, we should use calls if we have a slightly bearish bias.

Even if a stock is not trading exactly at a strike, the choice between puts and calls should not be ignored. With a stock trading at $98, establishing a calendar spread with the 100 put options would be slightly bullish, as we would look for the stock to rise to or slightly above $100. Using the 95 calls would be slightly bearish, because we would look for the stock to drift down to or slightly below $95.

CALENDAR SPREADS: PROS OR CONS

Pros:

- Risk limited on both upside and downside
- Benefit from option time decay
- Fine tuning possible through the use of puts or calls

CALENDAR SPREADS: PROS OR CONS (CONTINUED)

Cons:

- Profit potential can be, at best, only be estimated
- Position hurt by too large an up or down movement of the underlying up or down
- When short American-style options, possibility of early assignment exists

Recap

Calendar spreads are a limited risk strategy that will allow us to benefit from time erosion if the underlying value remains within a narrow range. Although the maximum risk is known, the profit potential can only be estimated.

Trading Strategy 4: Diagonal Spreads Using Leaps

For a conservative investor, one strategy that seeks to profit from option time decay is *covered writing*, selling calls on a share-for-share basis against owned stock. As discussed above in the section entitled Investing Strategy 3, if the underlying stock does not move, profits are realized as the short calls lose value, and; if the stock rises, the options are covered. For a trader, this strategy may have some appeal but it usually runs into a major obstacle: the capital required to purchase the stock.

Investment Strategy 4, discussed earlier in this chapter, was the purchase of in-the-money LEAPS calls as a stock substitute was discussed earlier as an investment strategy. In this trading strategy, LEAPS calls act as a stock substitute as part of a buy-write. Consider the following:

Medium Tech Fully Conductor (BIT):	$64
Jan 2001 50 calls (18 months):	$20
Sep1999 70 calls (2 months):	$ 1⅝

A plain vanilla covered write, as discussed in Investing Strategy 3 above, would consist of buying BIT at $64 and writing the September (1999) 70 calls on a share-for-share basis at $1⅝. This may be deemed acceptable for a conservative investor, but some traders may consider it too capital intensive.

As an alternative to purchasing the stock, a trader might consider purchasing the January 2001 LEAPS call at $20. And, against this long position, the trader could sell the September 1999 70 calls at $1⅝. The total cost to establish this position is $18⅜ compared to the $62⅜ cost of the plain vanilla covered write described above.

The reasons that deep-in-the-money LEAPS calls are chosen in this example because they possess the same given earlier in this chapter: low time premium, minimal time decay, and a relatively high delta. But what are the risks and potential returns of a diagonal spread using LEAPS calls?

In a plain vanilla covered write, the maximum profit potential is realized if the stock price rises above the strike price of the short call. In this case, the calls are assigned and the stock is sold. This is not the best outcome, however, for a diagonal spread involving LEAPS calls.

Consider what the outcome would be if the price of BIT stock rose substantially above $70 at September 1999 expiration and the September 70 calls were assigned. In this case, although the trader owns January 2001 50 calls, the trader does not own BIT stock that can be delivered against the assigned September 1999 70 calls. And, although it would be easy to exercise the January 2001 calls to get stock to deliver, this not likely to be an astute move. With fourteen months left to January 2001 expiration, these calls are likely to have a considerable amount of time premium, which would be lost if they were to be exercised. With the stock price above $70 at September, 1999, expiration, it is likely to be more profitable to sell the LEAPS calls to close the position—thereby capturing the time premium—and purchase stock to deliver against the assigned short calls. The disadvantage of using these two steps to meet the assignment is the number of transactions and, therefore, the increased commission costs. A trader who uses diagonal spreads must monitor the position as expiration of the shorter-term option approaches and compare the time premium in the LEAPS call with the transaction costs of selling that call and purchasing stock. The alternative with the lowest net cost, including time premium, will be the preferred choice.

What, then, is the maximum-profit scenario for the diagonal spread strategy? If the price of the underlying stock remains unchanged, then the short shorter-term calls expire worthless and the long LEAPS calls keeps the most time premium. In the example above, with BIT unchanged at $64 at September 1999 expiration, the September 1999 70 calls would expire worthless for a profit of $1⅝ and the January 2001 50 calls, assuming implied volatility and interest rates remain constant, would decline by approximately ½ to 19½. The net result, in this case, would be a profit of $1⅛, not including commissions, on an investment of $18⅜, not including commissions, in approximately 60 days. At this point, another short term option, the October or November series could be sold, once again with the goal of profiting from the accelerating time decay of short term options. Theoretically, up to 18 different one-month options could be sold against the 18-month LEAPS call in this example, but the success of continuously selling short-term options would depend on the neutral movement of the underlying stock price for that period of time.

And what could go wrong with the diagonal spread strategy? The same thing that could go wrong with a plain vanilla covered write could also cause a loss for a diagonal spread: the underlying stock could fall in price. The worst-case scenario is that the price of BIT stock drops substantially, causing the LEAPS calls to decline as well. If held to expiration, and the underlying stock price is below the strike price, the LEAPS calls could expire worthless and result in a loss of the entire 18⅜ paid to establish the original position.

DIAGONAL SPREAD:
WATCH IT!!

With plain vanilla covered writes, little day-to-day monitoring is required. The position is initiated, an eye is kept on the position, but nothing needs to be done if the stock rises above the strike price of the short call.

With a diagonal spread, however, monitoring is definitely the order of the day. What would happen if the stock price ran above the strike price of the short call? The short call could be assigned. Is this a problem? In theory this is not a problem, because the LEAPS call could be exercised to purchase the stock for delivery. But, in practice, exercising the LEAPS call is unlikely to be the best way to meet the assignment, since because there may be some time premium remaining in the long LEAPS call. Consequently, when faced with an assignment on the short call it is likely to be more profitable to close the LEAPS call by selling it (to capture the remaining time premium) and to purchase the stock in the open market to be delivered. If this sounds messy, it is.

Traders want to avoid assignment on the short calls in a diagonal spread. And, while there is no guaranteed method of avoiding assignment, the chances of being assigned can be reduced by monitoring the position and closing out the short calls by repurchasing them when assignment risk is deemed to be high.

Assignment risk can be gauged by looking at the time premium left in the short calls. When there is some time value in an option's price, the risk of assignment is relatively low. If time value is at or close to zero, however, the risk of early assignment increases.

Should the risk of early assignment become great, a trader may "roll out" the short call by closing the current position by repurchasing it in the market and selling another call with a later expiration.

Recap

A diagonal spread with LEAPS calls is created by purchasing a longer-dated LEAPS call and, simultaneously, selling a call with a shorter-term expiration and a higher strike price. The goal of the strategy is to leverage percentage returns from selling covered calls relative to traditional covered writing. The risk is that percentage losses can be leveraged as well if the price of the underlying stock declines.

Diagonal spreads are neutral market strategies, because the maximum profit potential occurs when the price of the underlying stock remains within a relatively narrow range.

DIAGONAL SPREAD:
PROS & CONS

Pros:

- Takes advantage of options' time decay
- Greater leverage than covered writing
- Limited potential risk

Cons:

- Substantial downside risk
- Requires monitoring
- Possibility of early assignment

SUMMARY

Options can be used to target either investment-oriented or trading-oriented goals. The primary focus of an investing strategy is to buy, sell, or manage a position in the underlying stock, and options are a tool used to achieve that goal. In contrast, options also are the focus of a trading strategy.

This short chapter did not attempt to enumerate all of the possible uses of options; only six investing-oriented strategies and four trading strategies were discussed.

For investing purposes, options can be used in an attempt to buy stock at a price below the current market price (writing puts) or in an attempt to sell stock above the current market price (writing covered calls). The buy-write strategy is a neutral-market strategy which attempts to earn a "high" current income. By combining the covered call and cash-secured put, an investor can either add to an existing stock position or reduce it, and collect option premiums at the same time. Options can also be used as an alternative to "doubling up" on an unprofitable stock position. The goal of the option strategy discussed in this regard is to lower the break-even point on an unprofitable stock position without increasing risk, which doubling up would do.

LEAPS call options can be used as an alternative to purchasing stock on margin. Although the time premium in a LEAPS call can be compared to interest paid on a margin loan, a LEAPS call generally has significantly less total risk than buying stock on margin.

The potential opportunities and challenges of buying calls and puts are well known to many traders. The advantages are that the risk is limited and profit potential is leveraged relative to purchasing stock. The challenges are that the forecast has to be very specific with regard to price movement, time horizon, and implied volatility. Traders must forecast all three components when trading options.

Bull call spreads and bear put spreads are two-part option strategies which attempt to minimize the impact of time decay and changes in implied volatility. The disadvantage is that profit potential is limited.

Calendar spreads involve the purchase of one call (or put) and the simultaneous sale of another call (or put) with the same underlying and same strike price but with an earlier expiration. Calendar spreads attempt to profit from option time decay while limiting risk at the same time. Stocks must trade in a narrow range near the strike in order for the maximum profit potential to be realized.

Diagonal spreads using LEAPS call options is a trader's strategy, which is sometimes compared to covered call writing, a strategy for investors. Diagonal spreads typically require a lower investment than covered writing and, therefore, percentage profits are higher if the forecast is correct. The risk of a diagonal spread is the same as the risk for covered writing: that the underlying stock drops substantially in price. The best outcome, however, for a diagonal spread is if the underlying stock trades in a narrow range, whereas the best-case scenario for a covered write is if the stock price rises above the strike price of the short call.

Investors and traders alike can find a place for options in their activities if only they take the time to understand how options work and to learn the differences in thinking that is required.

CHAPTER 6

STRATEGIES FOR INSTITUTIONAL INVESTORS

James B. Bittman
Eric Frait
Andrew B. Lowenthal

A portfolio manager has many responsibilities beyond those related to specific equity selection. Some involve the broader issues of market timing, portfolio asset allocation, and trade execution. The need to deal with these responsibilities and the ever present pressure to reduce costs were among the driving forces that led to the creation of index options.

The use of index options has grown dramatically since the first index options were listed in 1983, and many index option markets have become extremely liquid. Because markets are liquid and able to handle sizable orders, index options are in fact, as well as in theory, a beneficial tool for the knowledgeable portfolio manager.

This chapter explains how a wide variety of option strategies can be used to manage risk and to increase returns. As strategies that can be used in different market environments are analyzed, the importance of having a market view is stressed. We first look at portfolio insurance, comparing index put options to equity puts, and discuss the advantages of each. We then look at five strategies that decrease market exposure and compare the results of each. The next subject considered is how to increase market exposure using index call buying strategies. Finally, other portfolio strategies, such as covered call writing, portfolio repair, and writing equity puts, are discussed.

TABLE 6-1

Put Options Compared to Insurance Policies

Insurance Policy	Put Option
Value of Asset	Underlying Value (Stock Price or Index Level)
Amount of Deductible	Option Strike Price
Duration of Policy	Time Until Expiration
Interest Rates	Interest Rates and Dividend Yield
Risk	Volatility
= Amount of Insurance Premium	= Amount of Option Premium

PORTFOLIO INSURANCE WITH INDEX PUT OPTIONS

Perhaps the most basic use of index options is to insure, or *hedge*, a portfolio against a broad market decline while at the same time allowing that portfolio to participate in any market advance. Table 6-1 shows the similarities between put options and standard insurance policies on a car or home.

Insuring a Simple Portfolio: Index Puts versus Equity Puts

It is easy to demonstrate how the purchase of index put options can protect the value of a well-diversified portfolio, the makeup of which generally matches the index on which the option is purchased. This concept is illustrated first in a simplified seven-stock, $1,003,100 portfolio that is assumed to track the performance of the S&P 100 Index. (The potential problem with this assumption for larger portfolios is discussed later.) Consider the seven-stock portfolio shown in Table 6-2. Closing prices were taken from *The Wall Street Journal*, May 1, 1998, 78 days from July expiration.

Insurance with Equity Puts

The question is, simply, other than liquidating the portfolio, what can a manager do to protect a portfolio from an expected short-term market decline? Buying puts, of course, is appropriate, but there are two types of put options from which to choose: individual equity put options and index put options. So the next step is to analyze the potential performance of both types and then to compare the results. A portfolio of equity puts that matches the equity holdings in the sample portfolio is presented in Table 6-3. Again, closing prices are from *The Wall Street Journal*, May 1, 1998.

Table 6-4 shows how this portfolio of puts would profit and how the stock plus puts portfolio would change in value if the stock market declines 15 percent. The assumption is that, at option expiration, each stock in the portfolio has declined 15 percent to match the decline in the overall market. At expiration, remember, there is no time premium left in option prices.

TABLE 6-2

A Seven-Stock Portfolio

Issue	# of Shares	Price	Value
AT&T	2000	60⅛	$ 120,250.00
PepsiCo	1800	39¹¹⁄₁₆	71,437.50
Chrysler	2400	40¼	96,600.00
Exxon	1900	73¹⁄₁₆	138,818.75
DuPont	2100	72¹³⁄₁₆	152,906.25
Merck	2500	120½	301,250.00
Texas Instruments	1900	64⅛	121,837.50
Total Value			**$1,003,100.00**

TABLE 6–3

A Portfolio of Equity Put Options That Matches the Seven-Stock Portfolio

Issue	Price	Put Options	Price	Quantity	Cost
AT&T	60⅛	July 60	2½	20	$ 5,000.00
PepsiCo	39¹¹⁄₁₆	July 35	½	18	900.00
Chrysler	40¼	July 40	2¹⁄₁₆	24	4,950.00
Exxon	73¹⁄₁₆	July 70	1¹¹⁄₁₆	19	3,206.25
DuPont	72¹³⁄₁₆	July 70	1⁹⁄₁₆	21	3,281.25
Merck	120½	July 120	5¾	25	14,375.00
Texas Instruments	64⅛	July 60	2¾	19	5,225.00
Total Cost					**$36,937.50**

TABLE 6-4

Effect of a 15% Decline in Each Stock

Put Option Portfolio				
Issue	**Stock Price (Down 15%)**	**New Option Price**	**Quantity**	**Value**
AT&T	51⅛	8⅞	20	$17,750.00
PepsiCo	33¾	1¼	18	2,250.00
Chrysler	34³⁄₁₆	5¹³⁄₁₆	24	13,950.00
Exxon	62⅛	7⅞	19	14,962.50
DuPont	61⅞	8⅛	21	17,062.50
Merck	102⁷⁄₁₆	17⁹⁄₁₆	25	43,906.25
Texas Instruments	54½	5½	19	10,450.00
Total put option value after 15% market decline				**$120,331.25**
Total put option cost				**$36,937.50**
Total put option profit				**$83,393.75**
Stock Plus Puts Portfolio				
Original Portfolio Value				**$1,003,100.00**
15% decline in value				**($150,465.00)**
Profit on put options				**$83,393.75**
Portfolio declined by				**($67,071.25)**

The $83,393.75 put option profit is the payoff from the "insurance policy." This profit directly reduces the total decline in value of the equity portfolio of $150,465. As a result, the net decline in value of the stock plus puts portfolio is $67,071.25 ($150,465 − $83,393.75). This is only a 6.7 percent decline compared to the overall market and uninsured portfolio decline of 15 percent.

Insurance with Index Puts

A second insurance strategy involves the purchase of index put options. On May 1, 1998, the S&P 100 Index (OEX) closed at a level of 536.48, and the July 535 put closed at 17½.

First, it is necessary to calculate the number of index puts required to insure the portfolio. OEX options represent a cash settlement value equal to $100 times the index level. This means that each option with a strike price of 535 represents a market value of $53,500 (535 index level × $100 index multiplier). In this example, the equity portfolio has a total beginning value of $1,003,100. Using the July 535 put options, insuring this portfolio requires the purchase of 19 puts ($1,003,100 portfolio value ÷ $53,500 value per put = 18.75 puts = 19). Obviously, some rounding is always involved in this calculation. At a price of 17½, the purchase of 19 options costs $33,250 (19 × $1,750), not including transaction costs.

The next task is estimating how this insurance strategy might benefit the portfolio. First, calculate the index level after a 15 percent decline. Second, determine the option price, assuming no time premium (a conservative assumption). Finally, calculate the put option profit and resulting benefit to the portfolio.

1. A 15-percent decline in the index from 536.48 results in a level of 456.01 (536.48 × .85).
2. At an index level of 456.01, the 535 put, at expiration, has a point value of 78.99 (535 − 456.01) and a dollar value of $7,899 (78.99 × $100).
3. With a dollar value of $7,899, each put has a profit of $6,149 ($7,899 value − $1,750 cost) for a total put profit of $116,831 ($6,149 × 19).

Table 6-5 summarizes the impact of a 15 percent market decline on the OEX Index, the equity portfolio, equity puts, and the index puts.

In this example, the payoff favors the index option purchase. The difference is $33,437.25, which is about 3.3 percent of the $1,003,100 portfolio. Although the index puts would have been the preferred

TABLE 6-5

Performance of the Index and Three Portfolios: Equity Portfolio, Equity Put Options, Index Put Options

	S&P 100 (OEX) Index	Equity Portfolio	Equity Put Options	Index Put Options
Beginning Value	536.48	$1,003,100.00	$ 36,937.50	$ 33,250.00
Ending Value	456.01	852,635.00	120,331.25	150,081.00
Change in Value	–15%	–$ 150,465.00	+$ 83,393.95	+$116,831.00

choice in this example, this is not always the case. There are other factors to consider, and a closer look at the two alternatives reveals several differences that become important as the portfolio in question gets bigger.

Consider first the issue of manageability. In the simple example given, the purchase of seven different equity put options is required, and all stock issues had July option series available 78 days out. In the real world, however, it is unlikely that all stocks in a diversified portfolio will have options available with matching expirations. Due to the way option expirations become available, there are times when some stock have options with 30-, 60-, 90-, and 150-day expirations and other stocks have options with 30-, 60-, 120-, and 210-day expirations. This means that matching a portfolio of equity puts with an equity portfolio inevitably result in the purchase of puts with different expirations. When different quantities of puts are added to the variety of expiration months, the management problem becomes obvious.

A second issue is the risk of portfolio disruption. When using equity put options to insure a portfolio, the risk of error is more than losing money. When equity puts are in-the-money, there is a great likelihood of automatic exercise at expiration. This means that stocks will be automatically sold. For taxable portfolios, the result can be significant. Extra commission charges from selling stocks and buying them back are other negatives.

There is also the possibility that listed put options are not available on some equities. Even though over-the-counter put options might be available, these options markets are generally not as liquid as listed options. The portfolio manager is left to contend with the problems of illiquid markets, such as wide bid/ask spreads and the possible difficulty of entering into or offsetting an option position.

ADVANTAGES OF INDEX OPTIONS

Fortunately, index options have features that offset some of the disadvantages of equity options. First, consider the issue of manageability. When insuring a portfolio with index options, one quantity of one option series is used. The problem of portfolio disruption is also absent. The cash-settlement feature of index options means there is no risk that individual equities will be sold by automatic exercise.

The likelihood of lower commissions is another advantage of index options. As explained above, one index option with a strike price of 535 covers a market equivalent value of $53,500. Compared to an equity put on 100 shares with a $50 strike, which only covers $5,000 in value, only 1 index option is required for every 10.7 equity options. This ratio, of course, is different for every portfolio, depending on the average stock price in the portfolio and the current index level. Nevertheless, commission discounts are likely to result from buying one larger quantity of index options, compared to buying several smaller quantities of equity options.

The relative advantages of index options apply to many situations and have fueled the growth of index option markets; however, these advantages do not apply in certain situations. For example, in times of market uncertainty, index options may not be cheaper than equity options. The possibility that index options could be more expensive than equity options may seem counterintuitive, because index options represent a diversified portfolio, and diversification generally implies less risk. Consequently, put options on an underlying with less risk should be cheaper. Yet, in times of market uncertainty, the implied volatility of index options can be higher than for a group of equity options, and the result can be a higher price for the index options.

In the simple example given above, the index put strategy requires a cash outlay of $33,250, and the equity put strategy requires $36,937.50. Consider also commissions and payoffs. Commissions for 19 index puts are obviously less than commissions for 146 equity puts on seven stocks.

In a real situation, the payoff issue is much more uncertain and depends entirely on how well the portfolio in question moves with the index on which the puts are purchased. As long as the two move in tandem, the index put payoff will theoretically equal the equity put payoff. However, portfolios that are weighted differently than the index on which options are used have a significant risk of behaving differently. If the manager of a specialized portfolio sees an imminent decline in its equities, there may not be a useful index option available.

PORTFOLIO INSURANCE: STRATEGIC CONSIDERATIONS

To implement a portfolio insurance strategy, a portfolio manager must evaluate alternatives. Using a general example, the following section illustrates how five strategies might perform and discusses considerations used in the selection of each.

The Alternatives

We look at five alternatives in terms of performance, then as a function of the portfolio manager's outlook, followed by the cost and risk of each alternative. We then compare the strategies using seven criteria.

Consider the investment alternatives of a portfolio manager with a bearish outlook. One choice is selling the portfolio. Although this has many wide-ranging implications, it is the ultimate insurance policy. When a portfolio is in cash, it has no risk if a market decline occurs. The second alternative is a variation on this strategy—selling stock index futures contracts. Although an in-depth discussion of stock index futures is not within the scope of this book, this strategy has attracted a considerable following.

Strategy alternatives 3, 4, and 5 involve the purchase of index put options. Alternative 3 is an index option strategy similar to the example with the seven-stock, $1,003,100 portfolio presented above. This strategy is known as *buying a portfolio equivalent of at-the-money puts*. Alternative 4 involves buying the same number of index puts as in the alternative 3, but with a lower strike price. This strategy is known as *buying a portfolio equivalent of out-of-the-money puts*. The final alternative is not obvious to many portfolio managers, because it involves buying a quantity of puts that is greater than the portfolio equivalent number in strategies 3 and 4. This strategy involves the purchase of a greater number of out-of-the-money puts. It is known as buying a *portfolio multiple of out-of-the-money puts*.

How the Strategies Perform

Figure 6-1 illustrates how the portfolio can change in value (vertical axis), given a change in the index level (horizontal axis) for each of the five alternatives. Lines 1 through 5 represent the theoretical outcome at expiration of the five strategies. Note that the strategies would have been implemented at some point prior to expiration.

PORTFOLIO INSURANCE STRATEGIES

Strategy 1: Selling the Portfolio

Converting a portfolio to cash eliminates the potential risk and the potential profit from market fluctuations. Consequently, the return, regardless of market movement, is a horizontal straight line above zero equal to the interest earned during the time period. In Figure 6-1, a 4 percent annual T-Bill interest rate is assumed, so the 90-day return is 1 percent.

FIGURE 6-1

Portfolio Insurance Strategies

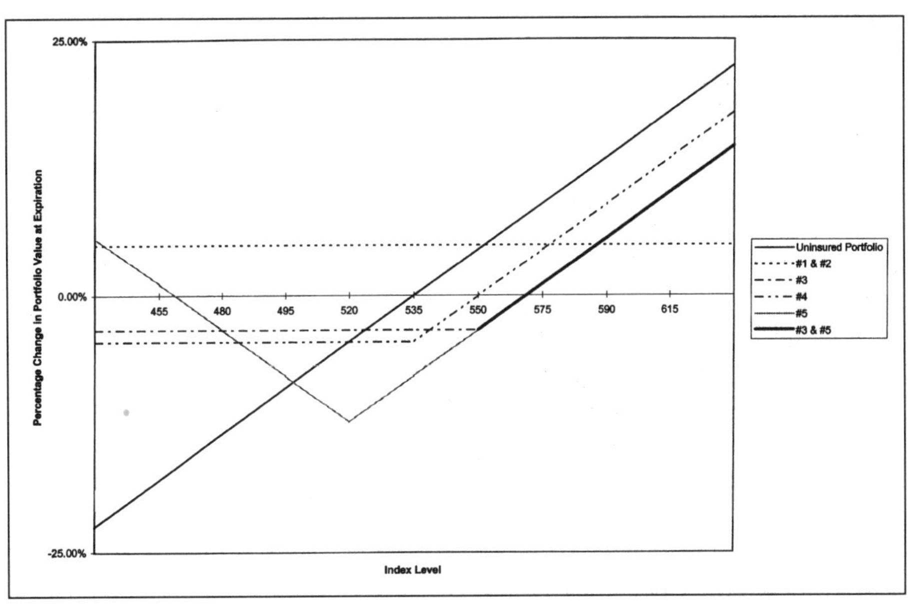

Strategy 2: Selling Stock Index Futures

It is not within the scope of this discussion to cover all the practical difficulties in the execution of this strategy. Suffice it to say that the goal of selling stock index futures against an equity portfolio is to eliminate the risk of a predicted market decline while still owning the portfolio. The negative is that profit potential from a market rise is also eliminated. As a result, at the end of the time period, the return from this strategy should, in theory, equal the T-bill rate of interest, because the dividends on an S & P 100 Index portfolio plus the futures premium equals T-bill interest. The horizontal line depicting strategy 2, therefore, is the same as that for strategy 1.

Strategies 3, 4, and 5: Buying Puts

The best way to explain these strategies is by using a model portfolio with a total value of $10 million. Assume that the portfolio is broadly diversified and that its performance matches that of the S&P 100 Index.

Specifically, assume that the beta of the portfolio is 1. In this example, the S&P 100 Index stands at a level of 535 when the insurance strategies are implemented. For this example, Table 6-6 contains strike prices and market prices of available 78-day put options.

The number of puts required to protect a portfolio is known as the *portfolio equivalent quantity of options*. This quantity is calculated by dividing the dollar value of the portfolio by the product of the index level and the index multiplier. For a $10 million portfolio with the OEX index at 535, the portfolio equivalent quantity of puts is 187; the calculation is shown at the bottom of Table 6-6.

Strategy 3: Buy a Portfolio Equivalent of At-the-Money Puts

For strategy 3, purchasing 187 90-day at-the-money 535-strike puts at 17½ each (from Table 6-6) costs $327,250. The balance of $9,672,750 remains invested in a diversified portfolio of equities that matches the performance of the S&P 100 Index (OEX). This means that 78-day "insurance" costs ap-

TABLE 6-6

Strike Prices and Market Prices of Available Put Options

S&P 100 at 535	
90-Day Put Strike Price	**Option Price**
535	17½
530	15¾
525	14¾
520	13⅛
510	11⅛
500	9⅛

Calculating portfolio equivalent quantity of options:

$$\text{Required Number of Puts} = \frac{\text{Portfolio dollar value}}{(\text{Index level} \times \text{Multiplier})}$$

$$\text{For a \$10 million portfolio: } \frac{\$10{,}000{,}000}{(535 \times 100)} = 187$$

TABLE 6-7

Strategy 3: Buying a Portfolio Equivalent of At-the-Money Puts

Index Level at Option Expiration	Index Percent Change	Equity Portfolio Value ($)	Put Option Value ($)	Total Portfolio Value ($)	Portfolio Percent Change
415	–22.43%	7,503,161.21	2,244.000.00	9,747,161.21	–2.53%
455	–14.95	8,226,357.48	1,496,000.00	9,722,357.48	–2.78
480	–10.28	8,678,355.14	1,028,500.00	9,706,855.14	–2.93
495	–7.48	8,949,553.74	748,000.00	9,697,553.74	–3.02
520	–2.80	9,401,551.40	280,500.00	9,682,051.40	–3.18
535	0.00	9,672,750.00	0.00	9,672,750.00	–3.27
550	+2.80	9,943,948.60	0.00	9,943,948.60	–0.56
575	+7.48	10,395,946.26	0.00	10,395,946.26	+3.96
590	+10.28	10,667,144.86	0.00	10,667,144.86	+6.67
615	+14.95	11,119,142.52	0.00	11,119,142.52	+11.19
655	+22.43	11,842,338.79	0.00	11,842,338.79	+18.42

187 put options with a strike price of 535 are purchased for $327,250 (187 × $1,750).

The balance of $9,672,750 remains invested in a diversified portfolio of equities that matches the performance of the S&P 100 Index (OEX).

* Dividends are not included.

proximately 3.8 percent of the portfolio. The results of this strategy are summarized both by line 3 in Figure 6-1, and by Table 6-7.

Table 6-7 and line 3 in Figure 6-1 demonstrate two important concepts about portfolio insurance. First, on the downside, the loss in portfolio value is limited to a maximum percentage, even though the market average could fall farther. On the upside, the portfolio remains intact and participates in a market rally. However, on the upside, the portfolio under-performs during the market rally because part of the portfolio is spent (and lost) on the put insurance.

A comparison of line 3 in Figure 6-1 and Table 6-7 reveals an apparent inconsistency. Line 3 in Figure 6-1 is horizontal and represents a constant maximum loss. The Portfolio Percent Change column in Table 6-7, however, indicates that a market decline of increasing proportion actually results in a smaller portfolio loss. How is this inconsistency explained?

In theory, as represented by line 3 in Figure 6-1, the maximum loss in the event of a stable or declining market is the cost of the put options, which act like insurance on the portfolio. In practice, however, paying for the put options reduces the amount invested in equities. The result is that 187 puts represent $10 million in market value, although only $9,672,750 is invested in equities. On the downside, the puts rise slightly faster than the decrease in the equities; and on the upside, the equities rise slightly less than the theoretical $10 million portfolio. It is possible to purchase a number of puts such that a maximum loss results in the event of a stable or declining market. That number is determined by the use of a linear programming optimizing technique that finds the number of puts equal to the portfolio value being insured. The issue here is that the cost of the put options is part of the total capital whose performance is being measured. Some writers have tried to avoid this issue by assuming that the cost of the put options equals the dividend income or that the puts are purchased with funds outside of the portfolio being insured. A complete discussion of portfolio performance evaluation is beyond the scope of this book, but portfolio managers must face this issue when using option-buying strategies.

Strategy 4: Buy a Portfolio Equivalent of Out-of-the-Money Puts

Out-of-the-money puts cost less than at-the-money puts. If out of-the-money puts are purchased, however, the equity portfolio is uninsured against a market decline between the current market level and the strike price of the out-of-the-money puts. Consequently, this portfolio insurance strategy is similar to purchasing a policy with a large deductible. In comparison, purchasing at-the-money puts is similar to buying a policy with no deductible.

As Table 6-8 indicates, the purchase of 187 puts with a strike price of 520 (15 points below the current market level) requires a cash outlay of $245,437.50. With these out-of-the-money puts, the maximum theoretical loss in the portfolio is approximately 5.20 percent. This occurs if the market declines to an index level of 520 or lower. At an index level of 520, the $9,754,562.5 invested in equities has declined to $9,481.070.09 and the puts—at a cost of $245,437.50—expire worthless for a total loss of $518,930 or approximately 5.20 percent of the original $10,000,000 portfolio. Below an index level of 520, the rise in value of the puts will, in theory, offset the decline in value of the equity portfolio, assuming the portfolio perfectly follows the index. On the upside, in theory, the insured portfolio under-performs the market by approximately 2.52 percent—the cost of the put options.

TABLE 6–8

Strategy 4: Buying a Portfolio Equivalent of Out-of-the-Money Puts to Insure a $10 Million Portfolio

Index Level at Option Expiration	Index Percent Change	Equity Portfolio Value ($)	Put Option Value ($)	Total Portfolio Value ($)	Portfolio Percent Change
415	–22.43%	7,566,623.25	1,963.500.00	9,530,123.25	–4.70%
455	–14.95	8,295,936.33	1,215,500.00	9,511,436.33	–4.89
480	–10.28	8,751,757.01	748,000.00	9,499,757.01	–5.00
495	–7.48	9,025,249.42	467,500.00	9,492,749.42	–5.07
520	–2.80	9,481,070.09	0.00	9,481,070.09	–5.19
535	0.00	9,754,562.50	0.00	9,754,562.50	–2.45
550	+2.80	10,028,054.91	0.00	10,028,054.91	+0.28
575	+7.48	10,483,875.58	0.00	10,483,875.58	+4.84
590	+10.28	10,757,367.99	0.00	10,757,367.99	+7.57
615	+14.95	11,213,188.67	0.00	11,213,188.67	+12.13
655	+22.43	11,942,501.75	0.00	11,942,501.75	+19.43

To insure a $10 million portfolio, 187 put options (90-day expiration) with a strike price of 520 are purchased for $245,437.50 (187 × $1312.50). The balance of $9,754,562.50 remains invested in a diversified portfolio of equities that matches the performance of the S&P 100 Index.

* Dividends not included.

Strategy 5: Buy a Portfolio Multiple of Out-of-the-Money Puts

This put-buying strategy has two significant differences from strategies 3 and 4. First, observe the difference in line 5 in Figure 6-1. With this strategy, the portfolio can actually show an increase in value with a market decline. This occurs because the quantity of puts purchased represents a larger market value than the equity portfolio. Table 6-9 shows, specifically, how this strategy performs at different index levels at expiration.

The second significant difference is that there are no strict guidelines as to how many puts to buy. The reasoning behind the choice of 249 puts (520 strike) options is as follows: First, $327,250 was the cost of the at-the-money puts in strategy 3, and that dollar amount was chosen to spend on the puts in strategy 5. This amount was chosen for strategy 5, because, in strategy 3, it was deemed an acceptable amount to risk in terms of under-performing the index if a market rally occurred. The quantity, 249, was determined by dividing the money available, $327,250, by the cost of each put, 13⅛, or $1,312.50, which was taken

TABLE 6-9

Strategy 5: Buying a Portfolio Multiple of Out-of-the-Money Puts

Index Level at Option Expiration	Index Percent Change	Equity Portfolio Value ($)	Put Option Value ($)	Total Portfolio Value ($)	Portfolio Percent Change
415	−22.43%	7,503,500.58	2,614.500.00	10,118,000.58	+1.18%
455	−14.95	8,226,729.56	1,618,500.00	9,845,229.56	−1.55
480	−10.28	8,678,747.66	996,000.00	9,674,747.66	−3.25
495	−7.48	8,949,958.53	622,500.00	9,572,458.53	−4.28
520	−2.80	9,401,976.64	0.00	9,401,976.64	−5.98
535	0.00	9,673,187.50	0.00	9,673,187.50	−3.27
550	+2.80	9,944,398.36	0.00	9,944,398.36	−0.56
575	+7.48	10,396,416.47	0.00	10,396,416.47	+3.96
590	+10.28	10,667,627.34	0.00	10,667,627.34	+6.68
615	+14.95	11,119,645.44	0.00	11,119,645.44	+11.20
655	+22.43	11,842,874.42	0.00	11,842,874.42	+18.43

Use $327,250.00 to purchase 249 put options (90-day expiration) with a strike price of 520 (327,250 divided by $1,312.50). The balance of $9,673,187.50 remains invested in a diversified portfolio of equities that matches the performance of the S&P 100 Index.

* Dividends are not included.

from Table 6-6. Out-of-the-money puts were purchased because the goal of this strategy is to take advantage of the leverage feature of options. Obviously, a portfolio manager would have a different market forecast to justify the selection of strategy 5 than for strategy 3. Consider how this strategy performs if the market remains steady or rises, if it declines slightly or if it declines dramatically.

If the market remains steady or rises, the portfolio insured with a multiple of out-of-the-money puts will under-perform the market by approximately 3.3 percent, the same as strategy 3. The under performance for strategy 5 is the same as for strategy 3, because the cost of the two strategies is the same, $327,250.

If the market declines slightly, this portfolio risks losing a maximum of 5.98 percent. This maximum possible loss occurs if the market declines to an index level of 520. At that index level, the equity portfolio has declined and the puts have expired worthless.

If the index declines below a level of 520, however, the leverage effect of the put options comes into play. At an index level of 495 and a market decline of 7.48 percent, for example, the puts have increased from the cost of $327,250 to a value of $622,500. The result, in this case, is only a 4.28 percent overall portfolio decline. If the market continues to decline, the leverage effect of the puts boosts portfolio performance to a profit position. At an index level of 415, a 22.43 percent market decline, for example, Table 6-9 indicates that the portfolio shows a 1.18 percent profit.

Choosing a Portfolio Insurance Strategy

For easy comparison, all three insurance strategies are summarized in Table 6-10.

In analyzing or choosing an option strategy, there are a number of important considerations. The first is the cost of the strategy; the second is the risk, and they are not always the same! Another important consideration is the portfolio manager's opinion of the market. Unfortunately, a vague opinion does not suffice. Because the most popularly traded options have a life of less than six months, market opinion must be specific in terms of direction, percentage of movement, and duration. A portfolio manager also must articulate the advantage that an option strategy can bring to the particular investment situation.

When the specific market forecast is formulated and the desired benefit is identified, the time frame should then be chosen. Some strate-

TABLE 6-10

Comparison of Results for Strategies 3, 4, and 5

Index Level at Option Expiration	Percent Change	Percent Change Strategy 3	Percent Change Strategy 4	Percent Change Strategy 5
415	–22.43%	–2.53%	–4.70%	+1.18%
455	–14.95	–2.78	–4.89	–1.55
480	–10.28	–2.93	–5.00	–3.25
495	–7.48	–3.02	–5.07	–4.28
520	–2.80	–3.18	–5.19	–5.98
535	0.00	–3.27	–2.45	–3.27
550	+2.80	–0.56	+0.28	–0.56
575	+7.48	+3.96	+4.84	+3.96
590	+10.28	+6.67	+7.57	+6.68
615	+14.95	+11.19	+12.13	+11.20
655	+22.43	+18.42	+19.43	+18.43

gies are implemented for the life of the option, perhaps as long as a year or more, and others are implemented for less than one month, at which point the option position is liquidated.

The portfolio manager should also take into account the implied volatility level of options. Although it is always better to buy something at the best possible price, the exact level of implied volatility is more important in some situations than in others.

Finally, the expected frequency of use for each strategy deserves discussion. Table 6-11 presents a summary of the analytical process in choosing between these strategies. A detailed explanation of each column in Table 6-11 follows.

Column 1: Selling the Portfolio

Selling the portfolio is, in effect, the ultimate insurance policy. Once in cash, there is no risk if the market declines. It is obvious, however, that selling an entire portfolio is not done lightly or frequently. In fact, this is a major policy decision that probably occurs only at the end of a market cycle. There would undoubtedly be strong fundamental economic considerations, perhaps the forecast of a prolonged recession. Consequently, the portfolio manager would not simply be bearish, he would be predicting a 25 to 40 percent market decline over one to three years. This strategy simply is not practical for a three-month or even a six-month period.

In the cost/risk area, however, this strategy has many ramifications. For a portfolio of any size, converting all equities to cash is not easy or inexpensive, and commissions are only part of the cost. The true full cost includes the spread between the current market price and the sale price of the equities. An increase in the supply of a given stock can drive down its price and, for large blocks of stock, this may be a significant discount. For taxable portfolios, this strategy has considerable cost consequences. For individuals with long-term holdings at a very low cost basis, this strategy is virtually impractical.

The ultimate test of conversion to cash is the price at which the portfolio is repurchased. The risk, therefore, is being wrong in the market forecast and missing a bull market move. If the repurchase occurs at a market level higher than at the time of sale, this strategy results in an opportunity loss, rather than a recognizable trading loss.

Column 2: Selling Stock Index Futures

A stock index futures contract is equivalent to some dollar value portfolio that matches the underlying index, and a portfolio manager seeking

TABLE 6-11

Portfolio Insurance Strategy Grid

	Strategy 1: Selling the Portfolio	Strategy 2 Selling Stock Index Futures	Strategy 3: Buying a Portfolio of At-the-Money Plus	Strategy 4: Buying a Portfolio Equivalent of Out-of-the-Money Puts	Strategy 5: Buying a Portfolio Multiple of Out-of-the-Money Puts
Cost	Commissions and impact on market	Discount or premium of futures to real market level	3.3%	2.45%	3.3%
Risk	Miss market rally	Miss market rally	Perform 3.3% worse than market	Maximum loss 5.19%	Maximum loss 5.98%
Specific Market Opinion	Long-term bearish	Bearing 4–12 months	Bearish, 78 days	Bullish, but worried	Short term very bearish
Benefit of Options	N/A	N/A	Market timing portfolio hedge	Disaster Insurance	Leverage profit from market decline
Time Frame of Strategy	1½–3 years	4–12 months	1–3 months	3–6 months	2–4 weeks; will liquidate if market does not begin to move
Implied Volatility	N/A	N\A	Should buy options at low end of implied volatility range	Relatively unimportant; do not want to buy extremely high implied volatility	Very important; must buy options at low end of implied volatility range
Expected Frequency of Use	Once in 5–10 years	Once every 2–3 years	Once per year	Once every 18–24 months	10 times in 20 years

N/A = not applicable

protection against a market decline can sell stock index futures contracts against a broadly based equity portfolio with the goal of getting out of the market. Although this is never perfectly achievable, the concept is that the loss experienced by an equity portfolio during a market decline is offset by the profit from the short stock index futures contracts.

This has the advantage of not disrupting the existing equity holdings, which is especially important to taxable portfolios. Other advantages are lower commissions and ease of execution. The ease of execution, however, has been called into question by periods of unusually high volatility in recent years. Nevertheless, during most periods of normal activity, this strategy can be implemented satisfactorily for portfolios with values that extend to hundreds of millions of dollars.

Because of the relative ease of entry and exit under normal circumstances, portfolio managers who employ futures contracts have more flexibility in acting upon bearish market forecasts than managers who only sell their equity holdings. If, for example, an institutional investor predicts a 10 percent market correction over three to six months, it generally would not be practical to sell the entire portfolio with the goal of buying it all back at a price 10 percent lower. Use of stock index futures contracts, however, could protect a portfolio during this period.

The risk of selling futures to neutralize a portfolio, as with selling all holdings, is missing a market rally. The cost, in addition to commissions, includes any discount or premium to the index at which the futures contracts trades are executed.

Column 3: Buying a Portfolio Equivalent of At-the-Money Index Puts

Index put options are available with maturities in excess of one year, but, for the sake of brevity, this discussion focuses on three-month option insurance strategies.

The market opinion required to make a portfolio insurance strategy appropriate must take into account the expected move relative to the price of the option. In the example discussed earlier, the 90-day at-the-money 535 puts were priced at $17.50 or 3.3 percent of the index value. This means that the market must decline 3.3 percent for the put position to break even at option expiration. Consequently, the portfolio manager must have a strong expectation that a market decline in excess of 3.3 percent is likely to occur. To understand why such an expectation is essential, look at the cost/risk of this strategy. The risk is under-performing the market by 3.3 percent—the cost of the put options. If the market remains at the same level or rallies, the put options expire worthless, and this insured portfolio under-performs the market by 3.3 percent.

Why would a manager be willing to risk under-performing the market by 3.3 percent during a 90-day period (approximately a 13.2 percent annual rate)? The manager is willing to take this risk because he or she has an extremely bearish view and hopes to beat the market performance by profiting from owning the puts. Clearly, the objective of the option purchase is to have profit and thereby to limit or insure against portfolio losses.

Because the puts in this case are at-the-money, they have the highest price change response to the level of implied volatility. In fact, during the period of 1986 to 1987, at-the-money put options traded as low as 3 percent of the index (11 percent implied volatility) to 15 percent of the index (45 percent implied volatility). The higher number occurred in the aftermath of the events of October 1987. Knowledge of implied volatility levels and an opinion about future levels are vital.

Column 4: Buying a Portfolio Equivalent of Out-of-the-Money Puts

What are the implications of buying puts with a strike price below the current market level? At first glance it may seem that there are only two differences between this strategy and buying at-the-money puts: a lower cash outlay (2.45 percent versus 3.3 percent) and a greater maximum portfolio decline (5.19 percent versus 3.3 percent). In fact, the other considerations for this strategy—market opinion, time frame, option objective, and implied volatility—are significantly different as well.

Market opinion, as with any equity investment decision, is the place to start. A manager who has a high expectation of a market decline is willing to pay more for portfolio insurance and buy at-the-money puts. One who is less certain, but has that nagging feeling that a market correction is within the realm of possibility, is willing only to pay a smaller amount.

To make this distinction clearer, consider the following two scenarios. The first scenario is after a two-and-a-half-year bull market, and a portfolio manager sees that gradually rising interest rates have not yet been digested by the market. Further, the latest research reports indicate that three major industries are facing labor negotiations and that walkouts in one or two are likely. This institutional investor feels that another major negative development is all it would take to send the market down, and his market experience and knowledge of the political and economic situation make him believe that such an event is likely in the next 90 days. In this scenario, the 3.3 percent cost of at-the-money puts is warranted.

In the second scenario, the market has been moving sideways for four months after a 60-day 8 percent rally. The research reports are favorable, and the portfolio manager's fundamental outlook is positive. Experience, however, suggests to this portfolio manager that all may not be as it seems. Bullish sentiment is high, the advance-decline line is sloping downward, and institutional cash holdings are at a low level. The portfolio manager is bullish, but is facing this quandary: Will the overbought condition be resolved by more sideways movement, or by a market correction? The manager is clearly willing to ride out a minor market correction, but he believes in the possibility of a short sharp correction. In this scenario, cheaper insurance is desired. The out-of-the-money puts are the best choice, because the portfolio manager is bullish in the longer term and does not want the expensive cost of at-the-money puts to significantly reduce performance relative to market averages.

What time frame is relevant for purchasing out-of-the-money puts? Because the certainty of the market decline in the second scenario is less than in the first, and because the cash cost is lower per unit of time for longer options, this strategy seems more appropriate for five-, six-, or seven-month options. The overall cost of the strategy is low and, with the timing in doubt, it is best to allow as much time as possible for the insurance to remain in effect.

The level of implied volatility and the objective for using options also needs to be addressed. A low level of implied volatility is always favorable when buying options. It is less important when buying "disaster insurance" with a longer time to expiration than when buying out-of-the-money options for a short-term trade. Although the change in option price has a nearly linear relationship to the change in implied volatility (because the overall cost of the out-of-the-money puts is relatively low), a slightly higher level of implied volatility does not significantly change the cost of this strategy as a percentage of the total portfolio. Of course, in extreme conditions such as the aftermath of October 19, 1987, this may not be true, but in normal times, the level of implied volatility does not greatly affect the decision to implement this tactic.

In this portfolio insurance strategy, the objective in purchasing the options is to have them expire worthless! Strange as it may sound, the portfolio manager actually hopes that these out-of-the-money put options expire worthless. After all, an honest homeowner does not hope that the house burns down so that the insurance policy will provide some compensation. Similarly, a portfolio manager who buys out-of-the-money puts does not hope that the market declines so that the portfolio will decrease in value by a lesser percentage than the overall market. This is a hard concept for many investors to digest. But there are times when the experienced money manager recognizes that a sharp market correction, even though not expected, has a probability of occurring that is high enough to justify the expenditure of 1½ to 2 percent of the portfolio on out-of-the-money puts for three to six months of insurance.

Column 5: Buying a Portfolio Multiple of Out-of-the-Money Puts

Strategy 5 has the possibility of actually profiting from a market decline. The cost of the options is 3.3 percent of the portfolio—the same cost as strategy 3. However, the risk is the greatest of any of the put buying strategies. If the overall market drops to an index level of 520 (equivalent to the strike price of the puts in this example), the portfolio loses 5.98 percent, which is worse than an overall market decline of 2.80 percent.

How often is this strategy useful? To quantify the answer, Table 6-12 shows that with a nearly 20 percent market decline, the strategy of buying a portfolio multiple of out-of-the-money puts will actually show a positive return for the entire portfolio. Because 90-day puts are used in the example, it seems reasonable to ask how often the Dow Jones Industrial Average (DJIA) declined by 15 percent or more in 90 days. If

TABLE 6-12

7 Times in 19 Years

	Start Date	DJIA	Finish Date	DJIA	Number of Days	Percent Decline
1	5/14/69	968.85	7/29/69	801.96	76	17.23%
2	4/9/70	792.50	5/26/70	631.16	47	20.36
3	10/26/73	987.60	12/5/73	788.31	40	20.18
4	7/24/74	805.77	10/4/74	584.56	72	27.45
5	6/23/81	1006.66	9/18/81	836.19	87	16.93
6	8/25/87	2722.42	10/19/87	1738.74	55	36.13
7	7/16/90	2999.75	10/11/90	2365.10	87	21.16

this has never happened, the discussion of this method would be no more than an academic exercise. This is a severe yardstick—a 15 percent or greater decline in 90 days. The number of such occurrences is shown in Table 6-12. This has happened seven times between 1969 and 1998, nearly once every three years. With a less strict test, say 12 percent in 100 days, the number of such occurrences is close to once per year. This knowledge should open the eyes of many portfolio managers to the possible benefits of this option strategy. Between 1969 and 1998 the Dow Jones Industrial Average fell more than 15 percent in less than 90 days seven times.

The objectives of this strategy are to take advantage of the leverage aspect of the options and to profit from the anticipated market decline. In order for a portfolio manager to take the risk of performing 10 percent worse than the market, he must be extremely confident of his market opinion. In fact, this strategy has no direct insurance analogy. It is more of a speculation on a market decline.

What, then, are the key elements to making this strategy successful? The answer to this question has three parts: the accuracy of the market forecast, the knowledge of implied volatility levels, and the choice of the appropriate time frame. As with any investment strategy, if the market forecast is inaccurate, the strategy has little chance of success.

Regarding implied volatility, although the level of implied volatility was not significant in strategy 4 (buying a portfolio equivalent of out-of-the-money puts), it is critical when buying a portfolio multiple of out-of-the-money puts. The implied volatility level is important for three reasons. First, because the quantity of options purchased is determined by the dollar amount (or portfolio percentage) allocated, more options can be purchased if the price (implied volatility level) is low. For this example, 3.3 percent of the portfolio was chosen because this is the same amount invested in at-the-money puts in strategy 3. Using this amount clearly demonstrates the effect of leverage when using out-of-the-money options. However, there is no specific rule for the amount allocated when using this strategy.

A second reason the level of implied volatility is important when implementing this strategy is that, if the puts are purchased at a high implied volatility level, then they will decrease in price rapidly should implied volatility levels decline. This could happen with or without the market declining as expected. Third, an increase in implied volatility, in addition to appreciation from market movement, greatly adds to an out-of-the-money options' increase in price. Consequently, the ideal time to implement this strategy is when implied volatility is low and expected to increase.

The best time frame is similar to the one chosen for any speculation—usually very short. Because success depends on a sharp market move (which, by definition, must be unexpected by the general market), it is likely that in two to four weeks this speculative strategy will prove to work or not. That is the point—to sell the options with a profit or no more than a small loss. This is a major difference from the other put buying strategies, in which the plan is to hold puts until expiration. With this strategy, however, the plan is to sell the options within two to four weeks of their purchase if the anticipated market decline has not begun.

The fact that this method calls for holding the options for no more than a month does not imply that options with four weeks left until expiration are those to purchase. Actually, options that have 90 days or longer until expiration are the preferred choice here, because option time decay is significant in the last 30 days of an option's life. The "extra element" is that option-buying strategies using these very short-term out-of-the-money options must be planned carefully. If 90-day put options are purchased, they will increase in price either from a sharp down move or from an increase in implied volatility. Perhaps both will happen. When using the 90-day or longer-term options, portfolio managers must have the discipline to liquidate the position at the point when either the strategy has worked successfully, or it has been judged to be unsuccessful.

Where the Insurance Analogy Breaks Down

The term "portfolio insurance" is not used in a strict sense to mean exactly the same thing it does to the buyer of homeowner's insurance. In fact, many attributes of homeowner's policies do not apply to portfolio insurance. First, the homeowner hopes that he or she will never have to make a claim on the policy. Second, mortgage lenders and many state laws require certain types of insurance. Third, homeowner's insurance generally costs less than 1 percent annually of the market value of the insured home. Finally, payoffs from can be several hundred times the premium.

Regarding portfolio insurance, there is certainly no requirement that portfolio managers purchase protection. The motivation behind buying put options is a negative market forecast and may entail a desire to profit from the put-option purchase. In the example above, of buying at-the-money puts (strategy 3), the total cost of insuring the portfolio for 90 days comes to 3.3 percent of the portfolio; that implies an annual rate of 13.2 percent, which is unquestionably too expensive to be used continuously. Although there are stories of options going up 1,000 percent, most success stories deal with options doubling or tripling rather than multiplying several hundred times.

The conclusion to be drawn from these differences is that portfolio insurance strategies are more properly labeled *market timing strategies*. They give the portfolio manager several alternatives to use for the purpose of increasing returns and managing market risk. Before the birth of these option products, the range of alternatives available was limited to owning stocks, investing in cash, or having a combination of the two.

MARKET TIMING

Although market timing has its critics, many professional investors recognize that a manager who changes the portfolio mix between equities and cash is essentially making market timing decisions. Almost by definition, unless a portfolio manager is 100 percent invested in equities all the time, he is a market timer. Consider a portfolio manager who, over the course of time, is "fully invested" at one point, "maintaining cash reserves for buying opportunities" at another point, and "largely in cash waiting for the market to bottom" at a third point. Many portfolio managers who make these statements also say, "I am not a market timer." But the motivation behind each of these statements is a market opinion and therefore a market timing opinion. "I am fully invested" means "I am bullish." In this case, the portfolio manager would not be using any of the portfolio insurance strategies discussed above.

"I am maintaining cash reserves for buying opportunities" means "I am long-term bullish but short-term bearish." This situation might dictate being fully invested and employing one of the put-buying strategies.

"I am largely in cash" means "I am bearish" and might indicate use of the put-buying strategy that profits from a sharp down move. Indeed, looking at portfolio insurance strategies in the light of market timing decisions can give the portfolio manager additional insight into the nature of these tools and their appropriate use.

DYNAMIC HEDGING REDUCES MARKET EXPOSURE USING PUTS

Index put options can be used on a very short-term basis (one to three weeks) to decrease a portfolio's exposure to the equity market during a temporary market decline or during the time it takes to sell individual equities. This strategy is appropriate in a situation in which a portfolio manager has decided to reduce the equity allocation and is concerned about the effect of a market decline during the two-week period that it will take to select and sell individual equities. To begin with, assume that the S&P 500 Index is at 1110. Anticipating a moderate decline in the market, a portfolio manager decides to reduce equity exposure by $7.5 million through the purchase of index puts. A 90-day, at-the-money S&P 500 put with a 1,110 strike is selling at $35¾ and has an estimated delta of −.45. The manager reduces equity exposure by purchasing 150 of these index puts [7.5 million/($100 multiplier × 1,110 index × .45 delta)]. The cost of this position is $536,250 ($3,575 premium per option × 150 options).

Two weeks later, the manager selects the equities to sell. If the index now stands at 1,100 and the S&P 500 put is $41⁷⁄₁₆, the put position has a value of $621,562.50, representing a profit of $85,312.50.

Assuming that the equities declined by the same proportion as the index, this profit offsets the $82,418 loss incurred on the sale of the equities.

This strategy is called *dynamic hedging* because the number of index options purchased depends on the delta of the option being used. Because an option's delta changes with market movement, the use of this strategy may involve the buying and selling of options during the hedging period.

In the previous example, if the market had moved to an index level of 1,100 before the individual equities had been chosen, the delta of the 1,110 put options would have risen to − 49. This would have changed the portfolio market value represented by the put options to $8,158,500 (1,110 × $100 × .49 × 150). Meanwhile, the equity portfolio would be $7,417,582 ($7,500,000 − $82,418). At the index level of 1,100, only 137 options would be required to hedge the equity portfolio [$7,417,582 ÷ (1,100 × $100 multiplier × .49 delta)]. Consequently, 13 of the options would be sold, and the portfolio manager would have a new position of delta equality—long $7,417,582 of equities and short $7,417,582—represented by index put options. Adjusting the option position could then be repeated until the time when the individual equities were sold, at which point the put options would also be sold.

This dynamic feature also works if only part of the equities are sold at one time. For example, if the equities are sold in three blocks, each block on a different day, the put options could also be sold in three blocks. The goal would be to maintain a short market position in options equal to the long market position in equities until both positions were sold. In this way, the equity market exposure of the portfolio is reduced during the period it took to implement the selling program.

Dynamic hedging is perhaps the most frequently misunderstood strategy. It is confused with insurance strategies, and its essentially short-term use is often overlooked. Successful use depends on knowledge of implied volatility levels and the option time decay curve. The portfolio manager must decide on the time period of the dynamic hedge and then carefully examine the possible returns from the options, given a change in market level and implied volatility. If these elements are carefully integrated, a manager can see how effective dynamic hedging can be.

INCREASING MARKET EXPOSURE WITH INDEX CALL OPTIONS

Because index call options replicate a dollar portfolio of the underlying index, it is possible to match a portfolio's total dollar investment by purchasing index call options. The advantages of doing this are similar to the advantages of buying call options on individual equities: for the limited risk of the premium paid, the call buyer can participate in a broad market advance.

There are four general investment-oriented ways to employ index call buying as opposed to speculative call buying. Three are variations on the same concept of buying calls with the goal of participating in a broad market rally while having limited risk on the downside. These three strategies are: (1) buying a portfolio equivalent quantity of at-the-money index calls, (2) buying a portfolio equivalent quantity of out-of-the-money calls, and (3) buying a portfolio multiple of out-of-the-money calls.

The fourth method of employing index call options is a short-term market timing tool used when increasing a portfolio's equity exposure. Proper use requires knowledge of implied volatility levels and the option's time decay curve, as will be discussed.

Figure 6-2 illustrates how the portfolio (cash plus calls) changes in value (vertical axis), given a change in the index level (horizontal axis) for each of the first three alternatives described above (lines 2, 3, and 4) relative to an index portfolio (line 1).

FIGURE 6-2

Index Call Buying Strategies

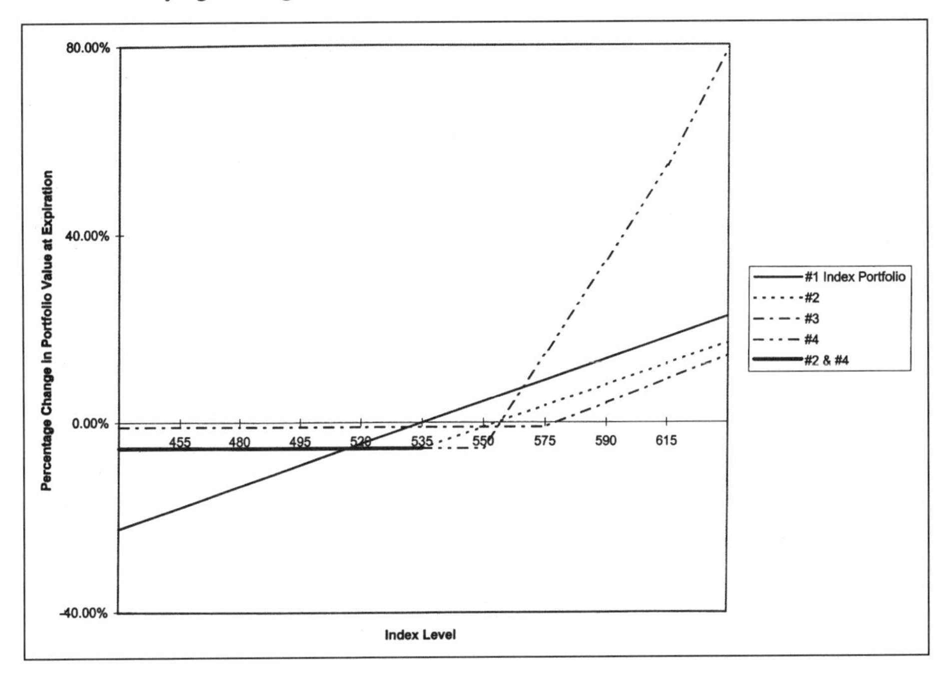

In the following examples, assume that the portfolio in question has $10 million in cash and the S&P 100 Index and related options closed at the prices indicated in Table 6-13.

Strategy 1: Buying a Portfolio Equivalent of At-the-Money Calls (The 90/10 Strategy)

A tactic traditionally known as the *90/10 strategy* enables portfolio managers to achieve two goals simultaneously: capital preservation and potential appreciation from a market rally. This method involves the purchase of index call options with a small portion of a portfolio's capital—say 10 percent—and the purchase of low risk cash equivalents with the remaining 90 percent. Hence the name 90/10. The maximum risk is the premium paid for the calls and, above the break-even index level (strike price plus call price), the index call options allow the portfolio to participate dollar for dollar in a market rally.

TABLE 6-13

Index Call Buying Strategies:
Available 90-Day Call Options

90-Day Call Strike Price	Call Option Price
535	25
540	21½
545	18¾
550	15¾
555	14½
560	12⅛
580	5¾

90-day Treasury bills pay 5.0 percent annual rate.

Calculating portfolio equivalent quantity of options:

$$\text{Number of calls} = \frac{\text{Portfolio dollar value}}{(\text{Index level} \times \text{Multiplier})}$$

$$\text{For a \$10 million portfolio: } \frac{10{,}000{,}000}{(535 \times 100)} = 187$$

From Table 6-13, one can calculate that 187 calls is the proper quantity required to replicate a $10 million portfolio. With the S&P 100 Index at 535, each option represents a market value of $53,500 ($100 × Index level of 535), and thus 187 options are required ($10,000,000 ÷ $53,500).

The at-the-money calls with a strike price of 535 are purchased at 25 each for a total cost of $467,500. The remaining balance of $9,532,500 is used to purchase 90-day Treasury bills that pay 5 percent annually, resulting in a maturity value of $9,651,656.25. Table 6-14 demonstrates how the portfolio of T-bills plus 187 at-the-money calls perform at various index levels.

The Percent Change column shows that the maximum loss on this portfolio is 3.48 percent, which is the premium paid for the index calls minus the interest earned on the cash investments. This occurs if the index level declines or remains constant and the calls expire worthless. With any market rally, however, this portfolio of at-the-money calls plus cash will participate in the market advance. If the market rallies to an index level of 615, a 14.95 percent advance, for example, this cash-plus-calls portfolio increases in value by 11.48 percent. It is important to note that the maximum loss of the portfolio is also the amount by which the portfolio will under-perform the market if the index level rises.

Strategy 2: Low-Cost Participation with Out-of-the-Money Calls

Participation in a market rally is also achieved by buying out-of-the-money calls. This method requires less cash outlay than buying at-the-money index call options, although participation in the rally begins at a later point in the market rise. Purchasing 187 options with a strike price of 580 at a total cost of $107,525 ($575 each × 187) is only 87 percent of the interest earned of $123,655.94 on the remaining cash balance of $9,892,475. Table 6-15 illustrates how this cash plus index call buying strategy performs at different index levels.

This strategy, as illustrated, has no potential for loss, because the interest earned is greater than the cost of the call options. However, in many cases there would be a loss if calls with a strike price closer to the current index level were purchased. The loss would be calculated by determining the difference between the cost of the calls and the interest

TABLE 6-14

Strategy 1: Buying a Portfolio Equivalent of At-the-Money Calls

Index Level at Option Expiration	Index Percent Change	Cash Plus Interest Earned ($)	Call Option Value ($)	Total Portfolio Value ($)	Portfolio Percent Change
415	−22.43%	9,651,656.25	0.00	9,651,656.25	−3.48%
455	−14.95	9,651,656.25	0.00	9,651,656.25	−3.48
480	−10.28	9,651,656.25	0.00	9,651,656.25	−3.48
495	−7.48	9,651,656.25	0.00	9,651,656.25	−3.48
515	−3.74	9,651,656.25	0.00	9,651,656.25	−3.48
535	0.00	9,651,656.25	0.00	9,651,656.25	−3.48
555	+3.74	9,651,656.25	374,000.00	10,025,656.25	+0.26
575	+7.48	9,651,656.25	748,000.00	10,399,656.25	+4.00
590	+10.28	9,651,656.25	1,028,500.00	10,680,156.25	+6.80
615	+14.95	9,651,656.25	1,496,000.00	11,147,656.25	$11.48
655	+22.43	9,651,656.25	2,244,000.00	11,895,656.25	+18.96

To replicate a $10 million portfolio, 187 call options (78-day expiration) with a strike price of 535 are purchased for $467,500 (187 × $2500). The balance of $9,532,500 is invested in Treasury bills and earns $119,156.25 ($9,532,500 × .05 × .25).

TABLE 6-15

Strategy 2: Buying a Portfolio Equivalent of Out-of-the-Money Calls

Index Level at Option Expiration	Index Percent Change	Cash Plus Interest Earned ($)	Call Option Value ($)	Total Portfolio Value ($)	Portfolio Percent Change
415	−22.43%	10,016,130.94	0.00	10,016,130.94	+0.16%
455	−14.95	10,016,130.94	0.00	10,016,130.94	+0.16
480	−10.28	10,016,130.94	0.00	10,016,130.94	+0.16
490	−8.41	10,016,130.94	0.00	10,016,130.94	+0.16
520	−2.80	10,016,130.94	0.00	10,016,130.94	+0.16
535	0.00	10,016,130.94	0.00	10,016,130.94	+0.16
550	+2.80	10,016,130.94	0.00	10,016,130.94	+0.16
580	+8.41	10,016,130.94	0.00	10,016,130.94	+0.16
590	+10.28	10,016,130.94	187,000.00	10,203,130.94	+2.03
615	+14.95	10,016,130.94	654,500.00	10,670,630.25	+6.71
655	+22.43	10,016,130.94	1,402,500.00	11,418,630.94	+14.19

To replicate a $10 million portfolio, 187 call options (90-day expiration) with a strike price of 580 are purchased for $107,525.00 (575 × 187). The balance of $9,892,475 is invested in Treasury bills and earns $123,655.94 ($9,892,475 × .05 × .25).

earned. In this example, the gain is 0.16 percent until the index reaches a level of 580, the strike price of the calls. Consider, for example, if the market rallies to an index level of 590, a 10.28 percent advance. In this case, this cash-plus-calls portfolio achieves a 2.03 percent increase, which is approximately 8.25 percent less than the rise in the index. This is calculated by adding the cost of the calls minus the interest earned to the difference between the strike price of the calls and the index level at the time of the call purchase. In this example, the cost of the calls minus the interest is negative $16,130.94, or −0.15 percent of the $10 million portfolio. The difference between the strike price of the calls (580) and the index level at the time of the call purchase (535) is 45 points, or 8.4 percent of the index level. Adding +8.40 and −0.15 returns approximately 8.25 percent, the amount by which the portfolio using this call buying strategy will under-perform the market in a broad rally.

Strategy 3: Leverage with Out-of-the-Money Calls

In this strategy, a quantity of out-of-the-money calls greater than the portfolio-equivalent number is purchased so that the portfolio's performance is increased by the leverage aspect of options.

Determining the quantity to be purchased is a subjective decision. One commonly employed practice is to calculate the dollar amount required to purchase at-the-money calls, as in Strategy 1, and then use that dollar amount to purchase as many out-of-the-money calls as possible from Strategy 2. In this case, $467,500, the amount used to purchase index call options in Strategy 1, will purchase 813 out-of-the-money 580 calls ($467,500 ÷ $575 − 813). The upside leverage payoff of this strategy is demonstrated in Table 6-16.

If the market rallies to an index level of 590, a 10.28 percent advance, this cash-plus-calls portfolio increases by 14.57 percent, outperforming the market by 4.29 percent. At an index level of 615, this portfolio outperforms the market by 27.56 percent. This portfolio will continue to outperform the index at an increasing rate, because 813 calls with a strike price of 580 represent the portfolio equivalent of $47,154,000 ($58,000 × 813). By comparison, a $10 million portfolio at an index level of 535 would have grown to only $10,841,121 at an index level of 580. Upside leverage does not come without a cost, however. For any decline in the index level, the maximum potential loss of Strategy 3, in this example, is 3.48 percent, the premium paid for the

TABLE 6-16

Strategy 3: Buying a Portfolio Multiple of Out-of-the-Money Calls

Index Level at Option Expiration	Index Percent Change	Cash Plus Interest Earned ($)	Call Option Value ($)	Total Portfolio Value ($)	Portfolio Percent Change
415	–22.43%	9,651,656.25	0.00	9,651,656.25	–3.48%
455	–14.95	9,651,656.25	0.00	9,651,656.25	–3.48
480	–10.28	9,651,656.25	0.00	9,651,656.25	–3.48
495	–7.48	9,651,656.25	0.00	9,651,656.25	–3.48
520	–2.80	9,651,656.25	0.00	9,651,656.25	–3.48
535	0.00	9,651,656.25	0.00	9,651,656.25	–3.48
550	+2.80	9,922.263.43	0.00	9,922,263.43	–0.78
575	+7.48	10,373,275.41	0.00	10,373,275.41	+3.73
590	+10.28	10,643,882.59	813,000.00	11,456,882.59	+14.57
615	+14.95	11,405,966.38	2,845,500.00	14,251,466.38	+42.51
655	+22.43	11,816,513.73	6,097,500.00	17,914,013.73	+79.14

Use $467,500 to purchase 813 call options (90-day expiration) with a strike price of 580 (467,500/$575). The balance of $9,532,500.00 is invested in Treasury bills and earns $119,156.25 ($9,532,500.00 × .05 × .25).

calls minus the interest earned on the cash investments. This is not the worst case for the portfolio manager whose performance is evaluated in comparison to that of the market and his peers. The worst case occurs when the market rallies to an index level of 580. At this point, the index has risen 8.4 percent and, with the 580 calls expiring worthless, this cash-plus-calls portfolio falls by .78 percent—the portfolio has underperformed the market by 9.18 percent.

Choosing the Appropriate Call Buying Strategy

Table 6-17 summarizes and compares the percentage changes in the portfolio for Call Buying Strategies 1, 2, and 3 at various index levels. Different payoff opportunities for each imply that various market environments and outlooks lead to the selection of each strategy.

Again, there are important considerations in selecting the appropriate strategy. The first is the cost of the strategy; the second is the risk. Risk is defined in terms of comparing the performance of the cash-plus-calls portfolio and the overall market performance. Market opinion, with all of the specificity described earlier, must be taken into account; and the benefit to be derived from buying calls must be clearly identified. Other considerations are the time frame for implementation and

TABLE 6-17

Comparison of Results in Index Call Buying Strategies 1, 2, and 3

Index Level at Option Expiration	Percent Change	Percent Change Strategy 1	Percent Change Strategy 2	Percent Change Strategy 3
415	−22.43%	−3.48%	+0.16%	−3.48%
455	−14.95	−3.48	+0.16	−3.48
480	−10.28	−3.48	+0.16	−3.48
495	−7.48	−3.48	+0.16	−3.48
520	−2.80	−3.48	+0.16	−3.48
535	0.00	−3.48	+0.16	−3.48
550	+2.80	0.26	+0.16	−0.78
575	+7.48	+4.00	+0.16	+3.73
590	+10.28	+6.80	+2.03	+14.57
615	+14.95	+11.48	+6.71	+42.51
655	+22.43	+18.96	+14.19	+79.14

the implied volatility level. Finally, no strategy can be used continuously, so the next section reviews how often a particular strategy might reasonably be implemented.

Analysis for Choosing Index Call Buying Strategy 3

The payoff table for this strategy (Table 6-16) shows the model $10 million portfolio significantly outperforming the general market in a major rally. Consequently, the market forecast must be extremely bullish for the very near term. Specifically, the portfolio manager must be looking for an increase in the index level from 535 to at least 575, a 7.5 percent market advance. Above that level, the cash-plus-calls portfolio increasingly outperforms the market. This move must be expected to begin in the very near future, probably within two to four weeks.

At this point in the discussion, it may be of interest to review a little history of market rallies. One may wonder, for example, how likely it is for the market to rally 15 percent in 90 days, the kind of market rally that would justify the selection of Strategy 3? Although it may come as a surprise to many investors, this kind of rally occurs, on average, nearly once per year. One rally of this magnitude occurred in August of 1982, when the Dow Jones Industrial Average rallied from 790 to 950 (a 20 percent advance) in just two weeks! Table 6-18 lists similar moves from 1969 to 1998. Between 1969 and 1998 the Dow Jones Industrial Average rose more than 15 percent in less than 90 days 24 times.

This strategy would be employed to take advantage of the leverage aspect of options. A portfolio manager who employs this strategy must be sufficiently confident of a short-term bullish market opinion that allows for a possible reward, despite the risk of under-performing the market. The time frame for this strategy is necessarily short. The portfolio manager's unusually bullish forecast must be prompted by something—a news event, a technical condition in the market, or some other pending development. Generally speaking, in such a situation the manager knows within two weeks to a month if the forecast is being realized. Consequently, the time to liquidate the option position is when the strategy is successful or when it becomes apparent that it is not working.

Choosing an option expiration is another matter. Expecting a major rally to begin in the next two to four weeks does not mean that the front-month options with 30 days or less to expiration are the right choice. This strategy is a short-term bullish volatility play, and two events can

TABLE 6-18

24 Times in 19 Years

	Start Date	DJIA	Finish Date	DJIA	Number of Days	Percent Increase
1	5/26/70	631.16	8/24/70	759.58	90	20.35%
2	7/7/70	669.36	10/5/70	776.70	90	16.04
3	10/29/70	753.56	1/26/71	866.74	89	15.02
4	11/18/70	754.24	2/16/71	890.06	90	18.01
5	11/23/71	797.97	2/16/72	922.94	95	15.66
6	8/22/73	851.90	10/26/73	987.06	65	15.87
7	10/4/74	584.56	11/5/74	674.75	32	15.43
8	12/9/74	579.94	3/7/75	770.10	88	32.79
9	4/7/75	742.88	6/30/75	878.99	84	18.32
10	12/5/75	818.80	2/25/76	994.57	82	21.47
11	3/7/78	746.79	6/5/78	863.83	90	15.67
12	8/12/82	776.92	11/3/82	1065.49	83	37.14
13	12/16/82	990.25	3/7/83	1141.74	81	15.30
14	1/26/83	1037.99	4/26/83	1209.46	90	16.52
15	9/20/85	1297.94	12/16/85	1553.10	87	19.66
16	1/22/86	1502.29	4/21/86	1855.90	89	23.54
17	12/31/86	1895.95	3/26/87	2372.59	85	25.14
18	5/20/87	2215.87	8/17/87	2700.57	89	21.87
19	10/19/87	1738.74	1/7/88	2051.89	80	18.01
20	11/16/88	2038.58	2/7/89	2347.14	83	15.14
21	1/9/91	2470.30	3/6/91	2973.27	56	20.36
22	9/5/96	5606.96	11/25/96	654.79	81	16.78
23	4/11/97	6391.69	7/8/97	7962.31	88	24.57
24	1/24/98	7700.74	4/21/98	9184.94	87	19.27

make it profitable: a market rally or an increase in implied volatility. With the front-month options so close to expiration, it is unlikely that they will benefit much from an increase in implied volatility. Also, if the market rally starts later than expected, these options can expire just before the market rallies through the strike price. Farther out options have a higher absolute cost, but they have the advantages of benefiting from an increase in implied volatility and a longer life to enable a bullish market forecast to materialize. A portfolio manager must weigh these tradeoffs when choosing which option expiration to buy.

Implied volatility is an important consideration when buying a large quantity of out-of-the-money options. If the strategy is to buy 90-

day options and sell them in two to four weeks if the rally has not started, the purchase decision is best implemented with a thorough knowledge of implied volatility levels. Changes in implied volatility can have a greater impact on the price of 90-day out-of-the-money options than time decay. If, instead, the strategy is to buy front-month options with three to four weeks until expiration and let them expire if the rally does not occur, implied volatility is not as important. In this second approach, the thinking would be to take a low-cost, short-term option risk. In this case, the total option price, not its implied volatility, is the most important factor.

Analysis for Choosing Index Call Buying Strategy 2

Strategy 2 (buying a portfolio equivalent of out-of-the-money calls) is employed for completely different reasons and has completely different criteria than strategy 3, yet both use out-of-the-money calls.

Examining payoff Table 6-15, one might ask: Why would a portfolio manager be willing to risk under-performing a market rally by 3.48 percent? The answer is that he or she has a bearish outlook and purchases calls for insurance. This strategy could be employed near the end of a bear market, when a portfolio manager is in cash and still bearish, but is looking for an acceptable price level at which to buy stocks. This manager is looking for the market to bottom in the next two to four months and is planning to be fully invested at that point. In the meantime, upside protection is needed in case the market rallies sharply and unexpectedly, as it so often does at the end of bear markets. By owning call options, the portfolio manager ensures participation in an upside rally.

Had this strategy been employed during one of the (approximately) 90-day, 20 percent market rallies in 1996, 1997, or 1998 (see Table 6-17), a cash-plus-calls portfolio could have increased by 14 or 15 percent. It may have under-performed the market by 6 percent, but it outperformed any portfolio that was in cash and out of stocks. The cost of options in one of these situations would have been approximately equal to interest income on cash reserves, and the options would have been an insurance policy against missing the big market rally which, in these years, actually occurred.

The optimal time frame for any conventional insurance policy is for as long as possible. With this strategy, a portfolio manager must determine over what time period the portfolio will become fully invested and plan the option purchase for that period. Generally, with this strategy, options are purchased with a view to carrying them until expiration. If the market does not rally while this portfolio is becoming fully invested, the options expire worthless—similar to any insurance policy expiring when no claim is made. The benefit to the portfolio, in this case, is the ability to buy individual equities at a lower market level.

Implied volatility is not a major consideration when buying out-of-the-money calls as insurance. Although changes in implied volatility affect the price of the options, because relatively few of the lower-cost out-of-the-money calls are involved, the total change in cost is not likely to be a meaningful percentage of the total portfolio. This strategy is designed as insurance against a major market move, so it is not likely that these options would be sold in the event of a quick market run-up or an increase in implied volatility. Doing so would eliminate the insurance before the cash reserves were invested in equities.

Purchasing out-of-the-money calls as insurance against a large rally is a strategy that can be employed whenever a portfolio has cash to invest and when the portfolio manager is short-to-medium-term bearish and waiting to buy individual equities. That might be as infrequently as at the end of each bear market cycle or as frequently as when new funds are received for investing in equities.

Analysis for Choosing Index Call Buying Strategy 1

Strategy 1, buying a portfolio equivalent of at-the-money calls, is similar to strategy 2 in that the portfolio participates in a market rally but under-performs the market—in this case by 3.48 percent. Also, at-the-money calls are more than twice as expensive as out-of-the-money calls.

What then must the market forecast be, and why purchase these expensive calls? The market forecast must be bullish, but the reason, again, is a perceived need for insurance.

There are two classic situations when this strategy is appropriate. The first is a major news event such as a presidential election. The portfolio manager predicts that the market will rally sharply after the election, but he or she realizes that the election results may cause the market to decline. In this situation, under-performing the market by 3.48 percent on the upside is a favorable trade-off relative to outperforming the market on the downside—a maximum loss of 3.48 percent versus a potentially significant market decline.

The second situation is one in which the money manager cannot afford to lose due to contractual obligations and, without the limited risk nature of call options, would otherwise be forced to buy only fixed-income investments. For example, assume a corporate treasurer must make a fixed pension contribution at some point in the future. If the money is available now and the treasurer is bullish, equities would not be appropriate due to the risk of a market decline. Any available funds, however, in excess of the fixed obligation (interest or excess accumulated funds) could be used to purchase call options, thereby ensuring participation in a market rally.

In this strategy, the market forecast is more important than the implied volatility level of the options. Obviously, buying options when implied volatility levels are low is always advantageous. But with this strategy, matching the market forecast to the break-even index level of the call option strategy is the determining consideration. For example, in a bull market, when new investment funds are coming to a portfolio manager, index calls may be the quickest and easiest way to commit the funds to the market. The considerations for index call buying strategies are summarized in Table 6-19.

DYNAMIC HEDGING USING INDEX CALLS INCREASES MARKET EXPOSURE

Using index call options to provide additional market exposure is a strategy with different implications than the insurance and leverage strategies just discussed. The goal here is to replicate market performance during the time it takes to shift a portfolio from cash investments into individual equity issues.

For example, assume that the S&P 500 Index is currently at 1111.77, and a portfolio manager, anticipating a market rally, decides to commit an additional $5 million to equities. Market exposure can initially be increased by purchasing S&P 500 Index calls. The plan is that stocks will be purchased as specific issues are selected and the amounts to be invested in each are determined. Assume further that the portfolio manager decides to use one-month S&P 500 calls with a 1,110 strike, that these calls cost $36½ each, and have a delta of .54.

To create an immediate $5 million exposure to the market on a point-for-point basis, 83 calls are bought [$5 million exposure ÷ ($100 multiplier × 1,111 index value × .54 delta)]. This represents an initial outlay of $302,950 ($3,650 premium × 83 calls).

If, one week later, the manager purchases $5 million of equities, the S&P 500 has advanced to 1,130 and the calls are trading at $40⅝, then the call position is valued at $337,187.50 ($4,062.50 × 83 calls). The calls can

TABLE 6-19

Index Call Buying Strategy Selection Grid

	Strategy 3: Buying a Portfolio of At-the-Money Puts	Strategy 4: Buying a Portfolio Equivalent of Out-of-the-Money Puts	Strategy 5: Buying a Portfolio Multiple of Out-of-the-Money Puts
Cost	**4.6%**	**1%**	**4.6%**
Risk	Under perform general market by 3.48%	Perform with the general market	Under perform general market by 3.48%
Specific Market Option	Bullish on pending developments, worried about sharp down move	Short-term bearish but worried about missing big market advance	Very bullish short term
Benefit of Options	Expensive insurance (No deductible)	Low-cost insurance	Leverage
Time Frame of Strategy	1 month maximum	Buy 3–6 month options, willing to let them expire	2–6 weeks, then close out if not successful
Implied Volatility	Should be on low end of range	Not an important consideration	Important consideration; Need low implied volatility
Expected Frequency of Use	As needed, depending on cash position of portfolio	As needed, depending on cash position of portfolio	Once every year, on average

then be sold for a net profit of $175,487.50 ($337,187.50 − $161,700) as the stocks are purchased. Assuming the selected stocks increased in proportion to the index, the outlay for the stocks will be $5,081,986, or $81,986 more than when the index was at 1,111. This increase in stock price is more than offset by the $175,487.50 profit in the option position, which provided immediate participation in the market advance.

A logical question is, what would have happened if the market declined? Quite simply, the loss on the index calls would have been largely offset by the decrease in the purchase price of the stocks. Remember, the calls were purchased so that market exposure was increased immediately, with results very nearly the same as purchasing a portfolio of equities—whether they rise or fall.

This strategy is sometimes called *dynamic hedging*, because the delta of an option changes as the index level changes. The mirror image of this strategy is to use put options to reduce market exposure. Consequently, maintaining approximate equality between the initially desired portfolio and the index options requires buying and selling options as the market fluctuates. This process is known as *adjusting* and explains why the strategy is dynamic.

In the example just given, assume that on the day the 83 calls are purchased the market rallies to an index level of 1,130, the options increase to 47⅛, and the delta of each one rises to .63. At this point, a $5,000,000 equity portfolio would have risen to $5,085,508.55 [5,000,000 × (1,130 ÷ 1,111)]; 83 calls with a delta of .63, however, replicate a $5,908,770 portfolio (83 × 1,130 index level × $100 multiplier × .63 delta). The proper number of calls for a $5,085,508.55 portfolio is 72 [$5,085,508.55 ÷ (1,130 index level × $100 multiplier × .63 delta)]. Consequently, at the end of day one, with the index up 19 points, 11 calls are sold so that the call option position remains in balance with the desired equity portfolio.

To make this successful, the time period must be chosen carefully, and the portfolio manager must have considerable knowledge of implied volatility levels. Also, this is a very short-term strategy, because time decay is a significant element. If the options are held too long, it is possible that time decay can take away all, or at least most of, the profit from owning them during a market rally. Depending on which options are purchased, this strategy is most effective in the two-to-four-week time frame.

Changes in implied volatility can also affect this strategy. The call option purchaser must have a basis for believing that implied volatility is not going to decline significantly during the period that dynamic hedging is implemented. Forecasting implied volatility requires knowledge and experience, but is not an impossibly difficult task.

OTHER PORTFOLIO STRATEGIES

Reducing Volatility by Writing Covered Calls

Covered call writing is a popular strategy among institutional investors. Selling call options against individual stocks reduces the variability of returns associated with stock ownership and enhances returns in stable or declining markets. Writing S&P 500 Index calls against a well diversified portfolio can provide the same benefits on a portfolio-wide basis.

For example, assume that the S&P 500 Index is at 1,110 and that a diversified $10 million portfolio approximately matches the index and yields 1.4 percent. Expecting the market to remain within a 10 percent range over the next three months, the manager decides to sell 90-day, at-the-money index calls with a premium of 36½. Dividing the value of the portfolio by the contract size times the index level, it is determined that 90 index calls can be sold against this portfolio [$10,000,000 ÷ (1,110 index level × $100 multiplier)]. Figure 6-3 illustrates the return profile for this position at expiration under different market conditions.

Table 6-20, column 4, lists changes in the call position under different market scenarios. Assuming that the stocks increase in proportion to the index, if the market advances 10 percent by expiration, the value of the portfolio will increase by $1 million. The value of the call option at expiration is 111, which is the in-the-money amount (1211 − 1110). Loss on the short call position is $74½, which is the in-the-money amount minus the premium received ($111 − 36½). Therefore, the total call position shows a $670,500 loss ($7,450 per contract × 90 contracts). If the call options expires unexercised, the seller keeps the total $328,500 premium.

Column 6 in Table 6-20 shows a three-month portfolio dividend of $35,000 (1.4 percent annual yield for three months on $10 million). The dividend is constant and is not dependent on market movements. The net change in the combined option/stock position is indicated in column 7, which is the sum of the value in columns 3, 4, and 6. The maximum profit from the combined position is indicated by the horizontal line in Figure 6-3. No matter how far the index advances, there is no potential for appreciation beyond the premium received.

The numbers in column 7 of Table 6-20 do not indicate a perfectly horizontal maximum profit line due to an arithmetic technicality. The

FIGURE 6-3

Writing Index Calls Against an Equity Portfolio (Reduces Volatility While Increasing Cash Flow)

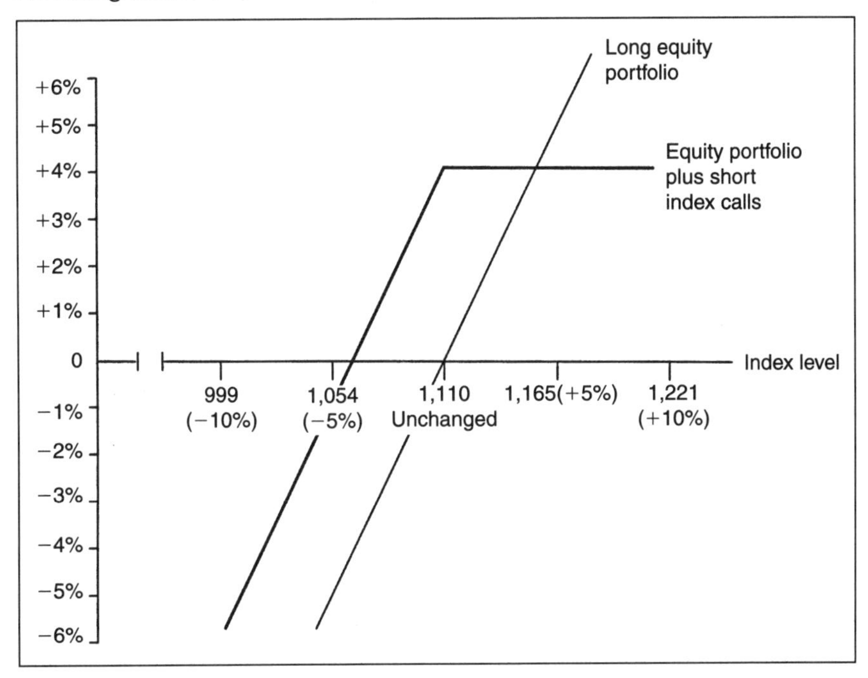

equity portfolio is assumed to be exactly $10 million at an index level of 1,110 and will behave proportionately with market rises and declines. The option position, however, will behave like a slightly larger portfolio if the index level rises. This happens because a 1,110 call priced at 36½ is similar to a starting index level of 1,146.50. The number of contracts, 90, equates to a $10,318,500 portfolio for index price rises above 1,146.50.

In the case of a market decline, the combined position outperforms a portfolio consisting of only stocks. If the market declines slightly, the value of a stock portfolio declines proportionately to the decline in the index, less dividends of $35,000, while the value of the portfolio with covered call options may actually increase.

The premiums earned by writing covered calls provide a cushion against loss in a declining market and extra income in a flat market. This is an effective strategy for institutional investors who (1) want to supplement the dividend income of a portfolio, (2) want to reduce the downside risk of a portfolio, (3) are willing to exchange upside potential for downside protection, or (4) believe that call premiums are overvalued.

TABLE 6-20

Writing Index Calls on a Diversified Portfolio

Range of Market Outcome (%)	S&P 500 Expiration Level	Change in Equity Position ($)	Change in Options' Position ($)	Value of Options' Position ($)	Dividends ($)	Profit/Loss Combined Portfolio ($)	Value of Combined Portfolio ($)	Percent Change (Annualized)	Profit/Loss Portfolio ($)	Value of Unprotected Portfolio ($)
+10%	1221.00	1,000,000	–670,500	999,000	35,000	364,500	10,364,500	+3.52%	1,035,000	11,035,000
+5	1165.50	500,000	–171,000	499,500	35,000	364,000	10,364,000	+3.51	535,000	10,535,000
0	1110.00	0	328,500	0	35,000	363,500	10,363,500	+3.51	35,000	10,035,000
–5	1054.50	–500,000	328,500	0	35,000	–136,500	9,863,500	–1.38	–465,000	9,535,000
–10	999.00	–1,000,000	328,500	0	35,000	–636,500	9,363,500	–6.80	–965,000	9,035,000

FIGURE 6-4

Writing Index Calls Against an Equity Portfolio (a Comparison of Strike Price Selection)

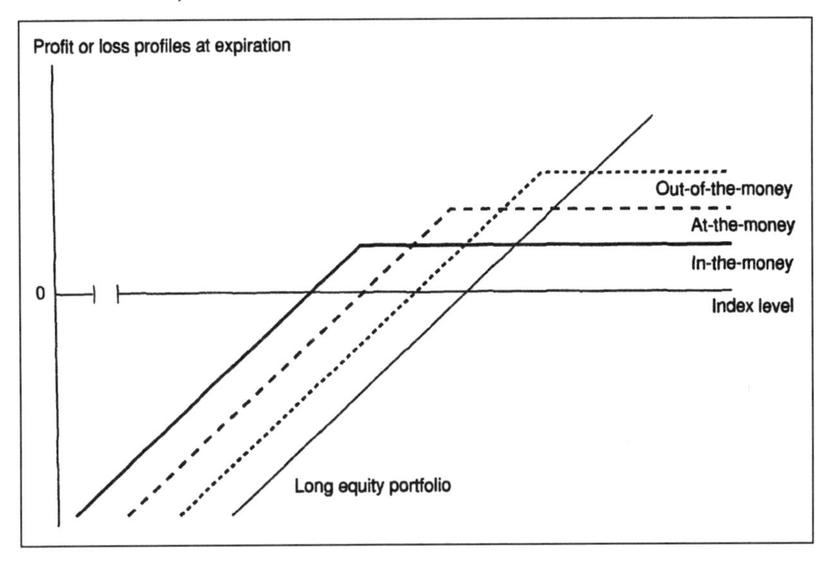

The risk/reward characteristics of a covered writing strategy change depending on whether in-the-money or out-of-the-money options are used. Some differences are graphically presented in Figure 6-4. A portfolio manager who writes in-the-money options exchanges upside potential for premium income that provides downside protection. The more a call is in-the-money, the more protection it provides. A manager who writes at-the-money or out-of-the-money options participates more fully in a market advance, but limits downside protection on the portfolio. Whether the option is in-the-money or out-of-the-money, however, covered call writing does not totally insulate a portfolio from a severe market decline.

The Fence Strategy

To build a fence around possible returns, combine two strategies previously discussed: buying an index put option for insurance and, at the same time, selling an index call to reduce volatility and enhance portfolio income. With an overall market decline, the long index puts limit the possible loss, while the returns from a market advance are limited by the sale of the index calls.

There is no real insurance analogy for the combination of these two strategies. This strategy is based on the portfolio manager's desire to lower the net cost of insurance (the put purchase) by giving up some upside profit potential (the covered call sale). A portfolio manager who expects the market to decline sharply, but feels that implied volatility levels are too high, might consider simultaneously selling a call option and buying a put option.

For example, assume that, with an S&P 500 Index level of 1,110, a $10 million portfolio can be insured for 60 days by buying 90 index put options with a strike price of 1,110 [$10,000,000 ÷ (1,110 × $100)]. At a price of 35¾ each, the total cost is $321,750 (90 × 3,575). The index level at which the put options breakeven is 1,074.25 (1110 − 35¾). This represents a market decline of 3.2 percent (1,110 − 1074.25) ÷ 1110).

The cost of the puts could be reduced by selling index calls with a 1,135 strike price for $21¼ each. The total premium received from the sale of 90 calls would be $191,250 ($2125 × 90). This would lower the net cost of the strategy to 14½ per put or $130,500 total.

Sale of the calls also raises the downside break-even index level to 1,095.50 (1,110 − 14.50), which represents a market decline of 1.3 percent. Although this level of downside protection is attractive compared to the break-even level of the put purchase alone, this benefit is not achieved without a cost. That cost is the opportunity cost of limiting the upside potential of the portfolio to an index level of 1,135 strike of the sold calls. Above that level, the short calls lose as the market (and the portfolio) rises.

The risk/reward profile for the combined strategy of buying a put and selling a call is presented in Table 6-21. This table describes the returns the investment manager could expect from a $10 million portfolio protected by the fence strategy (purchase of puts strike 1,110 and sale of calls strike 1,135). Figure 6-5 presents a picture of the return profile at option expiration.

In the fence strategy, the long put limits the downside risk, and the short call limits the upside potential. The fence is often seen as a low cost method of buying insurance, and it is frequently thought to be excellent if the insurance can be bought for "nothing." In some cases, calls that are closer to the money can be sold for a premium equal to or greater than the cost of puts, which are slightly farther from the money. Note, however, that focusing exclusively on the cost of the strategy ignores other important factors central to portfolio management. General issues of managing risk and reward are, by their very nature, subjective. Any in-

TABLE 6-21

The Fence Strategy with a Diversified Portfolio

Range of Market Outcome (%)	S&P 500 Expiration Level	Change in Equity Position ($)	Change in Options' Position ($)	Value of Options' Position ($)	Dividends ($)	Profit/Loss Combined Portfolio ($)	Value of Combined Portfolio ($)	Percent Change (Annualized)	Profit/Loss Portfolio ($)	Value of Unprotected Portfolio ($)
+10	1221.00	1,000,000	–904,500	–774,000	35,000	130,500	10,130,500	+1.29%	1,035,000	11,035,000
+5	1165.50	500,000	–405,000	–274,500	35,000	130,000	10,130,000	+1.28	535,000	10,535,000
+2	1135.00	225,225	–130,500	0	35,000	129,725	10,129,725	+1.28	260,225	10,260,225
0	1110.00	0	–130,500	0	35,000	–95,500	9,904,500	–0.96	35,000	10,035,000
–5	1054.50	–500,000	369,000	0	35,000	–96,000	9,904,000	–0.97	–465,000	9,535,000
–10	999.00	–1,000,000	868,500	0	35,000	–96,500	9,903,500	–0.97	–965,000	9,035,000

FIGURE 6-5

Fence Strategy

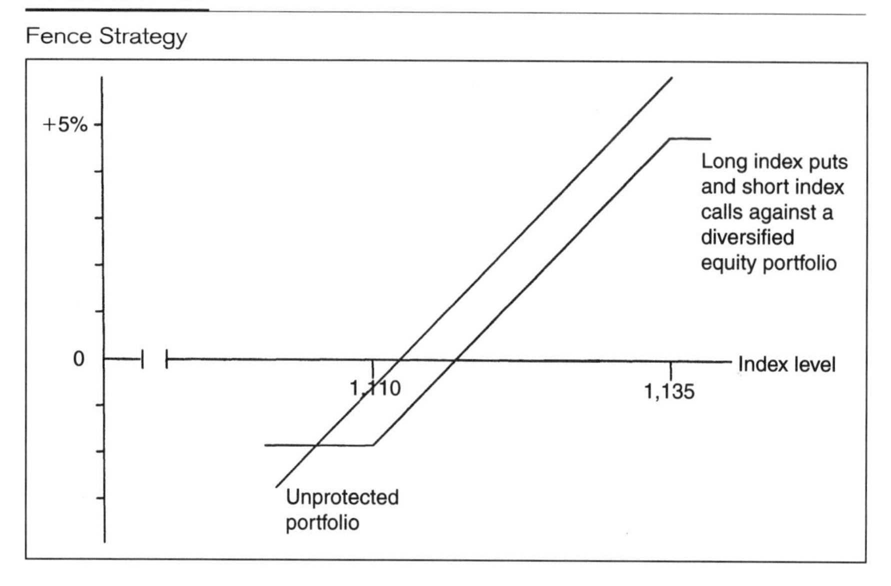

vestment decision is a decision that the prevailing market price is wrong in some sense. The decision to buy a stock presumes that the market is valuing a stock too low. Option trading decisions are no different in this basic sense. Portfolio managers buy puts as insurance because they believe the put premiums relative to current market levels and their market forecast represent a fair risk/reward ratio relative to owning a portfolio without insurance.

At a given decision point, a portfolio manager should consider the fence strategy as a viable alternative, but not because of its low cost alone. The first consideration should be the market forecast. The second consideration should be the portfolio's objective, and third should be the knowledge of current implied volatility levels.

The Portfolio Repair Strategy

A variation of the repair strategy for individual equity issues discussed in Chapter 5 can also benefit broadly based portfolios that have experienced a 10 to 15 percent decline. Assume that $10 million was invested in a diversified group of equities, the performance of which closely follows the S&P 500 Index when it is at 1,110. If the index subsequently declines 12 percent to 977, the portfolio would have a value of approximately $8.8 million. By using options on the S&P 500 Index, the break-even point on this portfolio can be lowered to an index level of approximately 1,048, instead of 1,110.

The portfolio repair strategy consists of two parts: first, buying one near-the-money call option and, second, selling two out-of-the-money call options. The goal is to initiate the combined position at a very low total net cost. With the S&P 500 Index at 977, the 980 calls are near-the-money calls. An $8.8 million portfolio equivalent number of these calls is 90 [$8,800,000 ÷ (980 × 100)]. At a price of $28 each, the cost of 90 calls is $252,000 ($2,800 × 90).

The second part of the strategy, selling twice as many index calls with a strike price of 1,050 at $13 each, brings in an option premium of $234,000 ($1,300 × 180). The total net cost of the strategy, then, is $18,000, not including transaction costs.

What has this accomplished? The long near-the-money index calls essentially double the market exposure of the equity portfolio. This means that the stock plus option portfolio can recoup losses in half the market rise required by the stock portfolio alone.

Portfolio appreciation, however, is limited to the level of the short index call options—in this case an S&P 500 Index level of 1,050. Above that point, at expiration, the short calls begin to appreciate in value and, therefore, negate additional gains in the long calls and the equity portfolio. How this strategy works is demonstrated in tabular form in Table 6-22.

In the example just given, note the use of S&P 500 Index options, instead of the more popular S&P 100 Index options. S&P 500 Index options were chosen because of their European-style exercise feature in which early exercise is not permitted. Because index options are cash settled, if American-style exercise index options which allow early exercise are used, then there is an additional risk in applying the repair strategy to a portfolio—a risk that does not exist when using equity options to repair a loss on an individual stock.

In the case of the options repair strategy on an individual stock, early exercise does not create additional risk, because early assignment of a short call can be met by delivering stock or by assuming a short stock position that can then be covered by the exercise of the long call, which has a lower strike price than the assigned call. For instance, assume the owner of 100 shares of a stock currently trading at $44 decides to repair an unrealized loss by purchasing one $45 call and selling two $50 calls with the same expiration date. Assume also that the stock then rose to $55 and, one week prior to expiration, assignment notices were received for both short $50 calls. The assignment of the first $50 call can

TABLE 6-22

The Portfolio Repair Strategy: Portfolio Values at Option Expiration

Equity Portfolio ($)	Index Level	Percent Gain	Long 90 980 Calls P/(L) ($)	Short 180 1050 Calls P/(L) ($)	Total Portfolio Value ($)
8,800,000	976.8	—	–252,000.00	234,000.00	8,782,000
8,900,000	987.9	+1.14%	–180,900.00	234,000.00	8,953,100
9,000,000	999.0	+2.27	–81,000.00	234,000.00	9,153,000
9,100,000	1010.1	+3.41	18,900.00	234,000.00	9,352,900
9,200,000	1021.2	+4.55	118,800.00	234,000.00	9,552,800
9,300,000	1032.3	+5.68	218,700.00	234,000.00	9,752,700
9,400,000	1043.4	+6.82	318,600.00	234,000.00	9,952,600
9,500,000	1054.5	+7.95	418,500.00	153,000.00	*10,000,000
9,600,000	1065.6	+9.09	518,400.00	–46,800.00	*10,000,000
9,700,000	1076.7	+10.23	618,300.00	–246,600.00	*10,000,000
9,800,000	1087.8	+11.36	718,200.00	–446,400.00	*10,000,000
9,900,000	1098.9	+12.50	818,100.00	–646,200.00	*10,000,000
10,000,000	1110.0	+13.64	918,000.00	–846,000.00	*10,000,000
10,100,000	1121.1	+14.77	1,017,900.00	–1,045,800.00	*10,000,000

*Numbers are rounded

be met by delivering the owned (or long) 100 shares of stock, and the assignment of the other $50 call can be met by exercising the owned $45 call. Between the time that the assignment notices are received and the time when the $45 call can be exercised, there is no market exposure, because the short $50 call became a short stock position when it was assigned. If the short stock position fluctuated between the receipt of notice of assignment of the short $50 call and the exercise of the long $45 call, the long $45 call also fluctuated and offset any gain or loss of the short stock position. Thus, early assignment of short equity calls does not create a risk for an investor who uses the stock repair strategy.

Early assignment of cash-settled index options, however, does create a risk when the portfolio repair strategy is being used. Because cash is delivered against an assigned index call, early assignment leaves the long portfolio and the long index call position with no offsetting position. The cash payment that meets the early assignment is never adjusted as the market fluctuates during the next session, so the portfolio manager is left with a market exposure that is long twice the desired amount until the time when the long calls can be sold. And, if the market opens substantially lower, it is possible that the long calls would be sold at an index level below the level where the short calls were settled. The result could be a substantial loss. This is known as *overnight risk*, because assignment notices are not received until the morning after exercise occurred. Cash settlement, however, is based on the closing index level on the day of exercise.

To avoid overnight risk, or *early exercise risk*, use European-style index options, such as the S&P 500 Index options. Early exercise is not permitted with these options. No early exercise privilege means these options are generally lower in price. Sometimes the lower prices affect the relationship of the prices of the at-the-money and the out-of-the-money options. And, as a result, it might be necessary to use different strike prices than optimally desired or to use an expiration farther out than with American-style options.

Selling Equity Puts: A Portfolio Management View

Selling equity puts is, without question, the most controversial of all option strategies. It is viewed by some investors as highly risky and very speculative. Some critics have gone so far as to blame sharp market down movements in part on a large open interest of uncovered, or *naked*, put options. But is all this true?

Chapter 3 demonstrated that selling puts has exactly the same payoff diagram as covered writing. In Chapter 2, the discussion of put–call parity also demonstrated that the two strategies are identical.

Nevertheless, selling puts continues to be vilified while covered writing is accepted as a safe, conservative strategy. During an Options Institute class in July 1993, an opinion poll was taken of 32 participants. Twenty-two said that selling puts was "risky," and twenty-eight said that selling puts was "inappropriate for institutions." Twenty-five said selling puts was "inappropriate for individuals." In contrast, 29 people said that covered writing is "conservative." Only five thought that covered writing was "inappropriate for individuals."

Why the difference? There seem to be three reasons. First, options are complicated. People may believe that the two strategies are different, because they have not been introduced to the concept of put–call parity. They may have also heard about or read about the "risk of naked put selling."

Second, for some inexplicable reason, there is a tendency to analyze a covered write from the point of view of, "How much can I make?"

In contrast, the tendency is to analyze selling puts from the point of view of, "How much can I lose?"

Third, the use of margin changes the profit or loss potential of an option strategy as a percentage of capital. And, typically, put sellers establish positions with less account equity to support a position, *i.e.*, on margin, than do covered writers who tend to pay for the positions fully in a cash account. Consequently, stories about losses from selling puts on margin—so-called naked puts—tend to be more dramatic than stories from covered writers who can brag that "The $2 call I sold expired worthless," even though the stock may have declined $10.

This discussion is a good introduction for considering equity put selling from a portfolio management perspective, because a comparison of the institutional investor and the individual speculator is most instructive. Their goals are different, the capital they manage is different, and the way they benefit from options is often different.

It is important to make a clear distinction between a portfolio manager and a speculator. A *speculator*, for the purposes of this discussion, has limited capital, uses margin, and concentrates on one or two stock or option positions at a time.

A *portfolio manager*, by comparison, does not use margin. Also, a portfolio manager is constantly dealing with several securities, probably at least 25 individual issues. Absolute quantity of capital is not the distinguishing feature, because there are many very large speculators. However, a portfolio manager has diversified his investments so that he does not consider himself to have limited capital.

With these clear distinctions in mind, consider how differently a speculator and a portfolio manager think about a short put position. The first obvious difference is the willingness to buy the underlying stock. The speculator has no intention of buying and owning the stock; so, if a put is assigned, the speculator was wrong. The speculator would sell the stock immediately. The portfolio manager, however, has sold puts on stocks that he is willing to own and that he would like to buy at a price below the market price on the day of the put sale. In fact, when a money manager sells puts on a group of stocks, he fully expects to get some delivered to him. Thus, assignment of short puts is not viewed by a portfolio manager as being wrong; it is viewed as buying a good stock at a good price.

There is a sharp difference in goals between the speculator and the portfolio manager. The speculator simply wants to collect the premium from the sale of the put option. The portfolio manager, however, is content with either outcome: collecting the premium or taking delivery of the stock. The result over time is income enhancement to the portfolio by collecting option premiums.

The very different risk profiles of the two investors is not so obvious. The speculator, as a result of selling puts on margin, stands the risk of losing a high percentage of capital. This is especially true if a speculator is "fully margined." A portfolio manager views the risk profile quite differently. First, selling puts for the portfolio manager is actually less risky than owning stock! Because selling a put is similar to buying stock at a lower price, the cash-collateralized put seller loses less than the owner of a stock when the market declines. Second, because the portfolio manager sells puts on stock considered desirable for the portfolio, he or she would be very disappointed if the market rallies and these stocks were not in the portfolio. Consequently, the risk for a portfolio manager is missing the big market rally—quite a difference from the speculator who does not care what happens to a stock when it is above the strike price of the sold put.

This difference in perceived risk implies a significantly different market opinion when initiating the put selling strategy. The speculator is neutral to bullish and does not want the stock to decline. Anything else is OK. The speculator is not wildly bullish, or he would buy stock or calls. The portfolio manager, however, has the goal of buying stocks cheaply. If a manager thinks prices will rally, he buys stocks now. To sell puts, therefore, the manager must forecast a neutral to bearish market. In the absence of options, a manager would place bids for stock under the market, expecting prices to dip so that the bids could be filled. Selling puts accomplishes the same result without a stock dipping to the desired price, as long as the stock stays below the strike price of the put.

For example, assume that a stock is trading at $50 and the appropriate $50 put can be sold for $2. A portfolio manager, who is short-term bearish but willing to buy this stock, places an order to buy the stock at $48. If the stock price drops to $48½ and rallies back to or above $50, the manager still owns no stock. If the stock drops to $45, the stock is purchased at $48 and experiences a $3 loss. If the stock rises to $55, the portfolio manager would have been wrong and made nothing. Selling the $50 put, however, makes two of these situations better and has the same outcome as the third. If the put is assigned, the portfolio manager has effectively bought the stock at $48. If the stock price declines to $45, there is a $3 loss exactly the same as buying the stock at $48. At any price between $48 and $50 at expiration, the portfolio manager is a buyer of stock at $48, gaining a profit. In the case of the rally to $55, he or she still collects the $2 option premium, nice consolation prize when the short-term forecast for neutral to bearish price action was wrong!

Viewed in its proper context, cash-collateralized put selling is a valuable income enhancement strategy for portfolio managers.

COVERED WRITING VERSUS SELLING CASH-SECURED PUTS: FINDING A DIFFERENCE WHERE THERE IS NONE

Starting from a cash investment and going to either a covered write or a cash-secured short put results in exactly the same payoff diagram. But if these strategies are used over time to enhance portfolio performance, there is a difference in timing as to when each should be implemented.

A portfolio manager who is fully invested, but is willing to sell some of the equities at higher prices, finds the strategy of selling covered calls attractive. This means that the manager is neutral to bullish and looking for a place to raise cash.

The portfolio manager most likely to sell puts, however, has cash and is looking for lower prices at which he hopes to buy stocks. Selling cash-secured puts is the appropriate strategy with this neutral to short-term bearish market forecast.

When considering the market cycle over the course of several months, the distinction is very obvious. A portfolio manager should be a covered call seller when willing to sell stocks and a cash-collateralized put seller when willing to buy stocks.

INTRODUCTION TO FLEX OPTIONS

In the previous examples, all of the strategies discussed involved the use of listed, standardized options (*i.e.*, those with strike prices, expiration dates, and other such factors set by the exchanges.) However, because the entire premise for the existence of options (as in *financial securities*) is to give "options" (as in *choices*), it would stand to reason that an ideal situation is when a user can select the *exact* strike prices, expiration dates, and other such factors he desires, instead of just choosing the *best available* from a roll of listed options. Instead of buying the SPX 1,000-strike call option expiring during the standard monthly expiration week, it may be more appropriate for the user to buy an SPX 1,013½ call expiring on December 3, if such an option fits the portfolio better. Even though it is not feasible to list options on every conceivable strike level and expiration date, several years ago the industry recognized that there was, indeed, a need for such "customized" options, and they responded to that need by creating *FLEX options*.

Consider a portfolio manager with, not $10,000,000 indexed to the S&P 500, but $500,000,000 indexed to the S&P 500. Although returns have been phenomenal, a couple of the larger investors in the portfolio are extremely nervous about the short-term prospects of the overall market. In fact, they are threatening to pull their equity if the manager does not reduce their exposure over the next three months. The portfolio manager knows that she can buy SPX put options for downside protection. Unfortunately, because of the added volatility in the market, put options are trading at historically high levels. Furthermore, she really does not have any cash on hand and does not want to sell part of her stock holdings and realize the capital gain. All this being said, she is very concerned that these investors will pull out their equity.

Examining her problem, the manager realizes that if she had the ability to customize expiration dates and strike prices, she could put on a *zero cost collar* for the next three months. This strategy involves buying an out-of-the-money put and funding that purchase by selling an out-of-the-money call for the exact premium of the cost of the put. Given her investors' time frame, she could do the collar for 90 days. This strategy allows her to reduce her downside exposure without any additional cash outlay. The manager can do this as an over-the-counter trade. Unfortunately for her, the way the fund is chartered, she can only invest in exchange-traded products. This was done to assure investors that at no time would they be exposed to counter-party credit risk. As she is explaining her quandary to a colleague, he asks her if she is aware of FLEX options, which trade on the CBOE.

In 1993, CBOE introduced Index FLexible EXchange (FLEX) options, and followed this up with the introduction of Equity FLEX options in 1995. FLEX options are designed to broaden institutional investor access to customized derivative products. Designed for initial transactions of at least $10,000,000 of notional principal in index products and the lesser of 250 contracts or $1 million underlying notional value in equity products, FLEX options, provide these important features: Option Clearing Corporation (OCC) contract guarantee and elimination of trading counter-party credit risk; contract terms that can be customized along four dimensions; price discovery in a competitive auction market; and the administrative convenience of listed options. In addition, FLEX are marked to the market independently on a daily basis by the OCC. Calls and puts on the Dow Jones Industrial, S&P 100, S&P 500, NASDAQ 100, and Russell 2000 Indexes, and over 250 equity classes are available for FLEX trading. These contracts are similar in structure to privately negotiated, "over-the-counter" call and put options and warrants.

Features Of Index FLEX Options

Index FLEX options allow users to select option contract terms. Users may specify any of the following terms:

- Underlying:Dow Jones Industrials (DJX), S&P 100 (OEX), S&P 500 (SPX), NASDAQ 100 (NDX), or Russell 2000 Indexes
- Option type:Call, Put, or Cap
- Expiration date: Up to five years from creation. The expiration date specified must be a business day and cannot fall within two days prior and subsequent to, or include the third Friday of a month (or the first preceding business day if the third Friday is a holiday)[1]
- Strike price: May be specified as an index level, a percentage, a numerical deviation from a closing index level, an intra-day value level, or any other readily understood method for deriving an index level, rounded to the nearest tenth of an index point (*e.g.*, 444.4)
- Exercise style: American, European, or capped
- Settlement values:Exercise (assignment) will result in the delivery (payment) of cash on the business day following expiration; there are two methods of exercise-settlement value determination:
  –Opening exercise-settlement value for these indexes is calculated using the opening price in the primary market for each component security on the specified expiration date of the FLEX option. If a security in the index does not open for trading on expiration day, the last reported sales price for that security will be used in calculating the exercise-settlement value.

[1] If for example, during a particular year, the third Friday of June is the 18th, the expiration dates for FLEX S&P 500 options could not include the June business days of the 16th, 17th, 18th, 21st, and 22nd.

–Closing exercise-settlement value is calculated using the closing price in the primary market of each component security on the specified expiration date of the FLEX option. If a security in the index does not open for trading, the last reported sales price for that security will be used in calculating the exercise-settlement value.

–NASDAQ 100 Index exercise-settlement value calculation is different from other CBOE index opening calculations. It is calculated based on a volume-weighted average of prices reported in the first five minutes of trading for each of the component securities on the last business day before the expiration date.

Features of Equity FLEX Options

- Underlying: Any equity class traded on the CBOE is eligible for FLEX. There is a OCC certification procedure which requires the submitting broker to notify the Exchange by 12:00 CST in order for that class to be eligible for trading on that day. Otherwise, the class will be available the next business day.
- Strike price: Any reference price in eighth increments for puts. Standard strikes for calls.
- Expiration date: A business day up to three years from trade date; however, it may not be the third Friday of the month (or the first preceding business day if the third Friday is a holiday) or two business days on either side of each of those days.
- Exercise style: American, European

Competitive Price Discovery

The FLEX trading process occurs on the trading floor of the CBOE, which means that customer transactions are effected according to the principles of a fair and orderly market employing trading procedures and policies developed by the CBOE. The Request For Quote (RFQ) is an indication of user interest for a particular option series to be created or that has been previously opened: a specific underlying, type of option (call or put), expiration date, strike price, exercise style, and settlement basis must be described. The RFQ may request either a market or a bid or offer. The RFQ sequence is as follows.

STEP 1: Requests for Quotes

- The submitting member brings a completed RFQ that specifies each of the variable contract terms to the FLEX Post.
- The submitting member presents the RFQ to the FLEX Post Official who reviews the RFQ to assure that the terms of the contract are within the necessary parameters.
- The RFQ is disseminated to all market vendors who carry FLEX data. This is to ensure that on- and off-floor members may participate in the response process by submitting their, or their customer's, best bids and offers on the series to be traded.

STEP 2: Responsive Quotes

FLEX market makers have an affirmative obligation to respond to a RFQ with a bid and an offer in a minimum of $10,000,000 in underlying value for Index FLEX products and 100 contracts for Equity FLEX options. Other market makers, member organizations, and floor brokers acting on behalf of customers may also provide a *responsive quote*, but are not bound by affirmative obligations.

- Responsive quotes must be communicated to the FLEX Post Official within the request response time, which will usually be between 3 and 10 minutes and will be disseminated to the market vendors
- Responsive quotes are made verbally at the FLEX Post at the end of the request response time. No transaction may take place until the request response time has expired.

STEP 3: Accepting the Best Bid or Offer

FLEX quotes are generated in response to an RFQ rather than from an order. The submitting member is under no obligation to accept the best bid or offer. If the best bid or offer is not promptly accepted, the quotes are no longer valid. Any other party interested in the same FLEX series must submit another RFQ. With the exception of those RFQs in which the submitting member has indicated a desire to "cross" an order, the following describes the transaction process:

- The submitting member decides whether to accept the best bid or offer, or request a BBO improvement interval

- Upon acceptance of a bid or offer by the submitting member, the FLEX Post Official will disseminate a last sale message as well as enter the transaction into the trade match system and generate a trade confirmation for the parties involved

SUMMARY

Portfolio managers must remember that options are like insurance policies. Just as equity options are insurance policies on individual stocks, so too are index options insurance policies on diversified portfolios that follow the index on which options are traded. Index put options can be used in a variety of ways to insure diversified portfolios, and index call options can be used to increase the market exposure of cash investments. FLEX options were created to expand even further the choices available for employing option strategies, and the market for FLEX options has matured to the point where the trading systems are nearly identical to those employed for traditional listed options.

Stated simply, index options offer portfolio managers a variety of strategic alternatives that are not available from other investment vehicles.

CHAPTER 7

HOW THE TRADING FLOOR OPERATES

Marshall V. Kearney

It is 9:00 AM EST, and you are sitting in your kitchen. In front of you are a cup of coffee, the business section of your newspaper, a pad of paper, and a pencil. You are preparing an order to buy call options on XYZ stock.

You tune your television to a business channel to get a feel for how the market will open. A young news reporter is broadcasting live from the trading floor of the New York Stock Exchange (NYSE). Many questions come to mind. Why is everyone walking so briskly? Will she be OK with all of that pushing and shoving? Are some of those people carrying orders to where options on XYZ are traded? Will one of them carry your order later this morning? Why are the people dressed so unusually—polka-dot ties, jackets with stars on the shoulders? And, oh my gosh, there is a button on a jacket with a picture of Elvis! Did that young man's mother dress him funny as a child? Can you assume that traders have poor taste in clothing, or is there a reason why they dress that way?

In this chapter, we explain how an options trading floor functions and how option orders are handled on these floors. Because this book is written by the staff of The Options Institute, the educational arm of the Chicago Board Options Exchange (CBOE), most of the examples address how business is transacted at the CBOE. Keep in mind that, while some of the terms used in this chapter are the same for all option trading floors, there may be slight or major variations at other exchanges. Questions regarding how a specific order is processed should be directed to your brokerage firm.

Trading floors are dynamic environments. The way orders are processed today is very different from the way they were processed just a few years ago, and it is reasonable to expect that change will continue at a rapid pace. Given advances in technology, there is speculation about whether trading floors will even exist in the future. But, regardless of whether or not physical trading floors are needed, option exchanges will continue to exist because they provide a financially secure, low-cost, centralized marketplace where interested parties can meet to negotiate the transfer of risk. Because traders of all types are customers of exchanges, the exchanges will constantly strive to provide the fast and accurate report of a trade and the secure transfer of cash and securities. This chapter explains how option exchanges work to deliver this service.

Questions about Elvis and how individuals dress will not be addressed. But, yes, people on trading floors do wear unusual jackets for a very good reason.

Because your order to buy or sell an option is frequently transacted with a professional market maker who is a member of the exchange, Chapter 8 discusses how market makers trade. That discussion will show how much the trading styles and objectives of on-floor market-makers differ from those of off-floor investors and traders. You will learn why off-floor investors and traders and on-floor market makers are not in competition with each other. Rather, they are in different businesses. They use different strategies, and they need each other to accomplish their individual objectives. Properly functioning markets bring people together who have different motivations and different market forecasts. Those people negotiate prices and, when a price is agreeable to both a buyer and a seller, then a transaction is made.

FROM PUT AND CALL DEALERS TO TRADING FLOORS

The organized securities business in the United States supposedly started under a buttonwood tree in New York City in 1792, when twenty-four brokers and merchants made an informal alliance. This was followed in 1817 by the founding of the New York Stock Exchange Board and, shortly thereafter, with the establishment of the American Stock Exchange (ASE). A little trivia may be of interest here. Many old timers in the securities business still refer to the ASE as "The Curb," because, in its early years, ASE traders literally conducted business outside, on the curbs of the busy Wall Street neighborhood!

In Chapter 1, The History of Options, you learned how, in the early 1900s, put and call dealers were the only professionals willing to make bids and offers for option contracts. A common frustration with that system was that these dealers were not required to make bids and offers all of the time. One complaint was that they would only do so if it were clearly to their advantage. Such a situation did not make for very liquid markets.

Another complaint about the early system was that, because there may have been only one put and call dealer making a market in XYZ options, the *spread*, or difference between the bid price and ask price, was very wide. The result was that *price discovery*, the process by which buyers and sellers arrive at a market price, was hindered. Too often, customers were left asking, "How do I know if the price quoted to me is the best price available, or even a fair price?" Concerns over fairness did not attract new market participants at a rapid pace.

All this changed on April 26, 1973, when the CBOE opened for business. The terms of option contracts—contract size, expiration dates, strike intervals, etc.—were standardized. Initially, call options on only sixteen stocks were listed, but this number quickly doubled and then doubled again. Put options were added soon after, and today, the listed options industry in the United States offers options on over 2,500 stocks and on over 60 stock indexes.

Other important differences between the system dominated by the put and call dealers and the modern option exchanges are (1) market maker obligations to make bids and offers continuously in all options, (2) world wide dissemination of bid and offer prices, and (3) the credit quality of the party on the other side of a trade. In the pre-CBOE environment, a broker had to find a put and call dealer who was willing to make a bid or offer for a particular option; that dealer then would write an option with unique terms; *e.g.*, the strike price being the current stock price and the expiration date being 90 days from today. Now, option contracts are standardized, and there are professional option market makers who compete in trading crowds on exchange floors. Standardization and competition tend to increase liquidity and narrow bid–ask spreads. In addition, today's market makers are obligated by exchange and SEC rules to make a bid and offer for all options. But that is not all. Price competition can also come from outside of an exchange. Because all option bid and offer prices are disseminated just like stock quotes, any market participant can "better the market," *i.e.*, enter a higher bid price or a lower offer price.

The problem of unknown credit risk has been eliminated from today's option market by the Options Clearing Corporation (OCC), a triple-A rated entity that guarantees the performance of all listed option contracts. If you buy a call option on XYZ stock through your broker, and if the person who sells it to you is a client of Tower of Babel Securities, then you do not have to worry about the credit worthiness of that individual or of Tower of Babel Securities. Technically, you purchased the call from the OCC, and the client of Tower of Babel Securities technically sold it to the OCC. The presence of the OCC and its credit worthiness have brought more participants into the options market, and this, in turn, has increased liquidity.

THE CBOE TRADING FLOOR

As discussed in Chapter 1, the CBOE was initially financed by the Chicago Board of Trade (CBOT), and its first trading floor was created from space previously used as the CBOT members' cafeteria and situated adjacent to the CBOT's trading floor. However, the CBOE soon grew to take over two full floors in the CBOT building, and, in 1984, the CBOE moved across the street to its current location, a seven-story custom trading facility.

Located at the south end of The Loop (Chicago's business district—given that name because of the shape of the famous elevated train (Ell), whose tracks surround the area), the CBOE building is connected by walkways to the Chicago Board of Trade, the Chicago Stock Exchange, and Financial Place, a 30-story office tower, where many trading firms and the OCC have offices. The present CBOE trading floor measures 50,000 square feet, comparable in size to a typical Wal-Mart store. Approximately 3,000 people work on the trading floor, and there are over 4,500 computer terminals there, more than in any other building in the world, including the Pentagon and the NASA Control Center in Houston. Over 50,000 miles of electrical wiring lie under the CBOE trading floor, enough to encircle the globe twice. The CBOE trading floor uses enough power to light the 110-story Sears Tower or the entire city of Rockford, Illinois (population 143,000). Because of the heat generated by the computer terminals and the people working on the trading floor, the heating system at the CBOE is only turned on when the outside temperature is ten degrees below zero Fahrenheit which, on average, is one day per year. For more information about the CBOE trading floor, visit the exchange's web site at www.cboe.com.

SEPARATION OF PRINCIPAL AND AGENT

There are two different types of traders on an option trading floor: *floor brokers*, who handle client orders only and may not trade for themselves, and *market makers*, or local traders, who trade with their own capital and who are required to make a bid and an offer for at least 20 contracts of every option in their trading pit. Market makers are not allowed to represent client orders.

This distinction is known as the *separation of principal and agent*. The floor broker is an agent for the off-floor investor or trader and has that person's best interests at stake at all times. Floor brokers may not trade for their own account because of possible conflicts of interest. In contrast, market-makers trade for themselves, but they have obligations to make bid prices and offer prices so that, at any time, an off-floor trader can make a trade "at the market."

BACK AT THE BREAKFAST TABLE

You have decided on a strategy, and you are ready to act.

The XYZ April 50 calls closed yesterday at 2⅞, and XYZ stock closed at $48¾. The reporter has indicated that there is no startling news this morning and that foreign markets are trading nearly unchanged from yesterday. You therefore believe that XYZ stock and XYZ options will open little changed from yesterday's close, and you are prepared to accept the risk of being wrong. Consequently, you decide to purchase ten of the XYZ April 50 calls *at the market*. With a market order, you are instructing your broker to buy these options at the best available price when your order is received in the pit. You pick up the phone and call your broker.

Technology has changed order processing so significantly in recent years that the "open outcry" trading floor of yesterday is hardly recognizable today. In order to fully understand the workings of a modern trading floor, however, it is instructive to review how the early options trading floor worked.

FLASHBACK

In the early days of the listed options business you could initiate an order to buy or sell an option only by calling your broker. Let's see, for example, how your order to purchase 10 XYZ April 50 calls at the market would have been processed in "the old days."

After receiving your verbal instructions, your broker wrote this information on a pre-printed two-copy form known as an order ticket. The order ticket was time stamped, and the top copy taken to the back office, or "wire room," for processing. The second copy was kept among your broker's records.

The back office first time-stamped its copy of the order ticket and then transmitted the information to your brokerage firm's booth on the trading floor by one of two methods: telephone or teletype. If by telephone, a back-office clerk at your local brokerage office would speak to a floor clerk at the exchange who would write the information on a trading ticket. If by teletype, the back-office clerk would type the information, and it would be transmitted to a receiving machine in the firm's trading floor booth, which would print out the "trading ticket." Whether received by phone or teletype, your firm's booth on the trading floor was in receipt of a trading ticket similar to that in Figure 7-1, which shows instructions to "buy 10 XYZ April 50 calls at the market."

FIGURE 7-1

Market Order

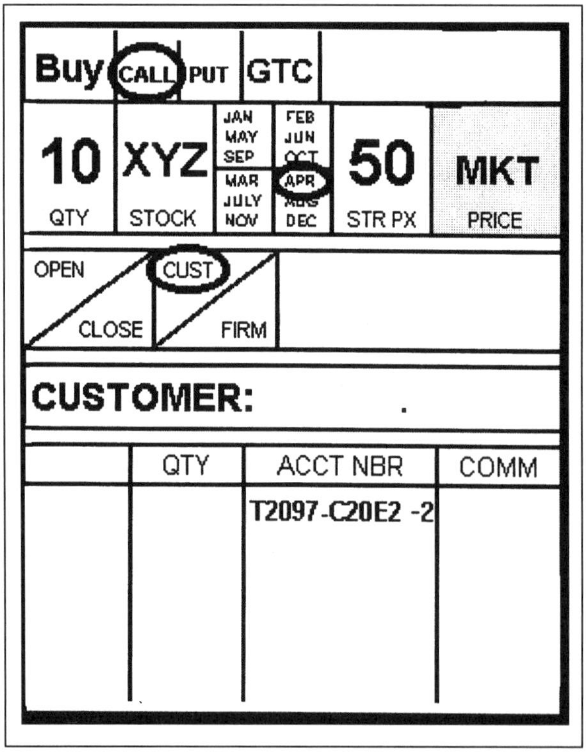

Buy CALL PUT GTC

10	XYZ	JAN MAY SEP / MAR JULY NOV	FEB JUN OCT / APR AUG DEC	50	MKT
QTY	STOCK			STR PX	PRICE

OPEN / CLOSE CUST / FIRM

CUSTOMER:

	QTY	ACCT NBR	COMM
		T2097-C20E2 -2	

The trading ticket had five copies. After being time stamped and checked for accuracy, the top copy was removed and kept at the booth. A runner then took the remaining four copies to the location at the exchange where XYZ options are traded. Trading locations are commonly referred to as "trading pits," because they are formed by circles of tiered steps so that traders can see and hear everything that is going on.

A trading pit is a centralized marketplace for all options on a particular stock. By Securities and Exchange Commission (SEC) regulation, all buyers and sellers must come to a centralized marketplace to conduct business. This assures buyers that they are getting the lowest available offer price from any seller at that particular moment, and it assures sellers that they are getting the highest bid price at that particular moment.

In the old days, the runner gave your order to a *floor broker* who is responsible for representing the best interests of you, the off-floor investor or trader. There are two types of floor brokers: *firm floor brokers* and *independent floor brokers*. Firm floor brokers are employees of a particular firm; they work for a salary and handle orders only for that firm. Independent floor brokers are self-employed; they are paid for each transaction they make, and they handle orders for many firms. Independent floor brokers are sometimes called "two-dollar brokers," which refers back to independent brokers on the New York Stock Exchange who received a $2 commission for handling a trade. If a firm does a sufficient amount of business in a particular trading pit, then it will have a full-time firm floor broker there. If, however, there is not enough volume to justify a full-time broker, then a firm uses an independent floor broker.

Upon receiving a paper market order, the floor broker first time stamped it, then read it and checked the *quoted market*, which appeared on a television screen above the pit. The quoted market is the most recently stated best (or highest) bid at which some buyer is willing to buy and the best (or lowest) offer at which some seller is willing to sell. Assume, for this example, that the quoted market is "2¾ bid, offered at 3."

The floor broker verbally checked the quoted market to make absolutely sure it is up to date. To do this the broker might yell, "Listen up, everyone! What are the XYZ APE 50s?" If this exclamation seems confusing, that is because it contains *floor speak*—terminology, slang, and verbal abbreviations used by pit traders. Let's review exactly what the floor broker means:

"Listen up, everyone!" means "I need to check the market on a particular option, and I may have an order that some of you may want to sell to or buy from." Because market makers make their living by buying and selling options, they will be eager to respond. The floor broker's opening exclamation has gotten the attention of everyone in the pit, and this increases the possibility of getting a better bid or better offer or both from someone who may have been thinking about something else at that particular moment.

"What are the XYZ APE 50s?" means "What is the highest bid and lowest offer right now in the April 50 calls on XYZ stock?" "APE" is obviously an abbreviation for April, and "50s" is obviously an abbreviation for the 50-strike. But did you notice that the word "call" does not appear? How do the market makers know that the floor broker is asking about 50-strike calls and not about 50-strike puts? In floor speak, no mention of calls or puts means calls. Had the broker said, "What are the XYZ APE 50 puts?" then the floor broker would have been asking about put options.

At this point you may wonder how long it takes to learn the language of floor speak? The answer is: not long. When money is involved, people tend to learn very quickly. Also, traders typically start out as assistants, or clerks, to other traders and an essential part of that job is to learn the language. But back to the floor broker and your order.

Note that, at this point, the floor broker has not indicated whether the order is to buy or sell, or even that there is an order. The floor broker's request for information makes all parties state their interests, and the lowest offer, which is what the floor broker wants in this example, is bound to appear.

If the best bid and best offer have not changed since the television screen was last updated, then the members of the trading crowd respond to the floor broker by yelling "2¾, at 3." In floor speak this means the best bid is 2¾ and the best offer is 3.

At this point the floor broker has a decision to make. Because the order contains instructions to buy 10 calls "at the market," the floor broker could turn to the market maker who first said "at 3" and simply say, "buy 10" which, in floor speak, means "I will buy 10 contracts from you at a price of 3 each." If more than one person were offering the XYZ April 50 calls at 3, then the floor broker could "divide up the order" by trading with more than one person. For example, the floor broker might say "3, 3, 2, and 2" while indicating four traders who had responded with offers of 3.

Alternatively, rather than simply buying 10 calls at the offered price of 3, the floor broker might attempt to get a better price by bidding in between the quoted bid price and quoted offer price. The floor broker might say, for example, "2⅞ bid for 10." If one or more of the traders who offered calls at 3 says, "OK, sell you at 2⅞," then the floor broker will have succeeded at buying the calls at a better price than the indicated offer. There is absolutely no guarantee, however, that any of the traders offering calls at 3 will "come down," *i.e.*, lower their price. In fact, there is some risk in not taking the best offer when it is made. In hectic market environments stock prices change every second or two and, as a result, bid and offer prices in options are adjusted continuously as well. Thus, there is a risk that when the floor broker bids 2⅞, something else changes—such as the stock price ticking up—that will cause the traders offering calls to raise their offer above 3. If this happened, it would be impossible for the floor broker to buy the calls at 3. Consequently, the floor broker has a big responsibility and some big decisions to make in representing your best interests.

Assuming the floor broker does buy 10 contracts for you at the indicated offer price, then, in floor speak, your order has been "filled at 3." The broker next writes down the essential information on the trading ticket: the quantity, the price, and the name of the individual or individuals with whom the trade was made. Figure 7-2 is an example of a filled trading ticket in which a total of 10 April 50 calls have been purchased at a price of 3, five from market maker XXX at firm #1 and five from market maker YYY at firm #2.

The filled trading ticket was time stamped again; the floor broker kept the second copy and gave the third copy to the exchange as a record of the trade. The CBOE had staff at each pit to keypunch the important information immediately after each trade occurs. (Just as the NYSE "ticker tape" disseminates information about stock trades, so too do the CBOE and other option exchanges immediately report option transactions.)

The remaining two copies of the trading ticket were then returned to your firm's trading booth. Typically, a floor broker is not able to leave the trading pit, because there would be other orders requiring attention. So the filled trading ticket with your trade information would be placed in a mail slot reserved for your brokerage firm, typically located at the edge of the trading pit. Every five to fifteen minutes your brokerage firm would send a runner out to look in all of their mail slots on the trading floor and retrieve any tickets placed there. Runners returned to the booth with a stack of filled trading tickets. At the booth, copy four of the

FIGURE 7-2

Filled Order

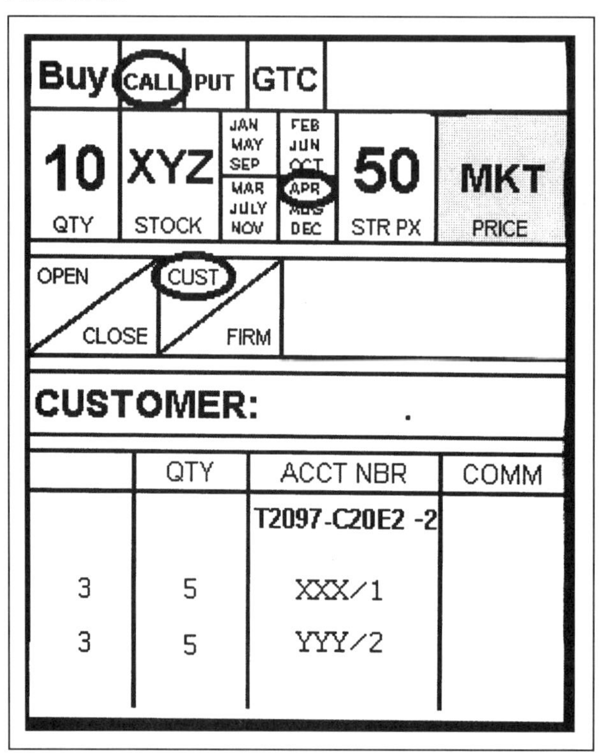

Buy CALL PUT GTC

10 QTY | XYZ STOCK | JAN MAY SEP MAR JULY NOV | FEB JUN OCT APR AUG DEC | 50 STR PX | MKT PRICE

OPEN / CLOSE | CUST / FIRM

CUSTOMER: .

	QTY	ACCT NBR	COMM
		T2097-C20E2 -2	
3	5	XXX/1	
3	5	YYY/2	

order ticket was removed and filed with the booth's records, and copy five was sent to the firm's back office at the exchange where trade information was typed into computers that communicated with the exchange's computers to match buys and sells and transfer funds. Eventually, a keypunch operator would get to your order, and a confirmation of your trade would be sent by teletype to your branch office. Your broker would be notified, and he or she would contact you with the fill information.

The back office process at an exchange is known as "clearing trades" and is similar to the processing of checks. Every check contains the name of a bank and an account number. When you deposit a check, your bank takes it to the indicated bank and that bank takes money out of the indicated account. Similarly, when your floor broker purchased call options for you, those calls were purchased from a specific individual who has an account at a specific brokerage firm. Money is taken out of your account at your brokerage firm and transferred to that individual's account at his or her firm. If that person owns calls, then they are taken out of that person's account and transferred to your account. If that person does not own calls, then a short call position is created in that person's account, and a long call position is created in your account.

How Long Did It Take?

At this point, two questions are typically asked: How long did the process take? And, why was the floor trading ticket time stamped so many times?

The time lag between your entering the order and it being filled could be from three to thirty minutes. The time lag between the order being filled and your receiving a report could be from five minutes to three hours. These indeterminate time lags depended on many things. How busy was your broker? How busy was your broker's back office? How busy was your firm's booth at the exchange? How many orders did the runner have to deliver before yours? How busy was the floor broker in the pit where XYZ options are traded? How long before your filled trading ticket was picked up by a runner? And how many tickets had to be keypunched before yours? So much for the "good old days." Although the system was sometimes inconvenient, it was a way of life.

Because the process took so long, time stamping a ticket at every step of the process provided a clear audit trail. If a problem were to occur, which it did on 2 to 3 percent of trades in the old days, then the time stamps might indicate where the problem occurred. For example, if your local back office received the order at 10:00 AM, but the booth at the exchange did not receive it until 11:00 AM, then the delay clearly occurred at the local back office. If, however, the time stamp at the booth is 10:01 AM and the floor broker's time stamp is 11:00 AM, then a delay occurred either at the booth or with the runner. Such an audit trail made it possible to improve the processing of orders and to assign responsibility when problems occurred.

This largely manual system was used until 1990. It was not bad by some standards and not speedy enough by others.

BACK TO THE PRESENT

In the middle 1980s an analysis of the orders received from individual investors, so-called "retail clients," brought out some interesting facts.

Approximately 35 percent of the orders from retail clients accounted for approximately 5 percent of the retail volume. In other words, there was a large number of transactions that involved a very small number of contracts. Upon reflection, the CBOE and brokerage firms arrived at a similar conclusion: these orders did not require human intervention, and everyone would be happier if the fill information from these transactions were reported quicker. This thinking, combined with the "10-Up Rule," (now the 20-Up Rule) made it possible to automate certain trades.

20-Up Markets and Maximum Spreads

Remember how illiquidity—the lack of bids and offers in sufficient size—was a problem with the early option markets made by put and call dealers? The CBOE originally implemented the "10-Up Rule" to avoid a similar problem in the listed options market. In 1998 the number was increased to 20-up in options on equities. This exchange rule obligates CBOE market-makers to buy a minimum of 20 options at the quoted bid price and to sell a minimum of 20 options at the quoted offer price. If, for example, the quoted market in the example above was 2¾ bid, offered at 3, and you are not a professional broker/dealer, then the exchange members of the trading pit are obliged to buy 20 contracts at 2¾ or to sell 20 contracts at 3. There are also rules about the maximum allowable spread between bid and offer prices. For example, when the bid price of an option is below 2, then the maximum bid–ask spread differential is ¼ of a point, *i.e.*, 1¾ bid, offered at 2 is allowable, but 1¾ bid, offered at 2¼ is "too wide." Market makers who violate the maximum bid–ask spread rules are subject to fines by the exchange's enforcement authority.

Automation

Given the 20-Up Rule, the desire to report fills faster, and no need for human intervention on certain orders, then, the thinking went, why not give retail clients an immediate, electronic confirmation that an order has been filled? Why not eliminate the several-step human-handling process?

Obviously there was no compelling reason not to do this. In the mid-1980s the CBOE began work on a system known as RAES, the Retail Automatic Execution System. When either a market order or a limit order that can be filled at the current market price for 20 contracts in equities or 10 contracts in indexes or less is entered into RAES, the CBOE computers simply match the order with the quoted price and assign a market maker to take the other side. Assume, for example, that your order to purchase 10 XYZ April 50 calls at the market is entered on the RAES system. RAES determines that "3" is the quoted offer price, and your brokerage firm is told immediately that you purchased 10 contracts at that price. On the other side of the trade, the RAES system randomly selects a market maker who has volunteered to accept such orders. At the same time that your purchase of 10 calls is reported to your firm, a printout is also produced in the pit where XYZ options are traded. The market maker selected to take the other side of the trade knows within 30 seconds that he or she has sold the 10 calls that you purchased.

Other Automation

In today's world, rather than hand write an order ticket, your broker probably enters the information directly into a PC on his desk, and your order to buy 10 calls at the market is received almost simultaneously on a mainframe computer at the CBOE. This process is known as the Order Routing System (ORS). Figure 7-3 shows the routing alternatives available to the CBOE's ORS System.

To process your order, the computer must ask a few questions. Does the order contain a RAES-eligible trade? In other words, is it from a retail client, and is it for 20 or fewer contracts? If the answer to these questions is yes, then RAES kicks into action and, almost immediately, your broker is notified of a fill.

If the order is not RAES eligible, the computer must determine why. Is the order from a broker/dealer? Are there any contingencies or restrictions on the order such as Immediate or Cancel, Fill or Kill, or All or None? Another contingency could be that filling the option order is tied to the last sale price of the underlying stock.

If the order is not RAES eligible, the exchange's computer must decide what to do. Because different brokers have slightly different methods of operating, your broker will have instructions on file with the exchange so that the computer can direct how every type of order should be handled.

The Book

If a limit price order is "away from the market," *i.e.* the limit price to buy is below the current highest bid or the limit price to sell is above the cur-

FIGURE 7-3

CBOE Order Routing System

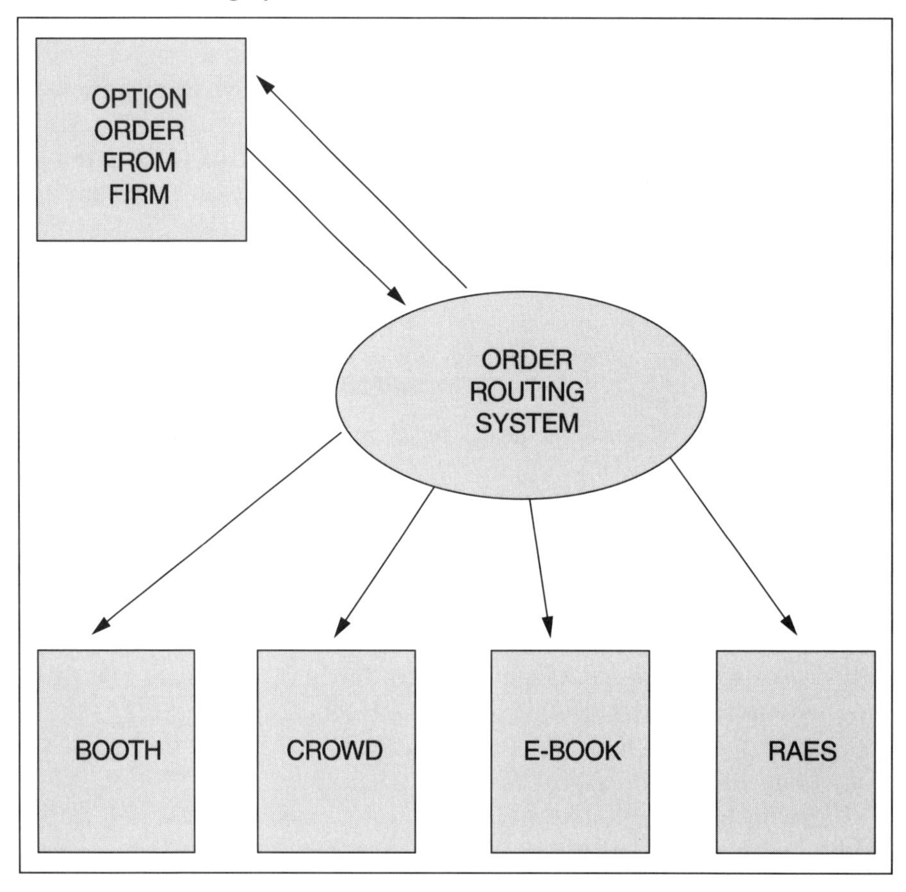

rent lowest offer, then your firm may direct the exchange to send such orders to the Public Limit Order Book. In floor speak, this is referred to simply as "the Book." The Book was established so that limit-price orders with no other contingencies from public investors and traders would have first priority to trades made at that price. Unlike RAES, there is no quantity restriction on orders placed in the Book. An order in the Book, for example, to buy 10 contracts at 2½ has priority over all other 2½ bids in that trading crowd. "Priority" means that if the quoted bid is 2½, then any seller at 2½ must sell to the Book first. No contracts can be sold at 2½ to floor brokers or to market makers in the trading crowd until the Book has been filled. The purpose of the Book is to offer off-floor public traders (as opposed to professional broker/dealers) the opportunity to be first in line for a particular trade if market conditions change to that price level. There is, of course, no guaranty that market conditions will change to that price level.

A limit-price order is not qualified for the Book if it is from a broker/dealer, or if it has contingencies other than price. Such orders may be sent electronically either to a floor broker in the trading pit, or they may be sent to a brokerage firm's booth on the trading floor. Sending an order electronically directly to a floor broker may be the fastest way of getting your order into the trading pit, but this is not always better. A floor broker may be very busy with other orders, so you might get a faster fill if the order goes to your firm's booth first. Regardless of where the order is sent, the need for a human runner has been reduced.

Processing a Non-RAES Order

Assume for this example that things are very busy in the pit where XYZ options are traded. Your brokerage firm's floor manager, therefore, has decided that non-RAES-eligible orders in XYZ options should be transmitted electronically to the booth. This gives the floor manager the opportunity to review current market conditions and to make a decision about the best way to represent your interests. The floor manager may still choose to send your order electronically to the floor broker in the XYZ pit, or the floor manager may decide to have your order represented by another floor broker who is not currently as busy as the floor broker in that pit. Let's see how this scenario might play out in today's highly automated market.

For this example, assume that your order is to "buy 30 XYZ April 50 calls *at a limit price* of $2\frac{7}{8}$" and that the current market in these options is $2\frac{3}{4}$ bid and offered at 3. Also assume that the XYZ pit is very busy and, therefore, your firm's floor manager has directed that all non-RAES XYZ orders be sent to the booth. When your order arrives at the booth, the manager determines that the firm's normal broker in the XYZ pit is very busy and decides to have the order handled by Carl LaFong, a firm floor broker in the pit next to the XYZ pit. To send your order to Carl, the floor manager does not need to send a runner. Rather, she can simply touch the same computer screen that notified her of the order, and, bingo! Your order appears on Carl LaFong's hand-held computer. The hand-held computer Carl is carrying is the size of a paperback book, and is actually a Pentium computer with infrared and microwave transmission capabilities within the CBOE building. Carl's acronym, initials which are displayed on a badge that he wears, are unique to him on the trading floor. Carl is not a small person. His acronym is TON.

Carl's screen immediately begins flashing, which indicates an order has just arrived. With the touch of his finger, your order appears on his screen. Seeing that it is for XYZ options, Carl walks hurriedly to XYZ pit and checks the monitors for the most recent bid and offer for the April 50 calls. Because your order is for more than 20 contracts, Carl must represent it in an open-outcry manner similar to that described above. To check the accuracy of the quoted bid and offer on the monitor, he yells, "XYZ APE 50 calls, how are they?"

Let's review what Carl did. He told the trading crowd that he wants a *two-sided market* in the XYZ April 50 calls. He is saying, "If I am holding a buy order, where can I buy some of these options? If it is a sell order, where can I sell some? I am not telling you, however, if I am a buyer or seller!" Remember, Carl's job is to represent you, the off-floor investor or trader. In this case, that means getting the lowest price, which cannot exceed 2⅞.

The trading crowd yells "2¾, at 3." Some traders were a little louder with the 2¾ bid, while others were more vocal with the 3 offer. TON declares in his baritone voice "I'll be a 2⅞ bid for a few." The market makers offering calls at 3 respond again, "At 3!" They do not know what instructions are on the order. TON feigns disinterest and repeats his 2⅞ bid. The traders that responded with the offer of 3 quickly scan their positions to see if they can sell some or all of those options. One trader might ask TON, "How many?" TON does not have to disclose the size or quantity of the order. In this case, however, he believes it is in your best interest to do so. TON replies by saying, "2⅞ bid for 30."

At this point, a runner wearing a distinctive (ugly?) dark blue jacket with stars on the shoulders walks up to TON. Hey, TON is wearing the same style of jacket! Among over 3,000 people on the trading floor, there are just twelve people wearing this jacket. That jacket helped the runner find TON. Unique clothing is a simple means of identification.

The runner whispers something to TON, and TON responds. Satisfied, the runner thanks TON and returns to his booth. What was the whispering about? Perhaps the floor manager wanted to find out the status of the order, the current bid and offer prices, and the possibility of filling the order at the offer price. Such information could be relayed to the off-floor investor or trader, and it might be used as part of the subjective decision-making process to "hit the offer," which means pay 3 in this example, or to "stand firm," which means to let the order remain unchanged. Because TON has no discretion to pay the current offer price of 3, he waits patiently.

Then something happens—TON's handheld computer beeps, and he reads some new instructions: "Buy 100 of the August 75 calls in Underwater Airways at the market." The Underwater Airways pit is across the aisle from XYZ, but TON cannot be two places at once. So what does TON do? He punches a button on his computer, and he yells "LZE has my 2⅞ bid." He then rushes across the aisle into the Underwater Airlines pit. What happened? Liz Ash, an independent broker in the XYZ pit with the acronym LZE was just handed your order electronically. When or if one of the market makers wishes to sell you your options, they will now make the trade with LZE.

Now, a floor broker with the acronym DCA enters the XYZ pit. DCA asks, "XYZ APE 50s, how are they?" LZE immediately yells back, "2⅞ for 30," and the market makers yell, "at 3." DCA then says, "sell you 20." This prompts a market maker with the acronym SLM to yell, "Sell you the balance." And LZE responds by saying, "Buy 20, DCA. Buy 10, SLM. I'm out of the APE 50s. Make them what you want."

Wow! That was a lot, and it was fast. What happened?

The most important point is that your order was filled. LZE purchased the 30 XYZ April 50 calls that you wanted to buy, and she purchased them at the limit price you wanted to pay.

When TON left the XYZ pit, he sent your order electronically to LZE who then became responsible for representing you. When DCA entered the XYZ pit, he asked for the best bid and best offer in the XYZ April 50 calls, but he did not disclose whether he was a buyer or a seller. When LZE responded by representing your order, it turned out that DCA had 20 to sell. We do not know if DCA had a market order or a limit order, and it does not matter; because, whatever order DCA had, it enabled him to sell 20 of the XYZ April 50 calls to LZE, which she purchased on your behalf.

Next—again, we do not know why—market maker SLM decided to sell the remaining 10 calls to LZE. LZE then confirmed the trades by stating each trader's acronym and the quantity purchased. When LZE said, "I'm out. Make them what you want," she meant that she had no more buy or sell orders in the XYZ April 50 calls, and, therefore, the market makers could re-establish the bid and ask prices.

After confirming the trade, LZE keys her side of the trade into her portable computer. This means she types in "buy" (the action), "20" (the quantity), "April" (the expiration), "50" (the strike), "calls" (the option type), "2⅞" (the price), "DCA" (the individual with whom she traded), and a number that indicates the brokerage firm that DCA was representing. She also types in the same information for the 10 options she purchased from SLM. LZE's typed confirmation is transmitted electronically to your broker's booth and its back office at the exchange.

As LZE was typing, DCA and SLM were also inputting their parts of the trade. When all three finish and press the "Submit" button, several things happen silently and electronically. First, DCA's, SLM's, and LZE's computers compare information. DCA and SLM sold, and LZE bought. If all information matches (quantity, strike price, expiration month, etc.), then the trade "clears." In layman's terms, this means that the OCC has confirmed that a trade has taken place. If the information does not match, then the computers of the people who disagree start flashing. This notifies them that there is a problem. They must either correct the information they input if they discover such an error, or they must talk to each other to resolve the difference.

Second, DCA's and SLM's computers report their trades into the exchange's *Time and Sales System*. Because the trade cleared, it is only necessary for one party to report it, and, by convention, the sell side of a transaction is responsible for reporting. The Time and Sales System is a record of every bid, every offer, every trade, and the times when they occurred. Time and Sales is a complete audit trail of every change in every market at the CBOE. If a problem with a trade is suspected, then this audit trail helps to determine if, in fact, there is a problem and who, potentially, might be responsible.

With this hand-held technology, there is no time lag in getting information onto or off of the trading floor. As of July 1, 1998, more than 80 percent of retail orders were handled or executed electronically. Today, from arrival at the CBOE to trade confirmation and dissemination, the average customer order spends approximately 48 seconds on the trading floor.

THE TRADING FLOOR DESCRIBED

A visit to the trading floor can be a bit intimidating at first, but it is fascinating and fun. In an average pit, options on ten to twenty different stocks are traded, and the number of people varies from ten to thirty market makers and two to five floor brokers. Figure 7-4 illustrates how a trading pit at the CBOE might be organized.

FIGURE 7-4

CBOE Floor Population

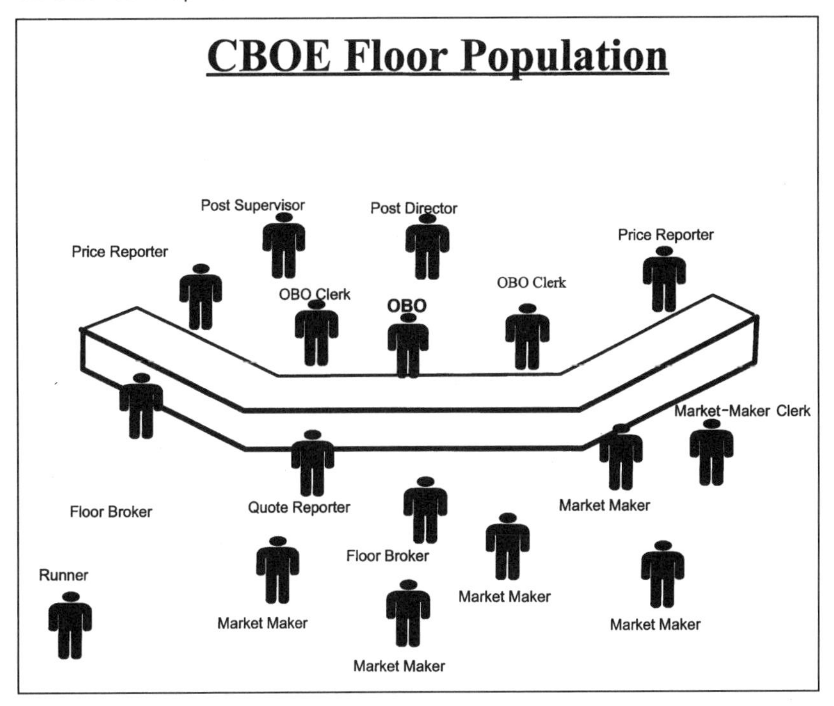

Market makers have the right to trade any options at the CBOE, but experienced market makers usually spend all of their time in one pit. They do this because the highest chance for success in market making comes from specializing on a small group of stocks and their options.

Stocks are not chosen for a given trading area, or pit, because they are in the same industry. In fact, you may find that options on Consolidated Lint, Underwater Airways, and Trans Atlantic Bridge all trade in the same pit. There are two reasons for this. First, when one stock in an industry has high volume, others tend to have high volume also. If the options on many related stocks traded together, then a small group of market makers could easily be overwhelmed when a major Wall Street research firm changed their opinion on that group.

The second reason that options on stocks in different industries are traded in the same pit has to do with the way the CBOE tries to insure that the best possible option markets are available to off-floor traders.

When the CBOE decides to list options on a new company, all trading crowds, or *pits* as they are called, are invited to apply for those options, and generally several trading crowds will apply. When the applications are received, the past performance of each crowd is evaluated. Has the crowd consistently made markets bigger than the minimum 20-Up Rule? And has the crowd made special efforts to make bids and offers on complicated spread transactions? The floor brokers for your firm and other firms participate in this evaluation, and the trading crowd with the best market making record will usually be awarded the new listing.

Big Pits

Some of the most interesting places to visit on the trading floor are the very active index option pits. These pits are where options are traded on the OEX (the Standard and Poor's 100 Stock Index), the DJX (the Dow Jones Industrial Average), the SPX (the Standard and Poor's 500 Stock Index), and the NDX (the NASDAQ 100 Stock Index).

The OEX pit, for example, is an oval that measures approximately 120 feet long and 80 feet wide. Approximately 350 traders, floor brokers, and clerks work there on an average day. Occasionally, however, that number of people more than doubles. In a crowd this size, people require a method of communication that is not needed in a smaller crowd. In these large pits, traders frequently "talk with their hands"; their language is that of hand signals.

In a large trading area, traders are often too far apart to hear each other. Is it me that guy wants, or is it the person behind me? Did he say September or December? Was it seven contracts or seventy? Hence the need for hand signals.

Earlier you were introduced to the verbal short hand of "floor speak." Now you will learn a little about the visual short hand of "floor sign language." Palms toward yourself means, "I'm buying contracts," while palms away from your body means "I'm selling."

Numbers from one to ten thousand are indicated by the number, location, and direction of fingers. Your index finger held vertically under your chin indicates "one." Index and second finger together, held vertically under the chin signifies "two." And so on up to "five." "Six" is indicated by the index finger held horizontally, with elbow held out, and touching the chin. The index and second finger held horizontally and touching the chin indicate "seven." And so on up to "nine."

Ten through fifty are indicated by fingers held vertically touching the forehead, and sixty through ninety are indicated by fingers held horizontally, with the elbow out, and touching the forehead. Hundreds are indicated by crossing arms, and thousands are indicated by hands behind the head. Figure 7-5 shows hand signals indicating various numbers.

There are twenty-six different hand signals for the letters of the alphabet, and there are customized (wacky?) signals which identify brokerage firms.

Traders in large pits also have some verbal short hand for times when they can actually be heard. "Labor Day" is used to indicate a September expiration, as in "I'm 3½ bid for the Labor Day 45 puts." Similarly, "Christmas" signifies the December expiration.

Opening Rotation

An aspect of trading floor activity that generates several questions is *opening rotation*. The purpose of an opening rotation is to insure that all opening trades in every option are conducted at the same price. Stocks are opened for trading in this fashion, but futures prices are opened in *ranges, i.e.*, at more than one price.

An opening rotation is conducted by the Order Book Official (OBO), an exchange official whose primary responsibility is to manage the Public Limit Order Book described above. There is an established order by which the options on a stock are opened. Generally options with the furthest out expirations are opened first, the in-the-money calls and out-of-the-money puts first, then the at-the-money options and, third, the out-of-the-money calls and in-the-money puts. LEAPS options are followed by earlier expirations, and front-month options are opened last.

While calls and puts are being opened in order during an opening rotation, no other trades can be made, not even in options that have been opened, until the entire opening rotation is completed. Also, the only option orders that are represented during opening rotation are those which arrived in the pit prior to rotation. Only when the OBO announces, "XYZ is open," can general trading begin.

During opening rotation, market makers are responsible for quoting a "two-sided market," that is a bid price at which they are willing to buy and an offer price at which they are willing to sell. The OBO then asks the floor brokers, "Is there any public?" This means, "Are there any

FIGURE 7-5

CBOE Hand Signals

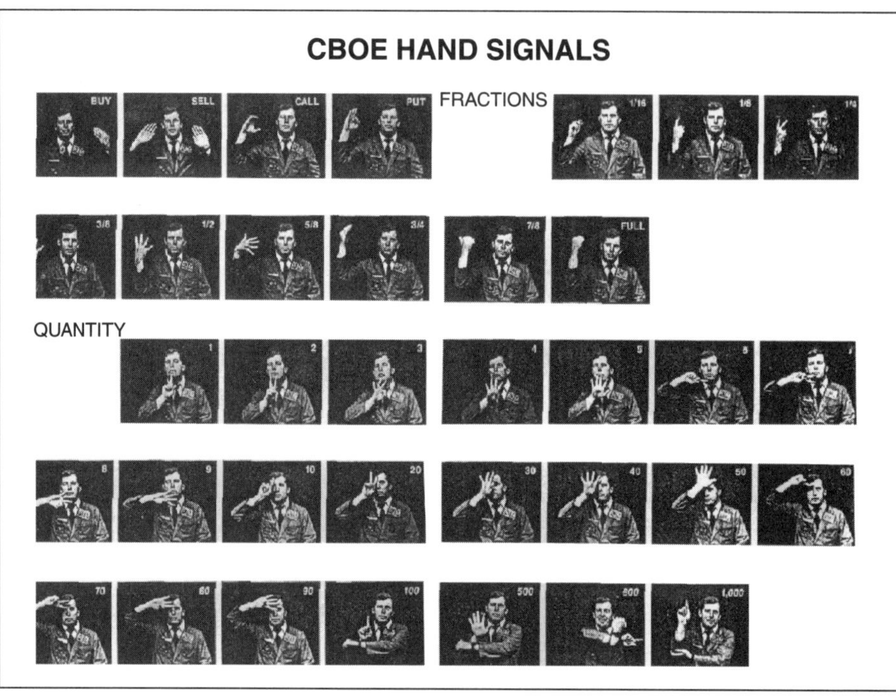

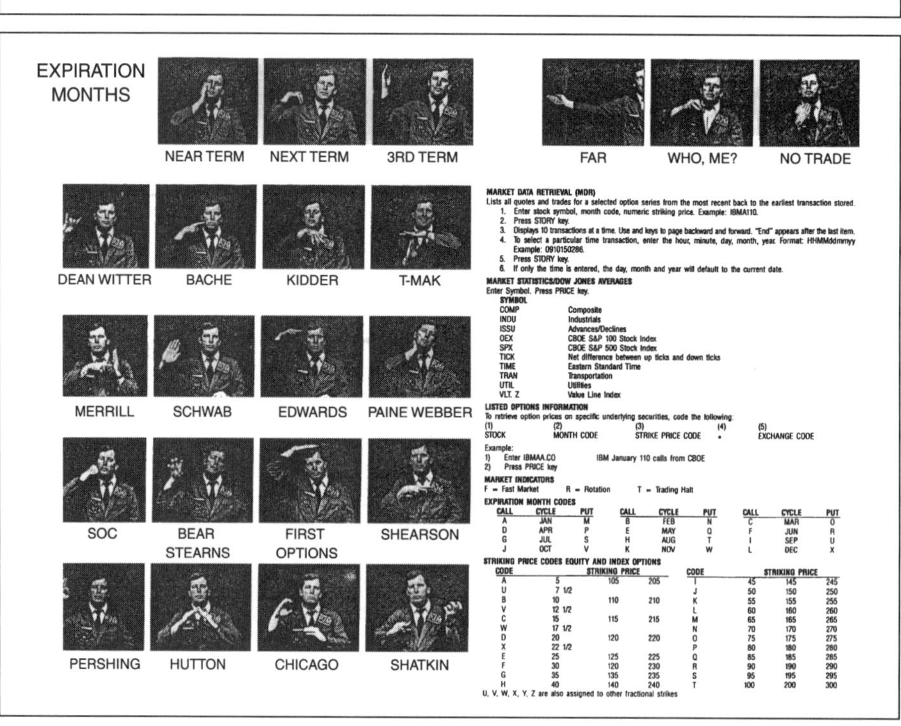

MARKET DATA RETRIEVAL (MDR)

Lists all quotes and trades for a selected option series from the most recent back to the earliest transaction stored.

1. Enter stock symbol, month code, numeric striking price. Example: IBMA110.
2. Press STORY key.
3. Displays 10 transactions at a time. Use and keys to page backward and forward. "End" appears after the last item.
4. To select a particular time transaction, enter the hour, minute, day, month, year. Format: HHMMddmmyy Example: 0910150286.
5. Press STORY key.
6. If only the time is entered, the day, month and year will default to the current date.

MARKET STATISTICS/DOW JONES AVERAGES

Enter Symbol. Press PRICE key.

SYMBOL	
COMP	Composite
INDU	Industrials
ISSU	Advances/Declines
OEX	CBOE S&P 100 Stock Index
SPX	CBOE S&P 500 Stock Index
TICK	Net difference between up ticks and down ticks
TIME	Eastern Standard Time
TRAN	Transportation
UTIL	Utilities
VLT. Z	Value Line Index

LISTED OPTIONS INFORMATION

To retrieve option prices on specific underlying securities, code the following:

(1)	(2)	(3)	(4)	(5)
STOCK	MONTH CODE	STRIKE PRICE CODE	•	EXCHANGE CODE

Example:

1) Enter IBMAA.CO   IBM January 110 calls from CBOE
2) Press PRICE key

MARKET INDICATORS

F = Fast Market   R = Rotation   T = Trading Halt

EXPIRATION MONTH CODES

CALL	CYCLE	PUT	CALL	CYCLE	PUT	CALL	CYCLE	PUT
A	JAN	M	B	FEB	N	C	MAR	O
D	APR	P	E	MAY	Q	F	JUN	R
G	JUL	S	H	AUG	T	I	SEP	U
J	OCT	V	K	NOV	W	L	DEC	X

STRIKING PRICE CODES EQUITY AND INDEX OPTIONS

CODE	STRIKING PRICE			CODE	STRIKING PRICE		
A	5	105	205	I	45	145	245
U	7 1/2			J	50	150	250
B	10	110	210	K	55	155	255
V	12 1/2			L	60	160	260
C	15	115	215	M	65	165	265
W	17 1/2			N	70	170	270
D	20	120	220	O	75	175	275
X	22 1/2			P	80	180	280
E	25	125	225	Q	85	185	285
F	30	120	230	R	90	190	290
G	35	135	235	S	95	195	295
H	40	140	240	T	100	200	300

U, V, W, X, Y, Z are also assigned to other fractional strikes market or limit orders from off-floor investors or traders that are close to being filled? During opening rotation, customer orders have priority over market makers and orders from broker-dealers.

As an example, consider an opening rotation in the XYZ June 45 calls in which the market makers' quoted bid is $7\frac{3}{8}$ and their quoted offer is $7\frac{5}{8}$. Also assume that floor broker A has an order to purchase twenty of these calls at the market, and that floor broker B has an order to sell five of these calls at a limit price of $7\frac{1}{2}$. The question is: at what price should these calls open?

Remember, the rules are: options open at a single price, and public orders have priority. Because the order to sell five calls at a limit price of $7\frac{1}{2}$ will not completely fill the order to buy 20 calls at the market, the OBO determines that "the 45 calls will open at $7\frac{5}{8}$." Floor broker A will then buy a total of 20 calls at $7\frac{5}{8}$. Five will be purchased from floor broker B, and the balance of 15 will be purchased from the market maker or makers who made the $7\frac{5}{8}$ offer. As this example illustrates, during opening rotations, market makers only participate in the overflow of volume after orders from off-floor investors and traders that can be filled are filled.

The opening rotation process is an efficient, orderly method of starting the trading day, but it is not perfect. One advantage of having an opening rotation is that both the off-floor buyer and off-floor seller trade at the same price, $7\frac{5}{8}$ in the example above. There is not an "opening range" where some of the off-floor buyers might pay $7\frac{3}{8}$ and some others might pay $7\frac{5}{8}$. Another advantage is that off-floor investors and traders have priority over on-floor market makers and broker-dealers.

A disadvantage of the opening rotation process, however, is that orders which miss the opening rotation deadline must wait until the completion of opening rotation before they are represented. And it is possible that market conditions could change significantly during this waiting period. Another disadvantage is that the RAES system is not available during opening rotations. And, finally, spread orders are usually not represented during opening rotations. *Spread orders* are multiple-part orders, and they are usually not represented during opening rotation because when options are being opened individually, it is difficult, if not impossible, to transact a multiple-part transaction.

Despite the speed at which opening rotations are conducted, increases in technology make it possible to envision even faster methods of conducting opening rotations. One idea, known as the Rapid Opening System (ROS), proposes to eliminate the verbal opening rotation and to open all options at the same time by using the computer to assign opening trades to market makers with the most competitive bids and offers, similar to the way that the RAES system assigns trades.

SUMMARY

The early days of option trading in which put and call dealers dominated the markets have been replaced by modern exchanges with competing market makers and advanced technology. Exchanges provide a centralized marketplace for all buyers and sellers to meet, and exchanges constantly strive to provide the fast and accurate report of a trade and the secure transfer of cash and securities. Also, the performance of all option contracts is guaranteed by the Options Clearing Corporation, a triple-A rated organization.

Separation of principal and agent means that floor brokers represent only the best interests of the off-floor trader whom they represent, and they cannot trade for their own account. It also means that market makers trade for themselves, and they cannot represent off-floor traders. They are responsible for making bids and offers in all options, and the 20-Up Rule for equity options requires them to purchase or sell a minimum of 20 contracts at those prices. Some index options have different RAES guidelines, and all of these guidelines are subject to change.

Open-outcry is still an important part of the modern trading business, because it seems to be the best way of transferring large quantities of risk. So, too, is the use of hand signals alive and well. They are primarily required in large trading crowds such as the OEX and SPX pits. But the original, multiple-step human-handling process by which off-floor orders were transmitted to the trading floor has been largely replaced by modern technology. Today over 75 percent of orders originated from off of the trading floor are transacted or delivered to the pit electronically. Exchange computer systems such as the Order Routing System and Retail Automatic Execution System allow efficient and individual handling of all orders. Although many advances have been made in recent years, floor technology continues to change at a rapid pace.

CHAPTER 8

HOW MARKET MAKERS TRADE

Marshall V. Kearney
James B. Bittman

In the last chapter, operations of the trading floor were viewed from the perspective of the off-floor investor and trader. In this chapter, we examine it from a different point of view, that of the market makers who stand on the floor all day, every day.

This chapter is not intended to turn you into a market maker. In fact, much of what you are about read will seem alien, because market makers buy and sell options for very different reasons than do off-floor investors and traders. Yet, as you will see, without market makers, life would be very difficult for off-floor investors and traders. It should also be noted that, without off-floor investors and traders, market makers would starve. The fact is, markets consist of participants with different motivations and different methods of attempting to earn profits. By coming together with the goal of fulfilling their own needs, each participant also helps to fulfill the needs of the others.

THE NEED FOR MARKET MAKERS

It should not be surprising to hear that, when an order from an off-floor investor or trader arrives on the floor, there is rarely an exactly opposite order from another off-floor investor or trader that is there waiting. If buy and sell orders that exactly matched each other did arrive on the floor at the same instant, then floor brokers could just make trades with each other and fill their customers' orders without market makers being present. Unfortunately, markets do not operate that simply.

When a floor broker represents a market order (an order to buy or sell at the best available price right now) from an off-floor investor or trader and there is no matching off-floor order to take the other side, then market makers are called upon. Market makers *must sell* to off-floor buyers (who are buying "at the market"), and they *must buy* from off-floor sellers (who are selling "at the market"). This is known as the market makers' *affirmative obligation to make a market*. For almost every option in their designated trading area, market makers must always have a stated bid price at which they are willing to buy a minimum of 20 contracts, and they must have a stated offer price at which they are willing to sell a minimum of 20 contracts.

Because of this affirmative obligation to make markets, it is often said that "market makers get what you don't want!" In other words, if you wish to buy an option at the prevailing market price, they must sell it. If you wish to sell an option at the prevailing market price, they must buy it. You might ask, "Why would anyone do such a thing?" The reason is that market makers are in a completely different business than off-floor investors and traders. This chapter is intended to give you a glimpse at how market makers think and how and why they make trades. Hopefully, then, you will appreciate that on-floor market makers are not in competition with off-floor investors and traders.

A "GREAT" MORNING

In the last chapter, we discussed how you woke up and decided to buy some calls on XYZ stock. Well, it just so happens that, when you were making your decision, one of the market makers in the XYZ trading crowd was also beginning his day. Let's call this market maker Godfrey Daniels, and, because all market makers have an acronym, or initials, by which they are known on the trading floor, let's give Godfrey the acronym GRT.

On this, a typical morning, GRT listens to the business news as he shaves, showers, and dresses. During the twenty minute train ride downtown, he scans *Investor's Business Daily* for news that might affect any of the twelve stocks on which he trades options. Note that XYZ is not the only stock on which options are traded in the trading area, or "pit," where GRT trades. Once downtown, GRT has a brisk eleven minute walk to his office, which is in a building adjacent to the options exchange.

At 7:30 AM CST, or one hour before the opening bell, when GRT arrives at his office, there is work to do. First, he must check that all of his trades from the previous day have "cleared." A trade is said to have *cleared* when a buy is matched to a sell and a trader's clearing firm gives a confirmation of the match. At that point the trader has a third party guarantee that the trade is good. A *clearing firm* is like a brokerage firm except that it specializes in providing back-office services to traders who are members of an exchange such as trade processing, record keeping, margin lending, and stock borrowing and lending. If there were an error, then GRT must meet with the person on the other side of the trade and resolve the difference. Second, he must review the clearing-firm–generated computer printout of his positions to make sure there are no discrepancies between that printout and his personal record. Third, for options on each of his twelve stocks, he must update the assumptions in the computer that he uses to calculate option theoretical values. There is definitely one day less to expiration. But his assumptions about interest rates and volatility may also have changed. Fourth, he evaluates the risk of his positions in each of his stocks, and, fifth, he formulates his strategy for each of his stocks for the day ahead.

Checking for Out Trades

In GRT's mailbox every morning are computer printouts that list every trade he made the previous day and summarize all outstanding option and stock positions. On a typical day GRT will make 50 trades that involve 900 option contracts and 10,000 shares of stock.

Modern technology has reduced the number of errors, or "out trades" as they are known on the floor. GRT typically has fewer than two out trades per month, but examples of errors that could occur are a "price out," in which the buyer and seller do not record the same price, or a "time out," in which different expirations are recorded. Typically, it will be obvious that one trader was in a hurry, wrote down the wrong information, and compounded the error by not accurately checking the trade. Consequently, such an out trade is a clerical error and not a real out trade.

Resolving Out Trades

There might, however, be a legitimate discrepancy such as a "quantity out" in which a buyer is convinced that he "bought 50" and a seller is convinced that he "only sold 15." In such a case, one or both traders would complete their orders on the opening that morning and make an accounting. In some circumstances, the two traders would then split the difference.

Assume, for example, that GRT and another trader, let's call him ERR, have a "50 versus 15 out trade at 4½." GRT recorded the purchase of 50 calls at 4½ and ERR recorded the sale of 15 calls at 4½. In this situation ERR is OK, because he actually sold what he thought he sold. GRT, however, does not have 35 calls that he thought he purchased, so he needs to buy them.

Ideally, GRT will be able to buy these calls at 4½ when the market opens today. If GRT can do this, then the out trade will be resolved without any loss. However, if the calls in question open at 5¼ and GRT purchases them at that price, then a loss will have occurred. GRT will actually have purchased the calls at a worse (higher) price than he recorded the day before. A loss of three-quarters of a point, or $75, per contract, on 35 contracts amounts to a total loss of $2,625, and GRT and ERR may have to split this amount. In this case, ERR will write a check to GRT for $1,312.50.

But not all out trades are losers. If the calls in question open at 4 and GRT purchases them at that price, then a profit will have occurred. GRT will actually have purchased the calls at a better (lower) price than he recorded the day before. A profit of one-half of a point, or $50, per contract on 35 contracts amounts to a total profit of $1,750, and GRT and ERR may split this amount. In this case, GRT will write a check to ERR for $875.

Fortunately, this morning GRT has no out trades. All of his trades from yesterday cleared.

Reviewing Positions

Modern technology has also made it relatively easy for market makers to compare their positions as they record them in their personal records with the clearing-firm–generated computer printout. But the task still requires attention to detail. GRT must go through each call and put, and each strike and expiration of each of his twelve stocks to make sure that the exchange and he are in complete agreement. Although his positions in the twelve stocks can vary dramatically from day to day, GRT can easily have a total of 5,000 open contracts in over 100 different options (calls, puts, strikes, and expirations). In addition, GRT could have positions in the underlying stocks that amount to more than 1 million shares (long or short).

Occasionally, GRT will forget to record a small trade in his personal records, but, generally, there are no discrepancies between his personal records and the clearing firm's printout.

Learning the Business

Some traders hire clerks to review their positions for them and to do other detail-oriented tasks such as checking trades during the day and adding and subtracting new positions as new trades are made. This is an excellent way for a new person to learn the business of market making, and it is an equally excellent way for an experienced trader to evaluate whether or not a new person has the potential, in their opinion, to become a trader. If a clerk exhibits the right skills and market sense, then an experienced trader who has sufficient capital to invest might decide to "back" that person. *Backing a trader* means forming a partnership and sharing profits or losses. A backer typically contributes capital, and a new trader typically agrees to work hard and share profits (hopefully) for three years. Trading on the floor of an exchange requires at least $100,000 in initial capital. Consequently, new people, especially young people, typically form a partnership with someone who has both the required capital and the experience to train them.

Updating Sheets

Until the development of hand-held computers, floor traders used to carry stacks of computer printouts that contained option theoretical values over a range of stock prices, and these were referred to as *sheets*. Market makers relied heavily on these sheets to make trading decisions, because it is impossible to carry that much information in one's head. But, like any information generated by a computer, the most applicable statement is, "garbage in, garbage out." For this reason, market makers spend much time and energy thinking about the assumptions that go into the formulas that produce option theoretical values.

Actually, the term "theoretical value" is a misnomer, because it implies that there is one absolute value that is an option's theoretical value. Market makers, however, are interested in "relative value." They do not care if an option is overvalued or undervalued in an absolute sense, but whether it is overvalued or undervalued in a relative sense; that is, relative to another option.

Market makers attempt to buy options that are "relatively undervalued" and sell others that are "relatively overvalued." Also, the quantities that are bought and sold have nothing to do with an opinion about market direction. Rather, purchases and sales are made in relative quantities based on the delta of each option. (Delta, remember, is the change in option price for a one-point change in price of the underlying stock. Consequently, by buying and selling options in ratios determined by their deltas, market makers create "delta-neutral" or "zero-delta" positions. As a result, they do not hope to profit from a forecast of market direction. Rather, market makers hope to profit by the options they purchase increasing in price relative to the options they sell.)

Updating one's sheets is a serious business, and GRT spends close to an hour every day reviewing the assumptions on which his option theoretical values are generated. Many traders do not leave this job to the morning before trading; rather, they do it before they go home at night. If it takes two or three hours, then they stay late. But the market opens at the same time every morning, and, if the job of updating sheets takes longer than expected, then a market maker who starts one hour before the market opens might be forced to start the trading day with inadequate or inaccurate information. And that could be costly!

Reviewing Position Risk

Risk, like beauty, is in the eye of the beholder. Nevertheless, risk means something different to market makers than it does to off-floor investors and traders. The risk (or risks) that concern market makers is unique to them, because their strategies are rarely, if ever, used by off-floor investors and traders. The primary risk of delta-neutral strategies such as option-to-option ratio spreads and stock-to-option ratio spreads is a change in implied volatility. The primary risk of arbitrage spreads (which are defined later) is a change in interest rates.

When GRT examines a position, he first looks at the "Greeks"—the delta, gamma, theta, vega, and rho values. These are computer-generated numbers which summarize how a position can be expected to change if market conditions change; *i.e.*, the price of the stock (delta and gamma), the time to expiration (theta), the implied volatility level (vega), or the interest rate (rho). GRT's biggest concern is delta. Typically, a market maker wants the position delta to be very close to zero. The *position delta* is the total delta of all options on a stock and the long or short shares of that stock. If the position delta is not zero, then, when the market opens, GRT will typically make a trade that returns the position delta to zero.

A zero delta is not GRT's only concern about market movement, however. Remember gamma (from Chapter 2)? It tells how much the delta of a position will change if the underlying stock price changes by one dollar. GRT must consider how his position delta will react to a change in the stock price. Notice that GRT's focus on gamma risk is not a concern about the *direction of movement*, but rather the *size of movement*. And how does GRT think about size of movement? It is not the dollar movement or even the percentage movement; GRT's concern is about the *standard deviation of movement*.

An in-depth discussion of standard deviations is beyond the scope of this book, but simply speaking, a *standard deviation* is a statistical range within which two-thirds of events are expected to occur. A "less than one standard deviation price change" is typical and a "more than two standard deviation price change" is unusual. In assessing risk, GRT must ask himself, "What are the chances of a two standard deviation price change occurring? How much would my position lose (or make) if this happens? Can I live with this risk?"

After evaluating gamma risk, GRT assesses the impact of time decay and the impact of a change in implied volatility on each position. He asks himself these questions: "If I do not trade today and the stock closes unchanged, will the passage of time help or hurt this position?" and, "Would an increase or decrease in implied volatility help or hurt this position?"

Note that, while a computer can calculate a numeric answer to these questions, an assessment of that numeric answer is subjective. Market makers must make their own assessments about risk. And every market maker must go through this process every day. If GRT thinks that one of his positions has an unacceptable level of risk, then he must formulate a plan for the day ahead that reduces his risk in that position.

FORMULATING THE DAY'S STRATEGY

Explaining how market makers think about or plan the day ahead is difficult to do in a few paragraphs, but, essentially, a market maker must do three things. First, he must price arbitrage strategies such as conversions, reversals, and box spreads. Second, he must know the implied volatility levels where he is willing to buy and sell options, and, third, he must decide whether he can add to or must subtract from existing positions.

Pricing Arbitrage Strategies

An *arbitrage strategy* is a multiple-part position that attempts to profit from price discrepancies in related, but different markets. The most basic arbitrage strategy involving options is the conversion. A *conversion* is a three-part strategy in which the underlying stock and a put are purchased and a call with the same expiration date and same strike price as the put is sold. The three parts are executed simultaneously (or nearly simultaneously) and on a share-for-share basis with each other. A profit and loss diagram of a conversion is presented in Figure 8-1. In this example 100 shares of stock are purchased for $50 per share, a 50-strike put is purchased for $2 per share and a 50-strike call is sold for 2½ per share. Dividends, commissions, margin requirements, and cost of funds (interest rates) are not included, but these real-world factors have an impact on this strategy, and market makers must include these factors in their calculations.

The solid horizontal line in Figure 8-1 appears to offer a riskless profit of ½ per share regardless of the price of the underlying stock at ex-

FIGURE 8-1

Market-Maker Strategy—The Conversion

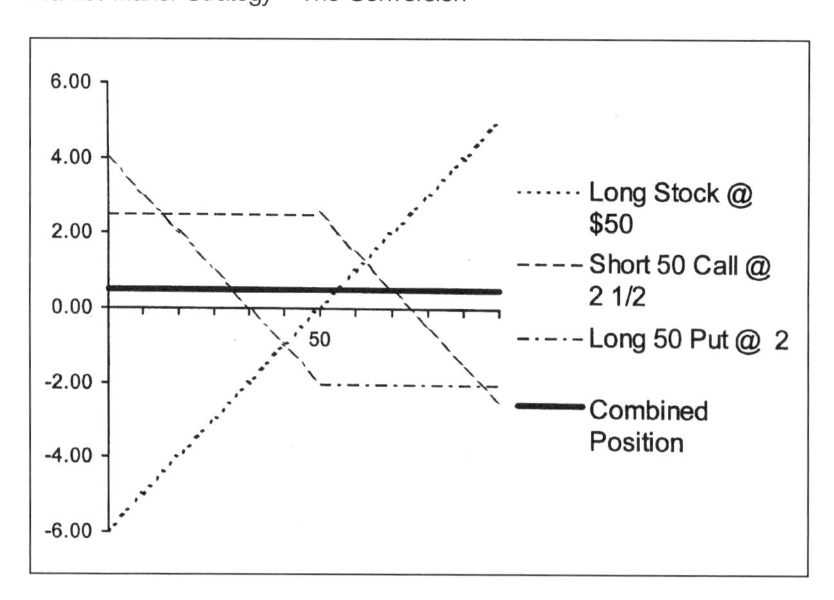

piration, and, conceptually, this is true. The original investment is \$49½ per share (\$50 paid for the stock + \$2 paid for the put − \$2½ received from selling the call = \$49½ net paid), and \$50 cash is received at expiration. Consider, for example, if the stock price is below the strike price at expiration, then the short call expires, the long put is exercised, the stock is sold, and \$50 is received for a profit of ½ per share not including transaction costs. Similarly, if the stock price is above \$50 at expiration, then the long put expires, the short call is assigned, the stock is sold, and \$50 is received for a profit of ½ per share not including transaction costs. And, finally, if the stock price is exactly \$50 at expiration, then both the long put and the short call expire, but the \$50 stock value represents an unrealized profit of ½ per share not including transaction costs.

Nothing is quite as good, however, as Figure 8-1 might indicate. If the stock closes exactly at the strike price at expiration then market makers face what is known as *pin risk*, an unknown and uncontrollable risk. With the stock price exactly at the strike price at expiration, then the market maker has a difficult decision to make. If the market maker does not exercise the long put and if the short call is not assigned, then the market maker has the risk of a long stock position over the weekend. If the market maker does exercise the long put and if the call is assigned, then the market maker has the risk of a short stock position over the weekend. Many a market maker has had a sleepless weekend from pin risk.

The most important observation to be made about Figure 8-1 is that this strategy is not feasible for off-floor investors and traders. First, transaction costs for off-floor investors and traders are too high. On-floor market makers pay commissions of one dollar per contract or less, because they buy and sell several hundred contracts per day. Second, the simultaneous, or nearly simultaneous, execution of two option positions and one stock position is a practical impossibility unless a trader is equipped with the right technology to do it. Third, the opportunity to establish a conversion rarely exists for more than a few seconds, because full-time, professional market makers are looking for these opportunities all day, every day. Fourth, and finally, the profit potential of ½ per share (or less after transaction costs) is generally only enticing to full-time market makers, and this small profit margin is only enticing to them because they hope to do it many times in the course of a month.

The conversion strategy is just one example that illustrates that market makers are in a completely different business than off-floor investors and traders.

Having introduced the concept of the conversion strategy as described above, let's see how GRT goes about "pricing his conversions." *Pricing a conversion* means adding transaction costs, borrowing costs, and a target profit to determine the amount by which the time premium of a call must exceed the time premium of a put with the same strike price and expiration date.

Using the above example, assume a stock price of $50, 90 days to expiration, a borrowing rate of 6 percent, transaction costs of 22 cents per share for each option, transaction costs of 6 cents for each share of stock, and a target profit of 25 cents per share. Borrowing $50 for 90 days at 6 percent amounts to 75 cents per share ($\$50 \times .06 \times 90 \div 360 = \$0.75$), and, adding the transaction costs, GRT determines that the time premium of the 50-strike call must exceed the time premium of the 50-strike put by $1½ in the following manner:

	Per Share
Borrowing costs	0.75
Stock purchase	0.06
Put purchase	0.22
Call sale	0.22
Target profit	0.25
Time premium of call over put	1.50

This calculation means that, if GRT can buy stock at $50 per share and if he can also buy the 90-day 50-strike put at $2 per share, then he can confidently sell the 90-day 50-strike call at 3½. If he does these three things, then he will have established a conversion at his target profit of 25 cents, or ¼, per share. His only concern will be pin risk.

Alternatively, if GRT can buy stock at $50 per share and if he can also buy the 90-day 50-strike put at $1¾ per share, then he can confidently sell the 90-day 50-strike call at 3¼. Again, he will have established a conversion at his target profit.

As a third example, if GRT can buy stock at $50½ per share and if he can also buy the 90-day 50-strike put at $1⅜ per share, then he can confidently sell the 90-day 50-strike call at 3⅜. Note that, in this example, the time premium of the call is 1½ greater than the time value of the put, because the call is ½ in-the-money. And, again, a conversion is established at GRT's target profit. (For a review of time value and intrinsic value, refer to Chapter 2.)

The conclusion to be drawn from these pricing exercises is that, for arbitrage trading, GRT does not need theoretical value sheets. He just needs to know the $1½ factor, and he needs to be able to do quick mental calculations. To beginners this may seem difficult, but market makers are trained from their first day on the job to do this. After a while it becomes second nature.

Knowing Implied Volatility Levels

Implied volatility is the volatility percentage which, if used in an option-pricing formula, returns the market price of an option as the theoretical value. Because option prices, like all prices, are determined by supply and demand, and because volatility is the only input to the option pricing formula that cannot be observed independently, then implied volatility is the common denominator of option prices. Market makers, therefore, must know what the implied volatility is when they buy or sell an option, because that is their basis for comparison after other factors such as stock price and time to expiration change.

Although it takes some practice, it is relatively easy for a market maker to become familiar with implied volatility levels. Remember, market makers usually trade options on the same stocks all day, every day. By noting implied volatility levels where bids and offers are made and where options are bought and sold, the numbers begin to stay in one's mind. It generally takes a market maker three to four months to become thoroughly familiar with the implied volatility of options on a new stock.

Adding To or Subtracting From Positions

The question market makers must ask themselves is, "How much risk can I assume?" And the answer to this question is the art of the market making business, because it determines how successful a market maker will be.

Market makers are generally described in terms of the number of contracts they trade. Beginning market makers typically trade 10 to 20 contracts per trade (a 10-lot trader). Experienced, but thinly capitalized market makers frequently trade no more than 50 contracts at a time (a 50-lot trader). Experienced and well capitalized market makers can trade 100, 200, and sometimes 300 contracts at a time. The off-floor source of such large orders is generally a mutual fund or pension fund. But regardless of the quantity of contracts involved, even if they are traded in a delta-neutral fashion, these positions still involve the risks of changing volatility and time decay.

At some point, a 10-lot trader must move up to 50 contracts or more. It is simply not possible to pay all the expenses of being a market maker and to earn a decent living by trading only 10 contracts at a time. So how does a 10-lot trader increase the quantity of contracts he trades? By making good judgments about risk.

If, for example, a 10-lot market maker decides that a particular option contract is "very undervalued relative to another option," then that is the time he may increase his size to 25 contracts. He purchases 25 of the option contract that he believes are relatively undervalued and sell 25 (or the proper quantity based on delta) of the other option contract. If that decision is validated with a profit, then the market maker will continue to cautiously make more 25-lot trades. If, however, that decision is proved wrong by a bigger than normal loss, then that market maker must lick his wounds and try again. Of course, a market maker can only lick his wounds and try again as long as his capital lasts. If a market maker depletes his capital with losing trades or an inability to cover expenses, then his career as a market maker comes to an end.

As GRT reviews his positions in the twelve stocks he trades, he makes personal and sometimes agonizing decisions about the risk of those positions. This is a totally subjective process, and it is different from the risk assessments made by off-floor investors and traders. Some of the language may be the same, but market makers are judging the risks of changing volatility and time decay of delta-neutral positions; their goal is to make a small profit on a high volume of trades. Off-floor investors and traders, in contrast, are judging market direction.

A TYPICAL MARKET MAKER

There may not be such a thing as "a typical market maker," but let's use GRT as a hypothetical example. GRT has traded in the same pit for 10 years. When he started, there were only three stocks in the pit, but, as the options market developed and options on an increasing number of stocks were listed, that number grew to the current twelve.

Like many traders at the CBOE, GRT began his career as a clerk for another CBOE market maker when he was 24 years old. He had majored in engineering in college, and, after graduation, he got a job with a small manufacturing company in Chicago. After two years at that company, he met some CBOE people socially, and the idea of working on the trading floor appealed to him. After interviewing with several market makers over three months, one finally agreed to hire him at one-half of his salary as a junior engineer. By taking a cut in pay, GRT showed the entrepreneurial spirit that is required to be a market maker.

As a clerk, he came in every morning two hours before the market opened to perform several clerical tasks for his boss. He checked trades, compared his records against the clearing firm printouts, prepared the daily theoretical value sheets, scanned the Dow Jones and Reuters services for news on the stocks his boss traded, and made coffee. During the day he updated position reports as his boss made trades, and he monitored the positions in the less active stocks when his boss was busy with the more active stocks.

At the end of each trading day he prepared position summaries for each stock for his boss to review at home in the evenings. In the process of doing this "grunt work," he learned about being a market maker. He learned to calculate arbitrage relationships between options and stock; he learned to evaluate position risk; and he learned how to make trading decisions. Most important, perhaps, he watched his boss make and lose money. Trading options as a market maker is a percentage business, and it is difficult for many people to adapt psychologically to the inevitable ups and downs. By walking in the shadow of an experienced market maker, GRT gradually absorbed the business. And he found that he liked it. Many people do not like it, but GRT did.

After eighteen months of clerking, GRT thought he was ready to trade. His boss, however, did not have the capital to back him, so he started talking to other market makers. After six months, another CBOE member, who traded in the same pit as his boss, decided to give GRT a try.

GRT and his backer signed a three-year partnership agreement. The backer deposited $100,000 in GRT's trading account and, in return, received a percentage of GRT's profits, based on a sliding scale. The backer received 60 percent of GRT's first $50,000 of net pre-tax profits, 40 percent of the second $50,000, and 20 percent of any additional net pre-tax profits. In the beginning, GRT leased an exchange membership for about $6,000 per month. Initially, GRT paid $1.10 per contract in clearing and exchange fees, and he traded 200 contracts per day on average in his first six months. At that rate, his expenses were nearly $10,000 per month. After three months, GRT was net down $8,000. After six months, he was back to even; and, at the end of the first year, his net pre-tax income was $25,000, $15,000 of which went to his backer. For his first twelve months of hard work and many sleepless nights, GRT kept $10,000 pre-tax for himself. His second year went much better; he took home over $50,000 pre-tax after paying all expenses and his backer's percentage share of profits.

When his three-year partnership agreement expired, GRT and his backer signed a new two-year agreement with more favorable terms to GRT. He had proven that he could make money, so he was in a stronger bargaining position. That agreement called for 25 percent of the first $50,000 of net pre-tax profits and 15 percent of all additional net pre-tax profits to go to the backer.

GRT continued to trade successfully, and, when his two-year agreement expired, he broke off on his own. This means he deposited his own money in a trading account, made his own seat-lease arrangements, and negotiated his own terms with a clearing firm. GRT has been on his own now for five years. He still leases a membership, and the rental rate has fluctuated between $6,000 and $10,000 per month. He does not have a clerk for two reasons. First, he feels it is an expense that he cannot justify, and, second, increases in technology have reduced the need for one. The daily reports he generated by hand for his former boss and the position-checking process that took over one hour every day are now both done automatically as he enters trades on a hand-held computer.

GRT's current goal is to purchase an exchange membership, and he estimates that he will have saved enough for the down payment within two years, about seven years after he started trading. CBOE memberships cost in excess of $500,000 and they have fluctuated in price between $350,000 and $750,000 between 1990 and 1998. During those years CBOE and options industry volume nearly doubled, and GRT believes in the long-term growth potential of the business. He therefore wants to "own rather than rent."

THE START OF GRT'S TRADING DAY

After checking his positions against the exchange's printouts and after scanning the news services for stories on his companies, GRT walks onto the trading floor and takes his usual spot among eight other market makers and two independent floor brokers, all of whom trade options on XYZ stock and the eleven other stocks in that pit.

Before the opening bell, GRT looks at the overhead monitors to see how the S&P 500 futures contract is trading. He also checks the price and direction of the futures contracts on the Dow Jones Industrial Average and the 30-Year U.S. Treasury Bond. The action in these futures gives GRT a feel for how the market in general, and the 12 stocks in his pit in particular, might open this morning.

The clock now reads 8:28 AM CST, two minutes before the opening bell. GRT looks around for Louie, his stock clerk. Louie is sitting at his usual booth just outside of the XYZ options pit. Through Louie, GRT uses a stock execution firm that specializes in buying and selling stock for market makers and other professionals. One of the ways that GRT manages risk is to trade stock against his option positions. This morning, Louie signals that indications from his associate on the New York Stock Exchange (NYSE) floor are that XYZ stock will open unchanged from yesterday's close.

If someone were to ask GRT for his opinion on the direction of XYZ's price today, he would respond as he would on most days, that he does not have a strong feeling either way. This is because GRT does not care whether the price of XYZ stock rises or falls. He is in the business of buying options that he believes are relatively undervalued and selling ones that he thinks are relatively overvalued. Occasionally, GRT may lean a little to the long side (a position with a positive delta) or to the short side (a position with negative delta), but he will never expose himself to the risk of an unforeseen event, such as an unexpected announcement, that could cause him a big loss. GRT knows that if he takes a large delta position, either long or short, then he would no longer be a market maker. He would then be a speculator. There is nothing wrong with being a speculator, because speculators are a necessary part of efficient markets, but speculating is not how market makers attempt to earn a living. GRT and other market makers simply want to buy at the bid price and sell at the offer price and thereby make the difference. That is the market maker way.

8:30 CST. The opening bell rings and everyone in the pit is concentrating on the market. The opening rotation, as described in Chapter 7, goes smoothly. It takes about four minutes to open all options on the twelve stocks traded in this pit. Eight of the stocks have about five orders each, and the other four have just a few each.

8:38 CST. Floor broker TON (you met him last chapter) walks into the XYZ trading crowd. He studies the computer screens that display the prices of XYZ options, and then he bellows, "XYZ APE 50 calls, how are they?" Without hesitation, GRT and some other market makers respond "2¾ – 3!" To review from last chapter, this floor lingo means that GRT is bidding 2¾ for at least 20 of the XYZ April 50 calls and is offering at least 20 at 3. TON responds, "I'll be a 2⅞ bid for a few." GRT and the others reiterate their 3 offer.

GRT quickly analyzes his XYZ position and comes to the conclusion that, with the stock and some XYZ options at their current prices, he has no interest in selling these calls at 2⅞. How did GRT arrive at this conclusion? He asked himself the same questions that market makers ask themselves over and over each day. And they are the same questions that his mentor taught him several years ago.

Question 1. "Am I *net long* or *net short* in my XYZ position, and do I need to get delta neutral?"

Net long or *net short* means a position with a non-zero delta. Because GRT is a delta-neutral trader as discussed above, it is unlikely that his position delta in XYZ is not zero. Because of gamma—the change in delta for a change in price of the underlying—however, it is possible that his XYZ position delta could become positive (long) or negative (short) as the price of XYZ stock changes. Today, however, XYZ is trading unchanged from yesterday's closing price, so GRT's position delta in XYZ is still zero. Consequently, GRT does not need to sell any XYZ April 50 calls to get delta neutral.

Question 2. "If I sell these contracts at 2⅞, do I have a reasonable chance to repurchase them at $2\frac{13}{16}$ in a short time frame?"

The motivation behind this question is a trading technique known as scalping. *Scalping* is a strategy that attempts to earn profits from buying and selling quickly. A "short time frame" could be from two to five minutes. If GRT could sell 10 contracts at 2⅞ and repurchase them at $2\frac{13}{16}$, then he would earn $\frac{1}{16}$ ten times or ⅝ total, or $62.50, before transaction costs of approximately $10. Scalping is an art. Initiating a position in the hopes of scalping a quick profit is based on instinct and market feel. If GRT sees that other stocks in XYZ's industry are trading lower, then he might be inclined to sell XYZ calls in the hopes that XYZ stock will also trade lower in the next few minutes.

Scalpers typically attempt to reduce risk by only entering a trade when they believe there is a good chance of "scratching" if the market moves contrary to their forecast. A *scratch trade* is a round-turn transaction, *i.e.*, a buy and sell, in which the buy price equals the sell price. The only cost to a trader in a scratch trade is the transaction fee. Because there are no sellers of these calls at $2\frac{7}{8}$, GRT does not believe that he can sell these calls at $2\frac{7}{8}$ and have a reasonable chance of scratching them if XYZ stock begins to rise in price. Consequently, GRT does not see a scalping opportunity right now, and he declines to sell these calls for this reason.

Question 3. "If I sell these April calls at $2\frac{7}{8}$ are there any other contracts I can buy?"

Market makers, remember, attempt to profit by purchasing options that are undervalued relative to another option that they can sell. This means that GRT is looking at *relationships between options.* One relationship of particular interest to market makers is a *time spread*, which is created by purchasing one option and selling another option of the same type with the same strike and same underlying but with a different expiration date. If GRT could, for example, sell the XYZ April 50 calls at $2\frac{7}{8}$ and purchase the XYZ May 50 calls at $3\frac{7}{8}$, then he would have purchased a time spread for a net cost, or debit, of 1. Because the deltas of these options are very similar, the net position delta of this time spread will be very close to zero, and, therefore, satisfy GRT's desire for a delta-neutral position.

In looking for a time spread opportunity, GRT's concern is not that the May 50 call is offered by other traders at $3\frac{7}{8}$, specifically. He is concerned about the *difference in price* between the April 50 call and the May 50 call. GRT's computer printouts calculate a theoretical difference based on the assumptions he entered in the computer, and he compares that difference to the difference in market prices currently available. Right now, his printouts indicate that the theoretical relationship between the April and May 50 calls is a $1\frac{1}{16}$ spread. Consequently, paying 1 for the time spread would be a small "advantage" relative to this theoretical calculation, but, in GRT's opinion, this is an insufficient advantage. His experience tells him that he needs at least an advantage of $\frac{1}{8}$, and preferably an advantage of $\frac{3}{16}$, to make it worth the risk.

This means that GRT would be very interested in establishing the time spread position if he could purchase May 50 calls for $3\frac{3}{4}$ and sell April 50 calls for 3. At those prices, he would be establishing the time spread for a net debit of $\frac{3}{4}$, a $\frac{3}{16}$ advantage to the $1\frac{1}{16}$ theoretical value. At that price, GRT believes that his chances of *unwinding*, or closing, the position in the near future, at a profit, would be good.

In addition to checking this time spread relationship, GRT also checks other options in XYZ, but finds that none of them offer any opportunity to get an *edge*, a term used by professional traders that means an advantage relative to a theoretical value, which was calculated using their inputs. If a spread position can be purchased at a price below a trader's theoretical value, then the difference is referred to as the "trader's edge." This edge is not the same thing as a profit, because a profit or loss is not realized until a position is closed.

Even though a position may be established with an edge, *i.e.*, at a price that is better than a trader's theoretical value, there is absolutely no guarantee that that position will be closed at a profit. Even a delta-neutral position, for example, has the risks of gamma (changing delta), vega (changing volatility), and theta (changing time). Because GRT sees no spreads that give him a sufficient edge relative to theoretical value, he declines selling the XYZ April 50 calls at $2\frac{7}{8}$ for this reason.

Question 4. "Where is the stock and where are the April 50 puts?"

Why does GRT care about the prices of XYZ stock and XYZ April 50 puts? TON wants to buy calls, not puts! Again, GRT is looking at the relationship between the stock and the options. This is how a market maker trades conversions. A quick look at the XYZ April 50 puts tells GRT that the time value of the April 50 calls, at $2\frac{7}{8}$, is not sufficiently higher than that of the April 50 puts to make a conversion profitable, and he declines selling the calls for this reason.

Question 5. "Is the implied volatility of the April 50 calls high enough?"

Similar to the logic of scalping, GRT could sell the XYZ April 50 calls at $2\frac{7}{8}$ and purchase shares of XYZ stock on a delta-neutral basis if he thinks he has a good chance of repurchasing those calls at a sufficiently lower implied volatility in a short time frame. A "sufficiently lower" implied volatility means enough lower to cover transaction costs and to make a profit. Because he observes every bid and every offer in XYZ options, he has a strong feeling for the implied volatility levels at which XYZ options generally trade and where he feels comfortable "buying volatility" and "selling volatility." Given the current price of XYZ stock and the price of $2\frac{7}{8}$ for the April 50 calls, GRT feels that the implied volatility is not sufficiently high to justify the risk of selling them delta-neutral against stock, and he declines selling these calls for this reason.

How long did this mental-five-question process take? Less than five seconds.

GRT is not interested in selling the XYZ April 50 calls at $2\frac{7}{8}$, but he is interested in selling them at 3, that is why he responded by saying,

"At 3." This lets TON know that he will not come down in price, *i.e.*, he is holding firm at his 3 offering price. Apparently, the other market makers are thinking along the same lines as GRT, because they all respond, "At 3" also.

For this example, TON has a market order for 20 XYZ April 50 calls at the market. He attempted to get a better price for his client by bidding 2⅞, but a market order requires him to get the best price when the order enters the marketplace. Consequently, when he determines that no one will sell him calls at 2⅞, he yells, "OK, pay 3."

GRT wants to sell some calls at that price, so he immediately yells, "Sold!" He hopes that TON recognizes that he was first, or at least tied for first. And, when TON, says, "OK, KNY, GYM , GRT, and III, 5 a piece," GRT knows that he was recognized, and that he sold 5 calls. He then immediately turns to LZE (the floor broker you met last chapter) and says, "LZE, buy 5 May 50s at 3¾."

LZE understands that this means GRT wants to buy 5 of the XYZ May 50 calls that she is offering for sale at 3¾, and she responds by saying, "OK GRT, sold you 5." GRT has now established a 5-lot time spread position. He has purchased 5 XYZ May 50 calls and sold 5 XYZ April 50 calls at a net debit of ¾ for each spread.

Having verbally committed to making these two trades, GRT then enters the necessary information into his hand-held computer. When TON and LZE enter their respective sides of the trades, GRT's computer will let him know that the trades have been confirmed.

We just saw the thinking process that GRT, a market maker, went through before he made a trade, and it is very different from the process that many off-floor investors and traders go through. GRT did not think in terms of direction, he thought in terms of relative option prices, and his goal was to create a spread position that was delta neutral. His goal is to make a small profit, essentially the spread between bid and ask, regardless of the whether XYZ stock moves up or down.

The example above shows that GRT and other market makers use strategies that are completely different from those used by off-floor investors and traders. Nevertheless, market makers and off-floor investors and traders need each other to accomplish their own individual objectives. If the off-floor investor or trader represented by the floor broker TON had not paid 3, then GRT and the other market makers would not have made a trade. On the other hand, if the market makers were not there to offer those calls at 3, then at what price would TON have been able to buy them? 3? 3½? 4? If TON had to rely on the presence of an offer from another off-floor investor or trader, would there have been one? Without market makers, what mechanism would assure continuous bid and offer prices? How liquid would the options market be? A strong argument can be made that, without full-time market makers and their affirmative obligation to make bids and offers for all options, the options market would operate in a much less efficient manner.

SPLITTING THE BID–ASK SPREAD

Off-floor investors and traders sometimes say, "I tried to split the difference between the bid and ask, but the market makers ignored me." Because this is a common complaint, it is worthy of discussion. First, an example of "splitting the difference" is presented, and, second, some hypothetical market scenarios are reviewed.

An example of splitting the difference between the bid and ask is an order to buy 10 calls at a limit price of 1⅜ when the current bid price is 1¼ and the current ask price is 1½. When your order to "buy 10 at 1⅜" reaches the pit, the market becomes 1⅜ bid and 1½ ask. What happens next depends on other events that may be occurring in the marketplace. Market makers will sell some contracts at 1⅜ if one or more of several things happen, and they will not sell any at that price if none of those things happen.

First, if puts with the same strike price and expiration are being offered at a price that makes a conversion possible, then market makers will sell the calls to you. They will also buy the puts and the underlying stock on a share-for-share basis and thereby create a conversion position as described earlier in this chapter.

Second, if some other option is offered at a price that is low relative to the call price of 1⅜, then market makers will sell you your calls and buy those other options on a delta-neutral ratio.

Third, if the stock price ticks down, then, by definition, the implied volatility of the calls in question, at a price of 1⅜, increases. If the market makers judge that this increase is sufficient to justify the risk of selling those calls delta-neutral against stock, then they will sell calls to you and purchase the underlying stock at the slightly lower price.

Fourth, if an order to sell calls at your price enters the pit from off of the floor, then you will purchase the calls being sold, because your bid is the highest bid at that moment.

In each of the four examples above, an event occurred that made it possible for you to buy the calls at your price. If however, none of these events occurred, then you would not be able to buy the calls at your desired price, because no one would be willing to sell them to you.

Does this suggest that "splitting the difference" is a bad idea? No. The success of any strategy depends on the market forecast being accurate. The implication of placing an order to buy at a limit price below the current offering price is that market conditions are forecast to change; so that the offering price will be lowered to the price where the buy order can be filled. If that forecast is correct, then the limit price buy order will be filled. If the forecast is incorrect, then the order will likely not be filled. But, regardless, the success of any decision to enter any type of order depends on the accuracy of the forecast.

SUMMARY

Market makers are an essential part of efficient markets. When an order from an off-floor investor or trader arrives on the floor, there is rarely an exactly opposite order from another off-floor investor or trader there waiting. Market makers, however, have an affirmative obligation to make markets. In equity options, for almost every option in their trading area, market makers must have a stated bid price at which they are willing to buy a minimum of 20 contracts, and they must have a stated offer price at which they are willing to sell a minimum of 20 contracts.

A market maker's day begins with checking trades from the day before, updating assumptions about option theoretical values, comparing clearing-firm–generated computer printouts to personal records, and formulating a plan for the day ahead. Modern technology has reduced the number of trading errors and the need for "sheets" of option theoretical values.

Market makers make trading decisions based on relative values. They try to establish positions that are delta-neutral, *i.e.*, positions that include long and short options and underlying long or short stock that have a combined delta of zero. The strategies that market makers use such as conversions and time spreads are not generally used by off-floor investors and traders.

Market makers do not compete with off-floor investors and traders. Rather, they use different strategies, and they make decisions in a different way. In fact, market makers and off-floor investors and traders need each other to accomplish their own individual objectives.

PART THREE

REAL-TIME APPLICATIONS

CHAPTER 9

INSTITUTIONAL CASE STUDIES

Gary L. Trennepohl

Development of the listed options market has dramatically enriched the set of portfolio management strategies available to institutional investors. Previous chapters have described numerous option strategies used to implement market views held by institutional and individual investors. For the institutional investor, options are best viewed as tools for portfolio risk management. They can be used to modify a portfolio's risk in ways not achievable with any other portfolio management technique. Understanding the proper role of options and other derivatives in portfolio risk management is a necessary component for successful institutional investment management in today's complex world.

Based on concepts presented in earlier chapters, the two case studies are designed for you to apply your knowledge of options as tools for portfolio risk management. Before institutional investors implement option-based investment strategies, they typically go through a two-step process. First is a learning process—investment managers learn how options can be used as strategic and tactical investment tools: What strategies are appropriate? When should they be used? What is different about options compared to other portfolio management techniques? Second is an approval process—the institutional trustees or directors must change the investment plan documents to allow for trading of options and other derivative assets. The approval process usually requires an extensive educational effort for the plan trustees and investment committee members about the costs, benefits, and risks of using options.

The trust plan managers making the presentation must be familiar with the characteristics of options and option strategies to earn the approval of their governing board.

The two case studies presented in this chapter were developed from discussions with a number of institutional investors and individuals who provide consulting services to institutional clients. These cases represent no particular institutional investor, but the situations described are based on events which actually occurred across several different companies.

In Case One, American National Pension Trust is beginning the option implementation process. Paige Hanson, the director of the trusts, is trying to educate her investment strategy group and herself about the proper use of options in portfolio risk management. Paige has been in institutional investment management over fifteen years, starting in the fund management business after receiving her MBA from Northwestern University in 1983. Her experiences during the decades of the 1980s and 1990s have convinced her that the new generation of professional fund managers must know how to use options and other derivatives in investment strategies. To help learn about the new techniques that are available, Paige also is studying for the Chartered Financial Analyst (CFA) designation, hoping to complete the third level exam next month.

American National Pension Trust is no stranger to the use of derivatives, having previously used covered writing strategies in the early 1980s and dynamic portfolio insurance techniques during the period of the 1987 market crash. Paige has experimented with a variety of option strategies using index options the past five years, and she is convinced that a rich set of portfolio management techniques based on listed options are available. She wants to institutionalize the use of options by the Pension Trust.

The analysis of this case should focus on development of the proper risk management strategies for the Pension Trust equity portfolio. What derivative strategies would you recommend in the current market environment? Because the Trust is invested primarily in large capitalization, domestic securities, some members of the investment committee are encouraging the Trust to consider small-cap stocks and even foreign securities to gain greater benefits from diversification. Can you develop option strategies that will control risk and facilitate investment in these securities?

In Case Two, the internal fund management of Norwalk Investment Fund is trying to gain approval from their trustees to change their plan documents and allow investment in derivative securities. Using the trustee meeting as a setting, the case involves members of the trustees and the fund management team discussing the pros and cons of investment in derivative securities. The purpose of this case is to make you aware of questions about options typically raised by board members, and to encourage you to formulate answers to these questions. Studying this case should enable you to avoid some of the problems and pitfalls faced by other institutional investors as they cleared the internal management hurdles to gain approval for the use of derivative securities.

You are encouraged to read each case study and formulate your own answers using the suggested outline of questions provided at the end of the case. As you develop an answer for each question, reference to other chapters in the book is encouraged. Afterward, you can compare your analysis to the suggested case solution. Keep in mind that the suggested solution is only one of many feasible answers that might have been developed; it is provided merely as a guide to the major points developed in each case study. Each investment manager has their own particular tradeoff for risk and return which will influence their answers. The great benefit of options is that they enable each manager to tailor a payoff pattern for their portfolio that provides the best fit for the manager's unique preferences.

CASE ONE: AMERICAN NATIONAL PENSION TRUST

Paige Hanson is the director of investment strategy for the $30 billion American National Pension Trust, the pension fund for employees of American National Energy Company. American National Energy is a large, integrated oil exploration and production company with headquarters in Chicago. The Trust employs more than eighteen external fixed income and equity managers who specialize in a variety of investment strategies. In addition, the Trust has fifteen internal managers whose responsibilities are divided among fixed income, equities, private investments, venture capital, and real estate.

Most fund managers employed by the Pension Trust typically trade well-diversified portfolios of U.S. stocks that track the major indexes quite well. In fact, one of their managers had popularized the idea of "indexing" early in the 1970s, when some sponsors believed that it was difficult to outperform the market, and they did not want to increase their risk exposure in excess of the market through active management. However, they were willing to place funds with specialized managers who had exceptional track records. For example, they recently committed a modest sum to a specialized manager in Dallas who had impressive performance over the past five years running a leveraged portfolio based on a proprietary, mathematical model.

Trust investment guidelines specify that the fund should target a 65 percent equity and 35 percent fixed income allocation, with a margin of 5 percent in either direction. The rising equity market over the past five years has moved the allocation to the upper limits on equities and motivated interest by some board members to raise the equity limit to 75 or 80 percent. This perspective bothered Paige because she had been in the business long enough to know that markets don't rise forever.

Recent headlines about the "Asian Contagion" made Paige more resolved about the need for a conservative approach in this market. One analyst suggested that financial crisis in Japan and China could lead to a 25 or 30 percent decline in U.S. markets, and a depression in Asia. These headlines impressed the more conservative trustees, who were convinced the U.S. markets were overvalued; they wanted the Trust to significantly reduce equity exposure.

Because of these conflicting opinions, Paige began to explore various option-related investment strategies appropriate for pension plan sponsors in the "irrationally exuberant" equity markets of the late 1990s. Her objective, taking the current market environment into account, was to design strategies to add value and reduce the risk of the trust's portfolios. Paige knew she would not be able to sacrifice much return potential in her product development plans, but at this time, she was more concerned about a market correction, than a strong market increase.

Paige knew she would have to explain the basic concepts of portfolio risk management to the Pension Trustees before she would ever be able to implement any strategies using derivative securities. A pension board meeting is scheduled three days from now, and they have requested that Paige discuss the strategies she is considering and describe their advantages and disadvantages for the Pension Trust portfolios. Paige has a lot to do in a short time.

Paige is highly regarded by her peers as a true "forward thinker." If someone is trying a new and successful investment strategy you could bet that Paige is one of the leaders of the group. She increasingly has become convinced that the pension fund manager's job should be viewed in the context of portfolio "risk management" rather than merely identifying stocks to buy and sell, or making overall asset allocation decisions which would change gradually over market cycles.

Portfolio Risk Management Techniques

Since the early 1980s, a variety of derivative products have been introduced to enable fund managers to more precisely control the risk of their portfolios, a process called *financial engineering*. Prior to 1983, an equity manager sensing a market decline in the making, could reduce risk only by selling stocks or shifting from "aggressive" to "defensive" securities. A plan sponsor could draw capital from equity managers and shift the funds to money market or similar funds, or move funds from more aggressive to more defensive managers. For sponsors, meeting infrequently, timing changes was more difficult. The new derivative products offer institutional investors the opportunity to move more capital, often more expeditiously, and usually with lower transaction costs.

Paige is familiar with one strategy, which gained popularity early in the 1980s, called *dynamic portfolio insurance*. Because no index option products existed when it was designed by Mark Rubenstein and Hayne Leland in Berkeley, California, it was implemented using stock index futures. The term "dynamic" was applied to the strategy because it required that positions in stock and futures be adjusted frequently (weekly or daily) to maintain the proper relationship between the two assets. Prior to the market correction in 1987, it was estimated that over $200 billion was managed under the dynamic portfolio insurance strategy. But the strategy performed poorly when the market collapsed during the week of October 19, 1987. Many plan sponsors dropped the strategy.

As she researched the topic, Paige found that portfolio insurance could be implemented using options instead of futures, a strategy called *static portfolio insurance*. For example, a plan sponsor might (1) buy an index put against a portfolio; (2) buy puts on individual stocks; (3) buy index calls and money market securities; (4) buy individual calls and money market securities, (5) sell stock index futures to create a synthetic put; or (6) sell the individual stocks, and go to cash. Paige thinks that the concept of portfolio insurance might be attractive to many plan sponsors, but she is sensitive to the problems that were encountered when it was implemented using futures. Perhaps options could provide a more effective vehicle for the strategy.

Simulation Study Results

Paige believes that evidence from a variety of research reports supports the strategy of portfolio insurance under certain market conditions. She is especially interested in a simulation study that illustrated how an "insurance" strategy of buying protective puts can modify the returns of unmanaged equity portfolios. The study used the S&P 500 Index and was designed to test the performance of a fund manger who is indexed to their target market and who purchases an at-the-money index put at the beginning of each quarter.

The put premiums are calculated by an option pricing model. (Paige wondered at the time if the standard Black-Scholes Pricing Model could be used to price both puts and calls, but she had to delay finding the answer to that question for a later date.) In both studies, it was assumed that the put was held to the end of the quarter and then sold for its intrinsic value, if greater than zero. Quarterly dividends are assumed to be reinvested in the index. The results of the simulation using the S&P 500 index are shown in Tables 9-1 and 9-2.

The simulation used European puts, which could only be exercised at the end of the period, instead of American puts, which could be exercised at any time throughout the period. However, American puts are more expensive than European. Paige wondered if it made any difference in the performance of the fund if different types of puts were used. Paige also questioned if performance would be different if individual puts were purchased against stocks held in the portfolio, rather than using an index product. She surmised the outcome of individual or index puts over time would be about the same, but had no theory or evidence to support this contention.

One thing Paige finds surprising is that no research referred to the strategy of covered call writing as portfolio insurance even though she knows many fund managers who offer covered call writing as a means to increase portfolio return while at the same time reducing its risk. She thinks that covered calls would appeal to some of the more risk-averse plan sponsors and should be considered as a possible strategy for risk management of their portfolios. She knows she will have to find an explanation before the board meeting in three days.

The Cost of Portfolio Insurance

Portfolio insurance, like any other form of insurance, is not free. If the plan trustees are uncomfortable with the level of volatility in the large-

TABLE 9-1

Quarterly Returns for S&P 500 Index 1971 to 1998 (2nd Q) and S&P 500 Index Insured with At-the-Money Puts*

(1)	(2)	(3)	(4)	(5)	(6)	(7)	(8)	(9)
Year	Q	End of Qtr S&P 500	Qtr Divd	S&P 500 Unhedged Qtr Return	Annual Return	S&P 500 At-the-Money Index Put	Insured Index+Put Qtr Return	Insured Annual Return
70	4	92.15				1.65		
71	1	100.31	0.75	9.67%		1.33	7.74%	
	2	99.70	0.78	0.17%		1.35	–0.54%	
	3	98.34	0.77	–0.59%		2.17	–0.57%	
	4	102.09	0.77	4.60%	14.22%	2.34	2.34%	9.03%
72	1	107.20	0.75	5.74%		1.37	3.37%	
	2	107.14	0.78	0.67%		1.57	–0.54%	
	3	110.55	0.78	3.91%		1.77	2.41%	
	4	118.05	0.84	7.54%	18.96%	1.76	5.85%	11.45%
73	1	111.52	0.77	–4.88%		1.99	–0.83%	
	2	104.26	0.83	–5.77%		2.77	–1.02%	
	3	108.43	0.83	4.80%		2.13	2.08%	
	4	97.55	0.95	–9.16%	–14.67%	3.58	–1.07%	–0.86%
74	1	93.98	0.83	–2.81%		2.66	–2.72%	
	2	86.00	0.89	–7.54%		2.39	–1.83%	
	3	63/54	0.92	–25.05%		3.12	–1.66%	
	4	68.56	0.96	9.41%	–26.31%	3.36	4.29%	–2.06%
75	1	83.36	0.80	22.75%		2.85	17.02%	
	2	95.19	0.93	15.31%		2.41	11.50%	
	3	83.87	0.92	–10.93%		2.27	–1.53%	
	4	90.19	0.93	8.64%	36.98%	2.19	5.78%	35.91%
76	1	102.77	0.91	14.96%		2.50	12.23%	
	2	104.28	1.00	2.44%		2.08	0.01%	
	3	105.24	1.01	1.89%		1.78	–0.10%	
	4	107.46	1.13	3.18%	23.81%	2.28	1.47%	13.77%
77	1	98.42	1.05	–7.44%		1.58	–1.12%	
	2	100.48	1.17	3.28%		1.73	1.65%	
	3	96.53	1.15	–2.79%		1.46	–0.57%	
	4	95.10	1.30	–0.13%	–7.19%	1.74	–0.16%	–0.22%
78	1	89.21	1.18	–4.95%		1.50	–0.58%	
	2	95.53	1.28	8.52%		2.06	6.72%	
	3	102.54	1.26	8.66%		1.71	6.36%	
	4	96.11	1.35	–4.95%	6.52%	2.78	–0.35%	12.47%

TABLE 9-1

Quarterly Returns for S&P 500 Index 1971 to 1998 (2nd Q) and S&P 500 Index Insured with At-the-Money Puts* (Continued)

(1)	(2)	(3)	(4)	(5)	(6)	(7)	(8)	(9)
Year	Q	End of Qtr S&P 500	Qtr Divd	S&P 500 Unhedged Qtr Return	Annual Return	S&P 500 At-the-Money Index Put	Insured Index+Put Qtr Return	Insured Annual Return
79	1	101.56	1.31	7.03%		1.56	4.02%	
	2	102.91	1.43	2.73%		1.43	1.17%	
	3	109.32	1.43	7.62%		1.61	6.14%	
	4	107.94	1.54	0.15%	18.50%	2.18	–0.06%	11.64%
80	1	102.09	1.46	–4.07%		2.68	–0.65%	
	2	114.24	1.56	13.43%		3.03	10.53%	
	3	125.46	1.56	11.19%		2.80	8.31%	
	4	135.76	1.58	9.47%	32.45%	3.11	7.08%	27.35%
81	1	136.00	1.58	1.34%		2.88	–0.93%	
	2	131.25	1.67	–2.26%		1.37	–0.87%	
	3	116.18	1.69	–10.19%		2.56	0.24%	
	4	122.55	1.69	6.94%	–4.88%	2.36	4.63%	3.00%
82	1	111.91	1.67	–7.32%		2.64	–0.55%	
	2	109.61	1.76	–0.48%		1.81	–0.77%	
	3	120.42	1.73	11.44%		4.22	9.63%	
	4	140.64	1.71	18.21%	21.50%	5.85	14.21%	23.56%
83	1	152.96	1.71	9.98%		4.10	5.58%	
	2	168.11	1.79	11.07%		3.11	8.18%	
	3	166.07	1.79	–0.15%		3.74	–0.77%	
	4	164.93	1.78	–0.39%	22.44%		–1.15%	12.03%
84	1	159.18	1.83	–2.38%		3.05	1.11%	
	2	153.18	1.87	–2.59%		3.99	–0.73%	
	3	166.10	1.88	9.66%		3.53	6.88%	
	4	167.24	1.93	1.85%	6.21%	4.45	–0.27%	6.99%
85	1	180.66	1.96	9.20%		4.06	6.37%	
	2	191.85	1.99	7.30%		4.04	4.94%	
	3	182.08	1.99	–4.06%		3.41	–1.05%	
	4	211.28	2.02	17.15%	31.69%	3.52	14.99%	27.01%
86	1	238.90	2.04	14.04%		4.46	12.17%	
	2	250.84	2.07	5.86%		6.61	3.92%	
	3	231.32	2.06	–6.96%		7.92	–1.77%	
	4	242.17	2.10	5.60%	18.61%	9.26	2.10%	16.92%

TABLE 9-1

Quarterly Returns for S&P 500 Index 1971 to 1998 (2nd Q) and S&P 500 Index Insured with At-the-Money Puts* (Continued)

(1)	(2)	(3)	(4)	(5)	(6)	(7)	(8)	(9)
Year	Q	End of Qtr S&P 500	Qtr Divd	S&P 500 Unhedged Qtr Return	Annual Return	S&P 500 At-the-Money Index Put	Insured Index+Put Qtr Return	Insured Annual Return
87	1	291.70	2.23	21.37%		6.61	16.90%	
	2	304.00	2.22	4.97%		9.25	2.64%	
	3	321.83	2.22	6.60%		11.69	3.45%	
	4	247.08	2.27	–22.52%	5.22%	9.43	–2.82%	20.63%
88	1	258.89	2.33	5.72%		12.96	1.84%	
	2	273.50	2.44	6.59%		13.23	1.50%	
	3	271.91	2.49	0.33%		10.95	–3.75%	
	4	277.72	2.56	3.08%	16.54%	7.54	0.91%	–1.41%
89	1	294.87	2.70	7.15%		6.60	4.32%	
	2	317.98	2.77	8.78%		6.56	6.40%	
	3	349.15	2.84	10.70%		7.15	8.46%	
	4	353.40	2.90	2.05%	31.66%	7.07	0.00%	20.37%
90	1	339.94	2.97	–2.97%		12.80	–1.14%	
	2	358.02	3.03	6.21%		9.71	2.36%	
	3	306.05	3.07	–13.66%		8.53	–1.81%	
	4	330.22	3.09	8.91%	–3.09%	12.47	5.95%	5.28%
91	1	331.75	3.07	1.39%		13.11	–2.30%	
	2	371.16	3.04	12.80%		10.46	8.51%	
	3	387.86	3.06	5.32%		12.70	2.44%	
	4	417.09	3.03	8.32%	30.47%	9.88	4.88%	13.90%
92	1	403.69	3.19	–2.45%		13.41	–1.57%	
	2	408.14	3.04	1.86%		8.86	–1.42%	
	3	417.80	3.36	3.19%		7.46	1.00%	
	4	435.71	3.03	5.01%	7.67%	9.49	3.17%	1.11%
93	1	451.67	3.12	4.38%		10.32	2.15%	
	2	450.53	3.13	0.44%		9.30	–1.56%	
	3	458.93	3.13	2.56%		12.76	0.49%	
	4	466.44	3.15	2.32%	10.02%	9.39	–0.45%	0.60%
94	1	445.75	3.18	–3.76%		13.60	–1.31%	
	2	444.28	3.18	0.38%		11.76	–2.27%	
	3	462.69	3.23	4.87%		9.58	2.17%	
	4	459.27	3.30	–0.03%	1.29%	10.11	–1.33%	–2.77%

TABLE 9-1

Quarterly Returns for S&P 500 Index 1971 to 1998 (2nd Q) and S&P 500 Index Insured with At-the-Money Puts* (Continued)

(1)	(2)	(3)	(4)	(5)	(6)	(7)	(8)	(9)
Year	Q	End of Qtr S&P 500	Qtr Divd	S&P 500 Unhedged Qtr Return	Annual Return	S&P 500 At-the-Money Index Put	Insured Index+Put Qtr Return	Insured Annual Return
95	1	500.71	3.43	9.77%		9.71	7.40%	
	2	544.71	3.34	9.45%		10.62	7.37%	
	3	584.41	3.39	7.91%		11.61	5.85%	
	4	615.93	3.45	5.98%	37.41%	12.54	3.92%	26.85%
96	1	645.50	3.52	5.37%		18.30	3.27%	
	2	670.63	3.57	4.45%		14.94	1.57%	
	3	687.31	3.61	3.03%		21.07	0.78%	
	4	740.74	3.72	8.32%	22.81%	24.61	5.09%	11.09%
97	1	757.12	3.67	2.71%		24.84	–0.60%	
	2	885.14	3.80	17.41%		23.34	13.68%	
	3	947.28	3.83	7.45%		40.66	4.69%	
	4	970.43	3.86	2.85%	33.27%	46.85	–1.38%	16.67%
98	1	1101.75	3.88	13.93%		45.77	8.69%	
	2	1133.84	4.11	3.29%	17.68%	52.11	–0.83%	17.78%
Cumulative Return '71 to '98 (2nd Q)					3321.71%			1183.61%
Geometric Avg. Annual Return				13.71%				11.48%
Annual Standard Deviation				16.73%				10.40%

*Source: The S&P 500 data from 1970 through 1983 is from a Salomon Brothers report by H. Nicholas Hanson. Data from 1983 to 1998 was updated by Marilyn Wiley and Gary Trennepohl.

cap market, or the amount of risk in small-cap or international stocks, they might consider the cost of the insurance too great. For example, if the market continues to rise, the cost of the option premiums would reduce the returns of the portfolio, causing it to under-perform the fund's benchmark portfolio.

As she considers how to reduce the cost of portfolio insurance, Paige remembers a discussion about a *costless collar* she had recently with representatives of a major investment banking firm. Although she doesn't recall all the details, the costless collar strategy involved selling calls against the equity portfolio and using the proceeds to buy index

TABLE 9-2

American National Pension Trust Cumulative Returns S&P 500 Index

Until Year end	From Beginning of Year													
	71	72	73	74	75	76	77	78	79	80	81	82	83	84
1971	0.14													
1972	0.36	0.19												
1973	0.16	0.02	–0.15											
1974	–0.15	–0.25	–0.37	–0.26										
1975	0.17	0.02	–0.14	0.01	0.37									
1976	0.45	0.27	0.07	0.25	0.70	0.24								
1977	0.34	0.18	–0.01	0.16	0.57	0.15	–0.07							
1978	0.43	0.25	0.05	0.24	0.68	0.22	–0.01	0.07						
1979	0.70	0.49	0.25	0.46	0.99	0.45	0.17	0.26	0.19					
1980	1.25	0.97	0.65	0.94	1.63	0.92	0.55	0.67	0.57	0.32				
1981	1.14	0.87	0.57	0.84	1.50	0.83	0.48	0.59	0.49	0.26	–0.05			
1982	1.60	1.28	0.91	1.24	2.04	1.22	0.79	0.93	0.81	0.53	0.16	0.22		
1983	2.18	1.79	1.34	1.74	2.72	1.72	1.20	1.37	1.22	0.87	0.42	0.49	0.22	
1984	2.38	1.96	1.49	1.91	2.96	1.89	1.33	1.51	1.36	0.99	0.50	0.58	0.30	0.06
1985	3.45	2.90	2.28	2.84	4.21	2.80	2.07	2.31	2.11	1.62	0.98	1.08	0.71	0.40
1986	4.28	3.62	2.88	3.55	5.18	3.51	2.64	2.92	2.68	2.11	1.35	1.47	1.03	0.66
1987	4.55	3.86	3.09	3.79	5.50	3.75	2.83	3.13	2.88	2.27	1.47	1.60	1.14	0.75
1988	5.47	4.67	3.76	4.58	6.58	4.53	3.47	3.81	3.52	2.81	1.88	2.03	1.49	1.03
1989	7.52	6.46	5.27	6.35	8.97	6.28	4.88	5.34	4.95	4.02	2.79	2.98	2.28	1.68
1990	7.26	6.23	5.08	6.12	8.67	6.06	4.70	5.14	4.76	3.86	2.67	2.86	2.18	1.60
1991	9.77	8.43	6.93	8.29	11.61	8.21	6.44	7.01	6.52	5.35	3.79	4.04	3.15	2.39
1992	10.60	9.16	7.54	9.01	12.58	8.91	7.01	7.63	7.10	5.83	4.16	4.42	3.46	2.65
1993	11.76	10.17	8.39	10.01	13.94	9.91	7.81	8.49	7.91	6.52	4.68	4.97	3.91	3.01

TABLE 9-2

American National Pension Trust Cumulative Returns S&P 500 Index (Continued)

Until Year end	From Beginning of Year													
	71	72	73	74	75	76	77	78	79	80	81	82	83	84
1994	11.93	10.32	8.51	10.15	14.13	10.05	7.92	8.61	8.02	6.62	4.75	5.04	3.98	3.06
1995	16.77	14.55	12.07	14.32	19.79	14.18	11.26	12.21	11.40	9.46	6.90	7.31	5.84	4.58
1996	20.82	18.10	15.06	17.82	24.54	17.64	14.06	15.22	14.23	11.85	8.70	9.20	7.40	5.86
1997	28.08	24.46	20.40	24.08	33.03	23.84	19.07	20.62	19.30	16.13	11.93	12.60	10.19	8.14
2Q1998	33.22	28.96	24.18	28.51	39.05	28.24	22.61	24.44	22.88	19.15	14.22	15.00	12.17	9.75
1971														
1972														
1973														
1974														
1975														
1976														
1977														
1978														
1979														
1980														
1981														
1982														
1983														
1984														
1985	0.32													
1986	0.56	0.19												
1987	0.64	0.25	0.05											
1988	0.92	0.45	0.23	0.17										

TABLE 9-2

American National Pension Trust Cumulative Returns S&P 500 Index (Continued)

Until Year end	From Beginning of Year													
	71	72	73	74	75	76	77	78	79	80	81	82	83	84
1989	1.52	0.91	0.61	0.53	0.32									
1990	1.44	0.86	0.56	0.49	0.28	–0.03								
1991	2.19	1.42	1.04	0.94	0.66	0.26	0.30							
1992	2.43	1.61	1.20	1.09	0.79	0.36	0.40	0.08						
1992	2.78	1.87	1.42	1.30	0.97	0.50	0.55	0.18	0.10					
1994	2.83	1.91	1.45	1.33	1.00	0.52	0.57	0.20	0.11	0.01				
1995	4.26	2.99	2.37	2.20	1.74	1.08	1.15	0.65	0.53	0.39	0.37			
1996	5.46	3.90	3.13	2.93	2.37	1.56	1.64	1.02	0.88	0.71	0.69	0.23		
1997	7.60	5.53	4.51	4.24	3.49	2.41	2.52	1.70	1.51	1.28	1.25	0.64	0.33	
2Q1998	9.13	6.69	5.48	5.16	4.29	3.02	3.14	2.18	1.95	1.68	1.65	0.93	0.57	0.18
1971	0.09													
1972	0.22	0.11												
1973	0.20	0.10	–0.01											
1974	0.18	0.08	–0.03	–0.02										
1975	0.60	0.47	0.32	0.33	0.36									
1976	0.82	0.67	0.50	0.51	0.55	0.14								
1977	0.82	0.67	0.50	0.51	0.54	0.14	0.00							
1978	1.05	0.88	0.68	0.70	0.74	0.28	0.12	0.12						
1979	1.29	1.10	0.88	0.90	0.94	0.43	0.25	0.26	0.12					
1980	1.91	1.67	1.40	1.42	1.47	0.82	0.60	0.60	0.42	0.27				
1981	2.00	1.75	1.47	1.49	1.54	0.87	0.64	0.65	0.46	0.31	0.03			
1982	2.70	2.40	2.05	2.08	2.14	1.31	1.03	1.04	0.81	0.62	0.27	0.24		

TABLE 9-2

American National Pension Trust Cumulative Returns S&P 500 Index (Continued)

Until Year end	From Beginning of Year													
	71	72	73	74	75	76	77	78	79	80	81	82	83	84
1983	3.15	2.81	2.42	2.45	2.52	1.59	1.27	1.28	1.03	0.82	0.43	0.38	0.12	
1984	3.44	3.07	2.65	2.69	2.76	1.77	1.43	1.44	1.17	0.94	0.53	0.48	0.20	0.07
1985	4.64	4.17	3.64	3.68	3.78	2.52	2.09	2.10	1.75	1.47	0.94	0.88	0.52	0.36
1986	5.59	5.05	4.43	4.47	4.59	3.11	2.61	2.62	2.22	1.88	1.27	1.20	0.78	0.59
1987	6.95	6.29	5.55	5.60	5.74	3.96	3.36	3.37	2.88	2.48	1.73	1.65	1.15	0.92
1988	6.84	6.19	5.45	5.51	5.65	3.89	3.30	3.31	2.83	2.43	1.69	1.62	1.12	0.89
1989	8.44	7.66	6.77	6.84	7.00	4.89	4.17	4.19	3.61	3.13	2.24	2.15	1.55	1.27
1990	8.94	8.11	7.18	7.25	7.42	5.20	4.45	4.46	3.85	3.35	2.41	2.31	1.68	1.39
1991	10.32	9.38	8.31	8.40	8.59	6.06	5.20	5.22	4.53	3.95	2.89	2.78	2.06	1.73
1992	10.44	9.50	8.42	8.50	8.70	6.14	5.27	5.29	4.59	4.01	2.93	2.82	2.09	1.76
1993	10.51	9.56	8.47	8.56	8.76	6.18	5.31	5.33	4.62	4.04	2.96	2.84	2.11	1.77
1994	10.19	9.27	8.21	8.29	8.49	5.98	5.14	5.15	4.47	3.90	2.85	2.73	2.02	1.70
1995	13.20	12.02	10.69	10.79	11.04	7.86	6.78	6.80	5.94	5.21	3.88	3.74	2.83	2.42
1996	14.77	13.47	11.98	12.09	12.37	8.84	7.65	7.67	6.71	5.90	4.42	4.26	3.26	2.80
1997	17.40	15.88	14.15	14.28	14.60	10.48	9.09	9.11	7.99	7.05	5.32	5.14	3.97	3.43
2Q1998	18.84	17.19	15.32	15.47	15.81	11.37	9.87	9.90	8.69	7.68	5.82	5.62	4.35	3.78
1971														
1972														
1973														
1974														
1975														
1976														
1977														

TABLE 9-2

American National Pension Trust Cumulative Returns S&P 500 Index (Continued)

Until Year end						From Beginning of Year								
	71	72	73	74	75	76	77	78	79	80	81	82	83	84
1978														
1979														
1980														
1981														
1982														
1983														
1984														
1985	0.27													
1986	0.48	0.17												
1987	0.79	0.41	0.21											
1988	0.77	0.39	0.19	–0.01										
1989	1.13	0.67	0.43	0.19	0.20									
1990	1.24	0.76	0.51	0.25	0.27	0.05								
1991	1.55	1.01	0.72	0.42	0.44	0.20	0.14							
1992	1.58	1.03	0.74	0.44	0.46	0.21	0.15	0.01						
1992	1.59	1.04	0.75	0.45	0.47	0.22	0.16	0.02	0.01					
1994	1.52	0.99	0.70	0.41	0.43	0.19	0.13	–0.01	–0.02	–0.03				
1995	2.20	1.52	1.15	0.79	0.81	0.50	0.43	0.25	0.24	0.23	0.27			
1996	2.55	1.80	1.39	0.98	1.01	0.67	0.59	0.39	0.38	0.37	0.41	0.11		
1997	3.15	2.26	1.79	1.31	1.35	0.95	0.85	0.63	0.61	0.60	0.64	0.30	0.17	
2Q1998	3.47	2.52	2.01	1.49	1.53	1.10	1.00	0.75	0.73	0.72	0.77	0.40	0.26	0.08

puts. The striking prices of the puts and calls could be adjusted so that the insurance premium (the cost of the puts) was exactly offset by the income received from the sale of the calls. The premiums and striking prices were related to a percentage of the current level of the market. For example, assume the market is at 100 percent; the striking price of the calls might be 108 percent and the strike for the puts would be at 95 percent. Paige wonders how the puts and calls were priced in this situation and what advantages and disadvantages accompanied this strategy.

As Paige considers the use of portfolio insurance strategies, she concludes that index puts, calls, and futures appear to offer the most cost-effective ways to implement risk management strategies. Index puts and calls are traded as listed contracts on various option exchanges or can be purchased using the OTC market, where an investment bank or brokerage firm creates an index option to fit the buyer's exact specifications.

Characteristics of Listed Index Option Contracts

Risk management strategies used by large pension funds often are based on index options, so Paige began noting information about these options which she thought would be interesting to the Board. She first gathered information about the different index option contracts available, and then began sorting out details about their characteristics.

A description of the listed index options contracts that Paige thinks would be most appropriate to consider for the Trust is given below. She knows she must describe to the Board the specifications of these contracts and how they are traded. Most important, Paige discovered that all index options are cash settled, usually based on the index value of the opening price on the Friday of the expiration week. In the usual sequence of events, trading in the contract would end on the Thursday of expiration week, the settlement price would be determined on Friday based on opening stock prices, and the contracts then expire on Saturday. Many investors liked the cash-settled feature because it reduced the turnover of the underlying portfolio—no stock would be called away. The cash settlement was equal to the difference between the option's strike price and the closing price of the index.

She also discovered that maturities on the contracts range from the nearest month up to nine months away for standard options, to almost three years for the LEAPS. Paige is unsure, however, if it would be better to use the short term options and roll forward positions as contracts expire, or to take a position for the long term using leaps, and hold it until expiration.

Paige thinks it is important to collect information about the characteristics of different index option contracts. With the broad range of index options available, it appears she will be able to find an index that is highly correlated to most of their equity portfolios.

Selected Index Option Contracts

Characteristics about the most popular listed index options available on the CBOE are shown below. Information about all CBOE listed options is available at their web site, www.cboe.com.

Standard & Poors 100 Index is a subset of stocks in the S&P 500, and is a capitalization-weighted index of 100 large companies from a variety of industries. The impact of a stock's price change is proportional to its total market value (shares outstanding × price per share). Contracts available on the S&P 100 include the standard OEX contract with maturities up to nine months, and longer term, reduced value LEAPS with maturities up to three years.

Option Contract:	OEX
Multiplier:	$100
Contract value:	$100 × S&P 100 index value *e.g.* $100 × 565 = $56,500 per contract
Style	American
Maturity	Four nearest months, plus one additional month from the March-June-September-December cycle.
Expiration date:	Saturday immediately following the third Friday of the expiration month.
Last trading day:	Trading in a series will ordinarily cease on the business day (usually a Friday) prior to the expiration date.

Standard & Poors 500 Index is a capitalization-weighted index of 500 stocks from a broad range of industries and is probably the most widely used benchmark for equity portfolios of institutional investors. Contracts available on the S&P 500 include the standard SPX contract with maturities up to nine months, and longer term, reduced value SPX LEAPS with maturities up to three years.

Option Contract:	SPX
Multiplier:	$100

Contract value:	\$100 × S&P 500 index value *e.g.* \$100 × 1,120 = \$112,000 per contract
Style:	European
Maturity:	Three nearest months, plus three additional months from the March-June-September-December cycle.
Expiration date:	Saturday immediately following the third Friday of the expiration month.
Settlement value:	The settlement value is calculated using the opening sales price in the primary market of each component stock on the last business day (usually a Friday) before the expiration date.
Last trading day:	Trading in a series will ordinarily cease on the business day (usually a Thursday) prior to the day on which the exercise-settlement value is determined.

Dow Jones Industrial Average is a price-weighted index composed of 30 of the largest, most liquid NYSE-listed stocks. It is probably the most familiar barometer of the stock market to investors as it been calculated in its current format since 1928. However, options and futures only began trading on the DJIA in 1997. Both standard options and LEAPS are available on the DJIA.

Option Contract:	DJX
Multiplier:	\$100
Underlying level:	1/100 of the DJIA, *e.g.*, if the DJIA is at 8700, the DJX equals 87.
Contract value:	\$100 × DJX level, *e.g.*, if the DJIA is at 8700, the DJX equals 87 and each contract is worth \$100 × 87 = \$8,700.
Style:	European
Maturity:	Three nearest months, plus three additional months from the March-June-September-December cycle.
Expiration date:	Saturday immediately following the third Friday of the expiration month.
Settlement value:	Calculated based on the opening prices of the component securities on the business day (typically Friday) prior to expiration.

Last trading day:	Trading in a series will ordinarily cease on the business day (usually a Thursday) prior to the day on which the exercise-settlement value is determined.

Russell 2000 Index is composed of the 2,000 *smallest* companies in the Russell universe of 3,000 stocks. It is considered one of the broadest measures of small-cap stock performance. It is a capitalization-weighted index of companies ranging in value from $40 million to $450 million, and company values are adjusted for cross-ownership.

Option Contract:	RUT
Multiplier:	$100
Contract value:	$100 × Russell 2000 Index value *e.g.* if the Russell 2000 is at 457, each contract is worth $100 × 457 = $45,700.
Style:	European
Maturity:	Three nearest months, plus three additional months from the March-June-September-December cycle.
Expiration date:	Saturday immediately following the third Friday of the expiration month.
Settlement value:	Calculated based on the opening prices of the component securities on the business day (typically Friday) prior to expiration.
Last trading day:	Trading in a series will ordinarily cease on the business day (usually a Thursday) prior to the day on which the exercise-settlement value is determined.

NASDAQ-100 Index is a capitalization-weighted index composed of 100 of the largest nonfinancial securities listed on the NASDAQ Stock Market. The index was created in 1985 with a base value set to 250 on February 1. After reaching a level of nearly 800 on December 31, 1993, the index level was halved on January 3, 1994. The NASDAQ-100 Index consists mostly of small, high-technology firms but includes industry giants Intel and Microsoft. It is considered a good measure of the market performance of small companies especially in the technology area.

Option Contract:	NDX
Multiplier:	$100

Contract value:	\$100 × NASDAQ Index value, *e.g.*, if the NASDAQ is at 1340, each contract is worth \$100 × 1340 = \$134,000.
Style:	European
Maturity:	Three nearest months, plus three additional months from the March-June-September-December cycle.
Expiration date:	Saturday immediately following the third Friday of the expiration month.
Settlement value:	Calculated based on the average price during the first five minutes of trading of the component securities on the business day (typically Friday) prior to expiration.
Last trading day:	Trading in a series will ordinarily cease on the business day (usually a Thursday) prior to the day on which the exercise-settlement value is determined.

Table 9-3 contains data from *The Wall Street Journal* that shows (1) the trading range for underlying indexes for which listed options are available, and (2) premiums for many of the contracts listed above. Paige recalls that only contracts on the S&P 100 were American-style options, while all the rest were European. She sees that the American style options seem to be relatively more expensive than equivalent European-style options. Should she consider the American or European style options in her strategies? Does the ability to exercise early (American style) have value to American National Pension Trust's equity managers? For the board meeting, Paige thinks it would be useful to construct payoff diagrams for some sample index option strategies using the data in Table 9-3.

OTC-Traded Index Option Contracts and FLEX Options

In the late 1980s some brokerage firms and investment banks began offering tailored option contracts to their larger customers. As the interest increased in portfolio risk management, fund managers began using these over-the-counter (OTC) options to implement risk management strategies.

Through discussions with other fund managers familiar with OTC options, Paige found that fund managers use these contracts because (1) listed contract position limits set by the clearing house or exchange were

TABLE 9-3

Selected Closing Index Option Premiums, June 30, 1998 from The Wall Street Journal

DJ INDUS AVG(DJX)

Strike		Vol.	Last	Net Chg.	Open Int.
Dec	66p	10	5 1/2	+ 1/8	705
Dec	72p	19	15/16	− 1/16	36,740
Dec	76p	311	1 1/8	− 1/8	46,846
Jul	80c	450	10 1/8	− 1/8	1,540
Sep	80p	200	3/4	− 1/16	14,109
Dec	80p	19	1 3/4	− 1/8	13,146
Jul	84p	500	1/4	...	3,808
Aug	84p	235	3/4	− 1/8	858
Jul	85c	160	4 7/8	− 5/8	266
Sep	85p	500	1 1/2	...	8,413
Jul	86c	21	4	− 5/8	569
Jul	86p	14	5/16	...	1,640
Aug	86c	4	5 1/4	− 1/2	109
Sep	86p	2	1 11/16	− 3/16	422
Jul	87p	636	9/16	+ 3/16	2,636
Aug	87p	402	1 1/4	− 1/4	4,025
Jul	88c	20	2 7/16	− 7/16	1,204
Jul	88p	373	3/4	+ 1/16	3,853
Aug	88c	4	3 3/8	− 3/8	310
Aug	88p	15	1 11/16	+ 3/16	573
Sep	88p	16	2 1/8	− 1/16	6,007
Dec	88c	4	7 1/4	+ 1 1/8	5,335
Dec	88p	310	3 3/8	...	2,605
Jul	89c	288	1 5/8	− 7/16	1,333
Jul	89p	323	1	+ 1/16	1,894
Aug	89c	4	3	− 3/8	152
Aug	89p	40	1 7/8	...	60
Jul	90c	277	1 3/16	− 5/16	7,540
Jul	90p	527	1 3/8	+ 1/4	11,497
Aug	90c	53	2 1/2	− 5/8	577
Sep	90c	132	3 5/8	− 1/4	2,273
Sep	90p	112	2 7/8	+ 1/8	2,359
Dec	90p	2	4 1/2	+ 1/4	1,429
Jul	91c	63	3/4	− 3/8	2,468
Jul	91p	194	2 1/16	+ 7/16	1,011
Aug	91c	3	2 1/16	...	436
Jul	92c	110	3/8	− 1/4	3,580
Jul	92p	39	2 11/16	+ 1/2	3,173
Aug	92c	30	1 1/2	− 5/16	952
Aug	92p	4	3 3/8	+ 1/2	983
Sep	92c	5	2 1/2	− 1/8	4,289
Sep	92p	12	3 7/8	− 1/4	2,263
Jul	93c	100	3/16	− 1/4	1,048
Jul	93p	55	3 1/2	+ 5/8	93
Jul	94p	5	4 1/8	+ 1/8	294
Aug	94c	20	3/4	− 1/4	1,113
Dec	94c	1	3 3/8	+ 3/4	17
Call Vol. 1,749			Open Int. 107,620		
Put Vol. 4,879			Open Int. 221,659		

DJ TRANP AVG(DTX)

Strike		Vol.	Last	Net Chg.	Open Int.
Sep	370c	1	4 7/8	+ 1/8	2
Call Vol. 1			Open Int. 819		
Put Vol. 0			Open Int. 3,217		

DJ UTIL AVG(DUX)

Strike		Vol.	Last	Net Chg.	Open Int.
Sep	280c	2	19 3/8	+ 7/8	3
Call Vol. 2			Open Int. 12		
Put Vol. 0			Open Int. 93		

NASDAQ-100(NDX)

Strike		Vol.	Last	Net Chg.	Open Int.
Jul	1100p	53	5/8	+ 1/8	462
Aug	1100p	10	3 3/4	− 1/2	52
Jul	1120p	10	5/8	− 5/16	283
Aug	1120p	20	4 7/8	− 1 5/8	27
Jul	1130p	3	3/4	− 1/4	27
Aug	1140p	10	6	− 2 1/2	24
Jul	1150p	3	1	− 1 7/8	554
Jul	1170p	23	1 9/16	− 3/16	134
Jul	1180c	13	163	− 2 1/2	59
Jul	1180p	19	1 5/8	...	480
Jul	1200p	5	2 1/4	+ 1/4	1,305
Aug	1200p	1	11	− 1/4	168
Sep	1200p	300	19	+ 1 1/2	1,419
Jul	1220c	4	124	− 2	1,417
Jul	1220p	460	3 1/8	− 1/4	2,751
Jul	1240c	152	103 1/4	− 2 1/4	1,532
Jul	1240p	401	5 1/4	+ 7/8	1,201
Jul	1260p	40	6 5/8	− 1/8	1,600
Aug	1260p	11	23	− 2	43
Jul	1280c	1	62	− 16	780
Jul	1280p	392	9 1/2	− 1/2	1,085
Aug	1280p	1	27	+ 3/4	25
Sep	1280p	5	35	− 4 1/4	7
Jul	1300c	2	55	+ 1 1/2	755
Jul	1300p	292	12 7/8	− 3/8	842
Aug	1300p	102	29 3/4	− 2 3/4	329
Jul	1320c	13	43	+ 3	1,517
Jul	1320p	283	18 1/2	− 1/2	643
Aug	1320p	20	38	− 5	10
Jul	1340c	393	25	− 2 3/8	1,249
Jul	1340p	101	27	+ 3/4	453
Aug	1340c	3	45 1/8	− 8 1/8	116
Jul	1360c	266	16 1/2	− 2 1/2	413
Jul	1360p	2	32	− 5	193
Aug	1360c	2	44	+ 2 1/2	53
Jul	1380c	89	9 3/8	− 1 1/8	521
Jul	1380p	6	48 1/2	+ 2	27
Aug	1380c	5	32	− 1	198
Sep	1380p	5	77	− 12 1/2	250
Sep	1420c	23	28	...	...
Call Vol. 1,178			Open Int. 14,093		
Put Vol. 2,581			Open Int. 20,309		

S & P 100 INDEX(OEX)

Strike		Vol.	Last	Net Chg.	Open Int.
Jul	460p	2,085	1/8	− 1/16	8,411
Jul	470p	190	1/4	...	9,090
Aug	470p	82	1 5/16	+ 1/16	2,911
Sep	470p	115	2 9/16	+ 3/8	616
Jul	480p	599	1/4	...	9,398
Aug	480p	5	1 5/8	...	2,589
Jul	490p	64	7/16	+ 1/8	11,234
Aug	490p	212	2 1/4	...	5,710
Sep	490p	3	4	+ 1/8	2,133
Jul	500p	630	5/8	+ 1/16	24,834
Aug	500p	121	2 7/8	...	1,909
Sep	500c	6	66 3/4	− 2	2
Sep	500p	2	5 3/4	+ 1/4	1,350
Jul	505p	604	3/4	+ 1/16	5,594
Jul	510c	250	47	− 5 5/8	654
Jul	510p	378	13/16	...	24,386
Aug	510p	218	4	+ 3/8	1,541
Jul	515p	584	1 1/8	+ 1/8	4,882
Aug	515p	77	4 3/8	+ 3/8	519
Jul	520c	477	37 1/2	− 2 5/8	3,450
Jul	520p	1,115	1 1/4	+ 1/16	9,884
Aug	520c	149	43 1/4	− 1 1/4	1,892
Aug	520p	114	5	+ 1/8	2,487
Sep	520p	320	7 3/4	+ 3/4	804
Oct	520p	49	10 3/4	− 1 1/2	75
Jul	525c	11	34 1/8	− 3 3/8	2,128
Jul	525p	694	1 11/16	+ 3/16	9,515
Aug	525p	15	6	+ 1/4	594
Jul	530c	4,206	28 1/8	− 4	10,445
Jul	530p	5,678	2 1/8	+ 1/4	17,995
Aug	530c	11	36	− 1/2	2,854
Aug	530p	79	6 7/8	+ 7/8	2,799
Sep	530c	9	41 7/8	+ 3 7/8	2,080
Sep	530p	205	10	+ 1 1/4	2,904
Oct	530p	75	13 1/4	− 1 3/4	782
Jul	535c	100	24	− 2 1/4	10,762
Jul	535p	1,492	2 3/4	+ 3/8	13,927
Aug	535c	1	33 3/8	− 3/8	686
Aug	535p	31	7 1/2	+ 1/4	289
Sep	535p	3	11 3/8	+ 5/8	1,386
Jul	540c	2,463	20	− 2 5/8	8,928
Jul	540p	3,087	3 1/2	+ 5/8	13,626
Aug	540c	29	27 3/4	− 2 3/4	5,701
Aug	540p	92	9 1/2	+ 1 1/4	5,036
Sep	540p	40	11 5/8	+ 1/4	2,516
Oct	540p	35	15	− 1 1/2	408
Jul	545c	713	15 1/4	− 2 3/4	6,526
Jul	545p	5,512	4 1/2	+ 5/8	10,179
Aug	545c	12	23 3/4	− 2 3/4	361
Aug	545p	65	10 5/8	+ 1 3/8	160
Sep	545c	1	29 1/4	− 2 3/4	9
Jul	550c	1,713	11 1/2	− 3	10,145
Jul	550p	5,763	6	+ 1 3/8	10,099
Aug	550c	16	21 5/8	− 1 3/8	3,092
Aug	550p	30	12	+ 1/4	1,813
Sep	550p	5	15 1/2	+ 3/4	477
Oct	550c	3,201	30 1/2	+ 1/2	56
Oct	550p	3,200	19	− 1 1/4	66
Jul	555c	7,600	8 3/4	− 2 3/8	6,732
Jul	555p	7,326	8 1/8	+ 1 3/8	5,402
Aug	555c	67	16 3/4	− 2 1/4	1,930
Aug	555p	7	13	+ 1/8	131
Sep	555c	17	22 1/2	− 2 3/4	1,196
Sep	555p	15	17	+ 3/4	180
Jul	560c	6,662	6 1/8	− 1 7/8	12,405
Jul	560p	2,108	10 1/2	+ 1 1/2	5,722
Aug	560c	113	15 1/4	− 1/4	4,390
Aug	560p	8	15 1/2	+ 1	320
Sep	560c	10	20 1/2	− 2 1/2	1,252
Sep	560p	120	19 3/4	+ 1 1/8	135
Oct	560p	4	22 1/4	+ 1 1/4	3
Jul	565c	3,457	3 7/8	− 1 5/8	12,714
Jul	565p	218	14 1/4	+ 2 1/2	672
Aug	565c	15	13	− 1/2	4,925
Jul	570c	3,746	2 7/16	− 1 1/16	10,279
Jul	570p	180	18	+ 3 1/8	502
Aug	570c	585	8 7/8	− 1 5/8	6,587
Aug	570p	48	20 1/2	+ 1/4	88
Sep	570c	12	14 1/4	− 2 7/8	981
Sep	570p	87	24	+ 1 1/2	66
Oct	570c	2	19 7/8	...	191
Oct	570p	85	28 1/4	...	5
Jul	575c	1,880	1 3/8	− 5/8	6,047
Jul	575p	45	21 5/8	− 3 3/8	49
Aug	575c	67	8	...	390
Jul	580c	1,287	3/4	− 7/16	6,558
Jul	580p	58	23 1/2	+ 1/8	346
Aug	580c	49	5 1/8	− 1 5/8	1,769
Aug	580p	1	29	+ 3 3/4	66
Sep	580p	32	30	+ 1 1/4	205
Oct	580c	201	15 1/4	− 1 3/8	166
Oct	580p	30	33	+ 1 3/8	5
Sep	585c	4	9	...	...
Jul	590c	1,902	5/16	− 1/8	6,027
Aug	590c	570	2 3/4	− 1 1/8	2,243
Sep	590c	6	7	− 1/2	2,230
Jul	600c	386	1/8	− 1/16	4,010
Aug	600c	73	1 9/16	− 5/16	3,481
Aug	610c	5	11/16	− 7/16	32
Call Vol. 42,194			Open Int. ... 178,696		
Put Vol. 44,091			Open Int. ... 243,532		

S & P 500 INDEX-AM(SPX)

Strike		Vol.	Last	Net Chg.	Open Int.
Jul	1025p	14	1 3/8	+ 5/16	9,231
Aug	1025p	14	6	+ 3/4	6,181
Sep	1025p	100	9 5/8	+ 1/8	15,437
Jul	1035p	150	1 1/2	− 1/2	680
Jul	1040p	88	1 1/2	...	735
Jul	1050c	19	85	− 10 1/2	608
Jul	1050p	145	2	+ 1/4	15,937
Aug	1050c	10	97	+ 14 1/2	24
Aug	1050p	4	7 1/2	+ 1/4	3,242
Sep	1050c	315	107	+ 20 3/4	19,835
Sep	1050p	352	13	+ 7/8	32,045
Jul	1060p	466	2 3/8	+ 1/8	2,229
Jul	1065p	85	2	− 1/4	25
Jul	1070p	509	3 3/8	+ 3/8	6,006
Jul	1075p	725	3 1/4	+ 3/8	15,162
Aug	1075p	154	10 1/2	...	3,460
Sep	1075p	26	17 3/4	+ 1 7/8	25,494
Jul	1080p	140	4	+ 5/8	4,718
Sep	1080p	10	17 1/2	− 1/4	16
Jul	1090c	10	50	− 11 1/2	2,482
Jul	1090p	208	5 3/8	+ 1 3/8	10,685
Aug	1090p	6	14	− 3/4	2,764
Jul	1095p	119	4 5/8	− 1/8	5,560
Sep	1095p	15	21 1/4	+ 1 3/4	3,131
Jul	1100c	129	41 7/8	− 6 5/8	17,117
Jul	1100p	468	6 1/2	+ 1	21,493
Aug	1100p	330	16 1/2	+ 1 3/4	10,033
Sep	1100c	40	68	− 4 1/4	27,027
Sep	1100p	168	23 1/4	+ 3/4	30,402
Jul	1105p	560	7 1/4	+ 2 1/2	963
Jul	1110c	155	32 1/4	− 6 3/4	9,608
Jul	1110p	137	7 5/8	+ 5/8	5,884
Aug	1110p	4	19 1/4	+ 2 1/4	434
Jul	1115c	52	29	− 7	3,954
Jul	1115p	450	9 5/8	+ 1 1/8	3,569
Aug	1115c	5	49 1/2	+ 2 1/2	165
Jul	1120c	5	25 3/4	− 3 5/8	8,533
Jul	1120p	620	10	+ 1 3/4	8,532
Aug	1120p	1	19 3/4	− 1 1/4	313
Sep	1120c	101	55 1/2	− 2 3/8	2,294
Sep	1120p	150	27 1/2	+ 1 1/4	3,151
Jul	1125c	58	21 7/8	− 4 7/8	6,901
Jul	1125p	208	12 3/4	+ 2 1/4	5,280
Aug	1125c	100	39 3/4	− 7 1/4	2,020
Aug	1125p	300	24 1/2	+ 2 1/2	1,977
Sep	1125c	1	49	− 5 1/2	18,946
Sep	1125p	58	31	+ 4	16,531
Jul	1130c	249	19 1/4	− 3 1/4	5,048
Jul	1130p	230	13 1/4	+ 2 1/4	1,810
Aug	1130p	13,288	26 1/2	+ 2 1/2	472
Sep	1130p	6	33 1/2	− 2 1/2	819
Jul	1135c	81	16 1/4	− 3 1/4	257
Jul	1135p	76	16 1/4	+ 5	1
Jul	1140c	1,862	13 1/2	− 3 1/2	10,923
Jul	1140p	698	18	+ 2 1/2	3,825
Aug	1140c	250	31	− 2	426
Aug	1140p	105	28 1/2	+ 3 1/2	696
Sep	1140c	245	40	− 2	446
Sep	1140p	62	38	+ 3 3/8	816
Sep	1145c	2,147	39 1/8	...	658
Sep	1145p	2,283	38 1/2	...	658
Jul	1150c	1,330	8 3/4	− 3 1/4	10,755
Jul	1150p	62	23	+ 2 1/2	1,951
Aug	1150c	563	27	− 3	4,109
Aug	1150p	1,710	33	+ 4	1,692
Sep	1150c	2,435	37	− 2	15,689
Sep	1150p	2,535	39	− 1/2	11,254
Jul	1155c	1	6 7/8	− 2 3/8	2,368
Jul	1155p	2	23	+ 2 1/2	2
Jul	1160c	1,131	5 1/4	− 2 1/4	6,275
Jul	1160p	22	30	+ 8	219
Aug	1160c	10	20	− 2	1,678
Jul	1165c	2	4 1/2	− 1 1/4	1,825
Sep	1165p	100	45 1/2	− 7 1/4	100
Jul	1170c	583	2 7/8	− 2 1/4	8,591
Jul	1170p	28	39	+ 9 1/4	158
Aug	1170c	12	15	− 4 1/2	699
Sep	1175c	2	23 1/4	− 5	9,378
Sep	1175p	7	53 1/2	+ 3	1,242
Jul	1180c	1,118	1 1/2	− 1 1/4	7,731
Jul	1180p	6	44	+ 4	443
Aug	1180c	250	12 1/2	− 2 1/2	644
Aug	1180p	1	51 1/2	+ 7 1/2	95
Aug	1190c	1	9 1/4	− 1 3/4	505
Aug	1190p	1	53	...	...
Jul	1200c	517	1/2	− 1/2	3,627
Jul	1200p	502	60 7/8	+ 7 3/8	29
Aug	1200c	164	7	− 1 1/8	3,672
Sep	1200c	229	15	− 2	15,773
Aug	1225c	4	2 1/2	...	...
Sep	1225c	139	8 1/2	− 1/2	3,623
Jul	1250c	448	1/8	+ 1/16	2,361
Jul	1250p	275	112	+ 11	1,009
Aug	1250c	305	13/16	− 13/16	66
Sep	1250c	292	4 5/8	− 3/8	12,481
Sep	1300c	950	1	− 1/8	3,530
Call Vol. 18,594			Open Int. ... 760,309		
Put Vol. 38,356			Open Int. .. 1,148,366		

LEAPS-LONG TERM

DJ INDUS AVG - CB

Strike		Vol.	Last	Net Chg.	Open Int.
Dec 00	76p	1	4 3/4	− 1/4	36031
Dec 99	80p	3	4 1/2	− 3/8	16979
Dec 99	84p	45	5 1/2	− 3/8	714
Dec 99	88p	6	6 7/8	− 1/8	2049
Dec 00	98p	2	10 3/4	+ 1/8	165
Dec 00	100p	4	12 1/4	...	1042
Call vol. 0			Open Int. 84,467		
Put vol. 61			Open Int. 157,910		

S & P 100 INDEX - CB

Strike		Vol.	Last	Net Chg.	Open Int.
Dec 99	60p	6	3/4	...	3109
Dec 98	75p	10	1/4	− 1/16	6088
Dec 98	100p	5	2 9/16	+ 3/16	4124
Call vol. 0			Open Int. 20,316		
Put vol. 21			Open Int. 116,989		

S & P 500 INDEX - CB

Strike		Vol.	Last	Net Chg.	Open Int.
Dec 00	70p	10	1 7/8	− 1/16	4608
Dec 99	75p	50	1 9/16	− 1/16	15203
Dec 99	77 1/2p	10	7/8	− 3/16	7341
Dec 98	80p	100	3/8	...	19240
Dec 98	85p	17	11/16	+ 1/16	13242
Dec 99	85p	2	2 5/8	+ 1/8	10121
Dec 99	87 1/2p	7	3	− 3/8	2306
Dec 98	90p	17	7/8	...	6376
Dec 99	90p	73	3 1/8	...	9855
Dec 98	92 1/2c	1	23 7/8	+ 7 5/8	1530
Dec 99	92 1/2p	7	3 3/4	+ 1/2	4540
Dec 98	95p	10	1 1/4	− 1/4	10323
Dec 99	95p	33	3 3/4	+ 1/4	20004
Dec 98	97 1/2p	2	1 1/2	− 3/8	765
Dec 98	100p	20	2	...	4658
Dec 99	100c	33	24	+ 3	15910
Dec 99	100p	10	5 1/8	+ 1/2	7047
Dec 98	105p	2	2 15/16	+ 1/16	940
Dec 00	105c	20	26 1/4	+ 7/8	201
Dec 00	105p	70	7 1/2	+ 1/8	9016
Dec 98	110p	3	4 1/8	− 1/8	1432
Dec 99	110p	2	7 3/4	...	1632
Dec 00	110c	50	22 3/4	− 3/8	5478
Dec 00	110p	50	9 1/8	...	12173
Dec 98	115p	12	6 1/4	+ 1/2	306
Dec 99	115p	3	9 1/4	+ 1/4	439
Dec 00	115p	43	10 3/4	...	14661
Dec 00	120p	14	12 7/8	− 3/8	12966
Dec 98	125c	2	8 1/4	+ 1/4	190
Dec 98	125p	2	4 7/8	− 2 1/4	36
Dec 99	125c	1	16 1/4	...	2
Dec 00	130p	6	16 3/4	+ 1/4	1078
Call vol. 107			Open Int. 86,946		
Put vol. 575			Open Int. 469,648		

RUSSELL 2000(RUT)

Strike		Vol.	Last	Net Chg.	Open Int.
Aug	425p	3	2 7/16	− 4 5/16	73
Jul	430c	500	26 1/2	+ 3 1/4	500
Jul	435c	400	22 1/2	+ 11 7/8	400
Jul	440p	23	1 11/16	− 1/16	31
Aug	440p	2	5 1/8	− 7/8	2
Jul	445c	6	13 1/2	+ 4 1/8	83
Sep	445c	10	24	+ 4	500
Sep	445p	5	7 1/2	− 2 1/2	500
Jul	450c	105	9 7/8	+ 3 1/8	371
Jul	450p	15	3 3/8	− 1	319
Aug	450c	1	15	− 1 1/4	19
Jul	455c	106	5 7/8	+ 1/8	53
Jul	455p	5	6 1/2	− 9 3/8	25
Jul	460c	11	3 1/8	− 1/2	69
Aug	460c	11	10 1/8	+ 2 1/4	52
Jul	465c	15	2	+ 1/8	371
Aug	465c	2	6	...	...
Sep	490c	5	2	+ 1/16	577
Call Vol. 1,172			Open Int. 16,150		
Put Vol. 53			Open Int. 22,935		

Tuesday, June 30, 1998

Volume, last, net change and open interest for all contracts. Volume figures are unofficial. Open interest reflects previous trading day. p-Put c-Call

not sufficient to hedge large portfolios, and (2) they had the ability to tailor the characteristics of strike, maturity, and size to match the portfolio being hedged.

Disadvantages to using the OTC options were often given as (1) counterparty risk—the risk that the "other side of the trade" would default on the position if exercised, and (2) the lack of a secondary market for the option. The only way to exit the option position prior to maturity is to trade it back to the brokerage firm at their determined price. Paige thinks she should be prepared to discuss the use of OTC options as an alternative risk management strategy during her presentation to American National's board.

In 1993 the CBOE began trading a product called FLEX options, designed to compete with over-the-counter options. It was an attempt to serve investors desiring customized options, but who wanted to trade with an organized exchange to reduce counterparty risk. FLEX options are similar to OTC contracts in that specifications of the contract are tailored to the needs of the buyer. They differ in that the contracts are traded by CBOE market makers, which enable the contracts to bought and sold at the exchange prior to expiration, thus addressing some of the concerns of buyers of OTC options. (For more information on FLEX options, see Chapter 6.)

American National Pension Trust Strategy Considerations

Because of the large size of the American National Pension Trust, Paige must consider the need for market liquidity and depth when designing a portfolio risk management strategy. She also recognizes that the internal and external fund managers used by American National are quite well-diversified with holdings in stocks, fixed-income instruments, real estate, and the money market. With their size and diversification, active management is difficult, and the Pension Trust management often has to diversify across managers and styles, rather than across the individual financial instruments themselves.

Paige reviewed the history of American National Pension Trust to determine if they had previous experience with option trading. She found that during the early and mid-1980s, the Trust had experimented with *overwrite programs* (selling calls against stock owned) and *buy-write programs* (buy the stock and write a call on the stock). Fund managers promoting these programs promised to reduce downside risk exposure for part of the portfolio at modest sacrifice of upside potential. Because sponsors tend to be risk-conscious in exercising their fiduciary responsibilities, limiting downside risk was attractive to them during periods of adverse equity market movements. However, after the start of the bull market in August, 1982, these strategies delivered poor performance for the American National Pension Trust because the written calls severely limited upside gains. Stocks which were appreciating were called away and significant opportunity losses were incurred.

The Past and Current Market Environment

The price appreciation from 1991 to mid-1998 in U.S. stocks represents one of the greatest bull markets in history. Not counting dividends, the Dow Jones Industrial Average has appreciated from 2,610 on January 1, 1991, to 8,952 on June 30, 1998. This is an increase of 343 percent over seven and a half years, not counting dividends. The S&P 500 has risen an almost-identical 348 percent over the same period, suggesting the market climb was not confined to the largest, blue-chip stocks.

This bull market, along with the Pension Trust's wise investment decisions, has created a situation in which the Pension Trust is significantly overfunded based on conservative actuarial assumptions. Pension funding costs, relative to payrolls, were declining sharply and Paige Hanson is in an enviable position regarding the fund's ability to meet its future payout demands. The overfunded pension situation causes Paige to be more cautious in her decisions, concerned more about protecting gains, rather than seeking aggressive investment strategies to overcome underfunded pension obligations.

Indeed, the period from fall 1997 to summer 1998 underscored to Paige the need for a consistent risk management strategy for the portfolio. At current levels, rather large daily price swings seem commonplace and the primary factor fueling volatility has changed from speculation in 1997 about "would the FED raise interest rates" to concern about the impact of the falling Asian economies on growth of U.S. companies. One headline suggested that the U.S. market could fall by 25 to 30 percent over the next year because of the problems in Asia.

By July many market observers were convinced the market was long overdue for a correction. Several of Paige's analysts predicted that the market will exhibit increased volatility during the remainder of the year. To the longer term, however, some were predicting the Dow will top 10,000 before the year 2000. Others believe a long-term top has been reached and are wary of a 5 to 10 percent correction. Because of the conflicting currents in the market, Paige and her management team have made the decision to accumulate a larger than normal cash position to have available in case a market correction occurs.

Paige knows that it is important to protect the current value of the Pension Trust's investments and she believes that options could be used in risk management strategies. She thinks that risk management is especially important if the fund decided to invest in smaller capitalization companies or emphasize particular industries, because either strategy would inject more risk into the portfolio than traditional investments of American National Pension Trust.

However, for any option-based strategy to work, two hurdles must be overcome. First, the Board must understand the meaning of portfolio risk management and second, they must be educated in the use of options and other derivative products used to manage risk. Paige knows that the education process is very important if the risk management strategies are to be accepted and implemented.

As the equity markets continues to exhibit increased volatility, some trustees increasingly question whether a significant market correction might be forthcoming. It is easy to draw parallels between this market and the disastrous correction of October 1987. Adding to Paige's concerns is the fact that traditional alternatives to reducing risk, such as moving to cash or bonds, are relatively unattractive because of relatively low interest rates. To Paige, this further motivates her to search for other ways to manage risk in the Trust's portfolios.

Designing a Strategy

As dusk falls on the evening before her Board presentation, Paige Hanson again gazes at the Chicago skyline and contemplates her presentation for tomorrow. In reviewing the market's performance over the past few weeks, it is apparent that equities have become more volatile. More economic data seems to suggest a weakening economy and increased problems in foreign markets, which could lead to a sharp market correction, or even signal the beginning of a bear market.

Given the performance of the market the past four months, Paige begins to sort out her thoughts on designing strategies to manage risk and return more effectively. She knows that listed equity options, index options, and portfolio insurance strategies provide attractive opportunities for enhancing return and reducing risk. Her problem is how to sort out the relevant information from the vast amount of data she has collected, and make a concise, hard-hitting presentation to the Board tomorrow.

Questions for Discussion

1. Define portfolio risk management and give examples of option strategies that can be used in typical risk management strategies. Demonstrate why portfolio insurance is a risk

management strategy. Explain why strategies such as covered call writing and market timing should not be considered portfolio insurance.

2. Using data in Tables 9-1 and 9-2, compare the returns of the S&P 500 over various periods. In what periods did the insured portfolio outperform the unprotected portfolio? What characteristic of the data seems to be necessary for the continuously insured portfolio to outperform the unhedged one?
3. Evaluate the data in Tables 9-1 and 9-2. Explain the implications of these results for static portfolio insurance. What criticisms would you make regarding the simulation? Why are such simulations termed "sample period dependent"? Is this type of data useful for making decisions about the implementation of portfolio insurance strategies?
4. Evaluate the advantages and disadvantages to American National Pension Trust of the different option contracts described in the case. How important are such characteristics as option maturity, style, and index composition? Should Paige consider individual equity options as well as index options? What about OTC options and FLEX options?
5. Using data in Table 9-3, develop a risk management strategy for a $10,000,000 portfolio using S&P 500 Index (SPX) options which reflects a market outlook that is (1) bullish, (2) bearish, and (3) neutral. Next, design a strategy using Dow Jones (DJX) options for gaining exposure to large cap stocks which assumes a (1) bullish, (2) bearish, and (3) neutral market view.
6. What strategy from question 5 do you believe is appropriate for today's market environment?

SUGGESTED ANALYSIS FOR AMERICAN NATIONAL PENSION TRUST CASE STUDY

Objectives of the Case Analysis

This case provides you with a framework for exploring portfolio risk management, and the opportunity for developing option-based strategies that can be used to control portfolio risk. It is assumed that you have read prior chapters in this book, with special attention given to Chapter 4, Option Strategies: Analysis and Selection, and Chapter 6,

Strategies for Institutional Investors. The case should enhance your understanding of options by allowing you to apply the concepts and ideas presented earlier. Your case analysis should focus on broad issues of portfolio strategies and risk management concepts rather than evaluation of detailed numerical data.

Portfolio Risk Management

Risk in a financial security, such as a share of stock, may be defined as the degree of uncertainty about the stock's value at any future date. A speculative stock will exhibit more variability in its price over time than a conservative stock, and for that reason is considered more risky. By quantifying that variability we can estimate an investment's risk. It also should be realized that the total risk or price variability in a stock can be decomposed into two components: firm-specific risk and market risk.

Firm-specific risk is the price uncertainty caused by events unique to the company itself. These include the demand for its product, labor costs, and executive management changes. *Market risk* entails factors that affect the prices of all securities. Events such as interest rate changes, a change in the economic outlook, the federal budget, and the federal deficit are market risk factors. When securities are combined into large portfolios, the firm-specific risk can be diversified away because random events cause some firms to prosper more than expected while others decline more than expected. However, the market risk cannot be altered. A portfolio containing more than 40 or 50 stocks will vary in price much like a broad market index such as the S&P 500 or the Russell 2000. The limit of risk reduction through diversification is the risk level of the market portfolio.

Investment return and risk can be visualized with a distribution of the expected (or past) returns for the investment. For example, Figure 9-1 shows a familiar bell-shaped histogram representing a normal distribution of returns, where the *mean*, or *expected return*, $E(r)$, of .10, divides the distribution exactly in half. Expected return is calculated as the sum of the product of each possible return times its probability, p, of occurring. When using past data, the mean of the distribution is calculated as the sum of the returns, r divided by the number of returns, n.

$$E(r) = \sum_{i=1}^{n} r_i p_i \text{ or when using past data, } \bar{r} = \sum_{i=1}^{n} r_i / n$$

A measure of risk in the distribution is the standard deviation of the returns, which is the average deviation from the expected return of each individual return weighted by its probability of occurring. Standard deviation is usually identified by the Greek letter sigma, Σ . The standard deviation of this distribution is .0537, and is calculated as,

$$\sigma = \sqrt{\sum_{i=1}^{n} p_i(\bar{r}-r_i)^2} \text{ or when using past data, } \sigma = \sqrt{\sum_{1}^{n} (\bar{r}-r_i)^2/(n-1)}$$

An investment which has a higher standard deviation has more risk, and one with lower standard deviation would be considered less risky. A virtually riskless investment such as a treasury bill will have a standard deviation near zero.

FIGURE 9-1

Symmetric Probability Distribution of Returns

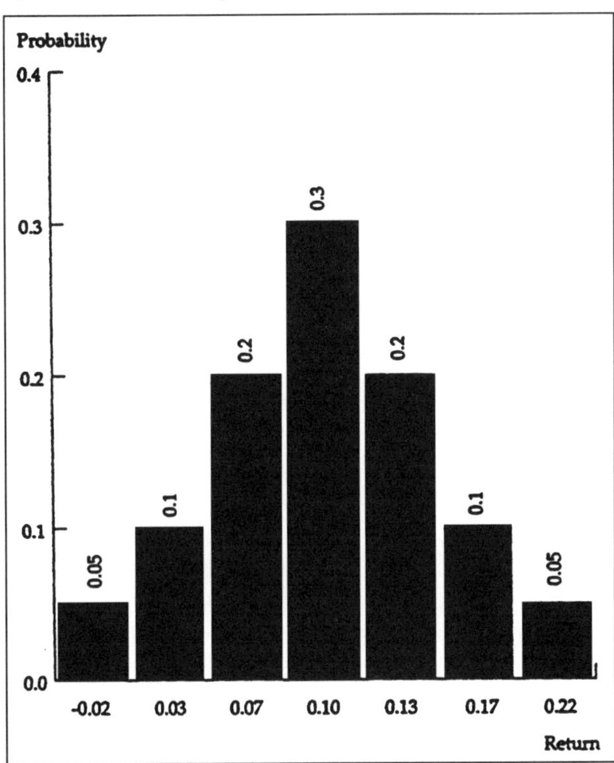

FIGURE 9-2

Return Distributions for a Stock Portfolio Compared to an 80% Stock, 20% Bond Portfolio

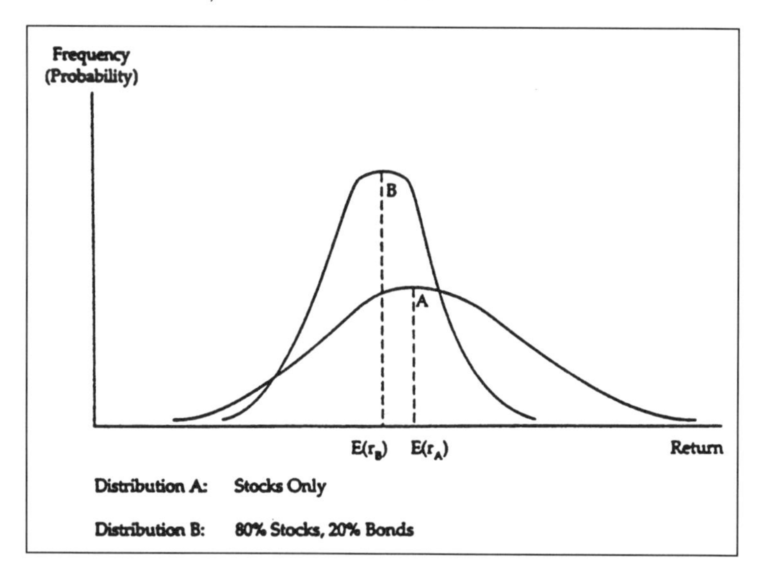

When using past data for large stock portfolios the distributions of historical returns are basically normal, meaning that the right side of the bell-shaped curve is a mirror image of the left side of the distribution. The same is true for bond portfolios, which have much lower standard deviations than stock portfolios, but still have normal distributions.

To understand how return distributions impound portfolio risk, consider what happens when you transfer money from a 100 percent equity position, to 80 percent equity, 20 percent bonds as shown in Figure 9-2. Note that the expected return of the distribution falls and the standard deviation is reduced. The distribution becomes more peaked, indicating that you are more certain of getting the distribution's expected return. At the limit, conversion of the portfolio to 100 percent treasury bills would produce a certain return of the treasury bill yield and would be shown as a vertical line on the expected return distribution.

Based on the concepts developed above, Paige can develop a definition of portfolio risk management to use in her presentation tomorrow, Molding the Expected Return Distribution of a Portfolio to Achieve the Objectives Desired by the Portfolio Manager. Using traditional investment strategies such as switching between stocks and bonds, portfolio managers have limited tools at hand to control the shape of the distribution. Stocks, bonds, and cash return distributions are basically normal, and any expected return distribution from combining these assets will have an approximately normal shape. Only the expected return and standard deviation can be adjusted.

To create distributions that are not normal, and may have more desirable characteristics to investors, Paige should consider *skewed* distributions. "Skewness" is a risk measure for a distribution that has *outliers,* or unusually large or small returns. The skewness for a normal distribution, such as Figure 9-2 equals 0.0, because the returns above and below the mean cancel each other in the calculation. However, when a return distribution has some large returns relative to the mean, and many returns near but slightly below the mean, it will exhibit positive skewness as shown in Figure 9-3A. If the outliers are below the mean, negative skewness will result as shown in Figure 9-3B. The calculation of skewness, σ^3, is similar to standard deviation, except the deviations from the mean are cubed, thus keeping the positive or negative sign intact, showing the direction of the deviation,

$$\sigma^3 = \sum_{i=1}^{n} p_i(\bar{r}-r_i)^3 \text{ or when using past data, } \sigma^3 = \sum_{1}^{n} (\bar{r}-r_i)^3/(n-1)$$

For non-normal distributions, the standard deviation does not fully capture all the risk or desirability of the investment. Most investors prefer distributions which have positive skewness because any surprises that occur will be returns greater than expected, rather than large losses. Distributions with negative skewness are usually not desired, because the large surprises in this situation will be larger negative returns.

It is important to think of risk management not only in the context of risk reduction, but as the opportunity to control risk in order to achieve desired returns. In cases where the portfolio manager is bullish, it may be desirable to inject risk into the portfolio to secure above average returns.

In the past, risk management tools were limited to changing the allocation between stocks and fixed income investments or cash. Risk could be reduced by lowering exposure to equities and could be increased by increasing equity exposure or by selecting more volatile stocks. As Figure 9-2 indicates, the distribution of returns using these techniques remains basically normal, only the standard deviation of the distribution is altered.

Some students in our seminars suggest that the expected return distribution from a futures contract is highly skewed, but it typically is not. Except for the leverage involved, a long position in a futures contract has

FIGURE 9-3A

Positively Skewed Distribution of Returns

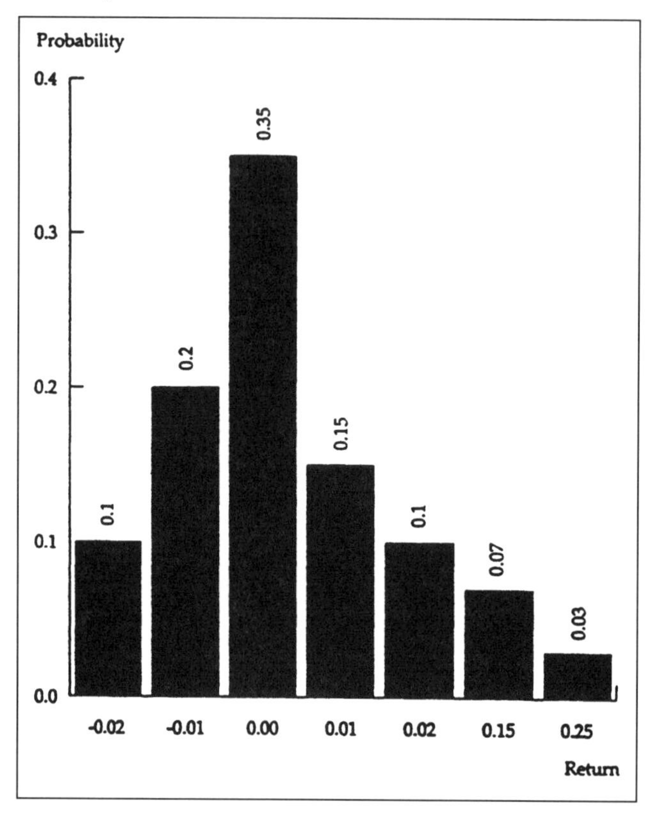

the same payoffs as a long position in the underlying asset. The high degree of leverage that can be used in a futures position produces more dispersion in the distribution of returns, and because the upside is theoretically unlimited while the downside is bounded by zero, some positive skewness may be observed. For example, if the S&P 500 appreciates, S&P 500 futures will appreciate by a relative amount; if the S&P 500 Index declines, the futures value also will fall. However, the leverage available in futures magnifies both their positive and negative returns.

Now we come to the reason options are so important for portfolio risk management strategies. Options produce what is called an *asymmetric payoff*. For example, when you buy a call, if the underlying stock rises, your gain is unlimited, while if the stock falls, your loss is limited to the price you paid for the call. Compare this to buying the stock itself, in which you will profit dollar for dollar as the stock rises, and lose dol-

FIGURE 9-3B

Negatively Skewed Distribution of Returns

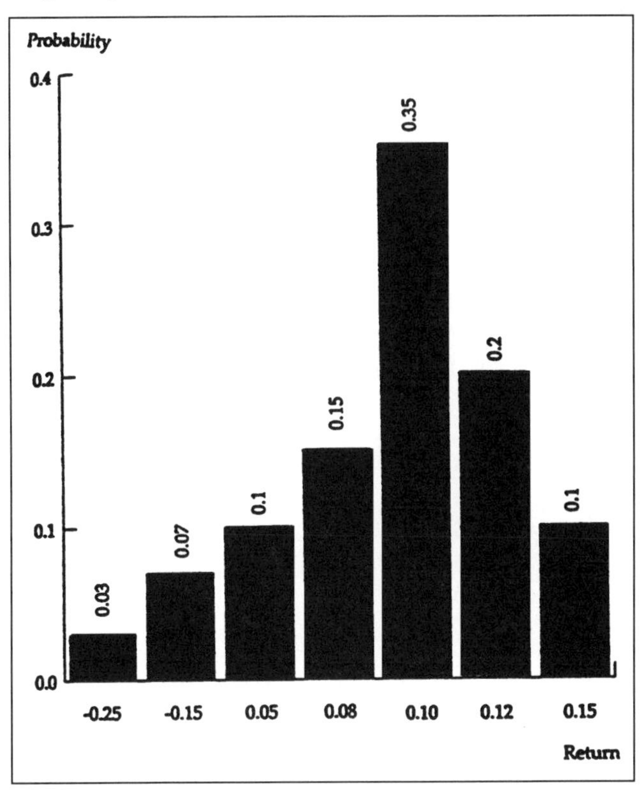

lar for dollar as the stock falls. Compared to stocks or futures, changes in option returns are limited on either the downside or upside, depending on whether you are long or short the option.

The availability of options provides a much richer set of tools for controlling portfolio returns. The unique benefit of options with regard to risk management is that options enable the portfolio manager to construct return distributions that cannot be duplicated with any other securities. Only options, because of their asymmetric payoffs, enable the portfolio manager to create distinctively skewed distributions. Why this is important to investors is illustrated in the examples below.

Using Options in a Risk Management Context

To understand how options can be used in portfolio risk management, consider the payoff diagrams for two popular stock and option strategies: covered call writing and use of protective puts.

Risk Management with Covered Call Writing

The profit/loss calculations for covered call writing (own stock and sell a call), and owning the stock itself are shown in Table 9-4 and the profit/loss lines are graphed in Figure 9-4. Assume that XYZ stock is at $70 and the 70 call, which expires in 70 days, can be sold for $5.25. The

TABLE 9-4

Profit/Loss Table for Covered Call at Option Expiration Stock Bought at $70, Option Strike at $70, Call Sold for $5.25

Stock Price at Expiration	Profit/Loss of Stock	Profit/Loss of Short Call	Profit/Loss of Covered Call Long Stock + Short Call
85	+15	(70 – 85) + 5.25 = –9.75	5.25
80	+10	(70 – 80) + 5.25 = –4.75	5.25
75	+5	(70 – 75) + 5.25 = .25	5.25
70	0	(70 – 70) + 5.25 = 5.25	5.25
65	–5	0 + 5.25 = 5.25	.25
60	–10	0 + 5.25 = 5.25	–4.75
55	–15	0 + 5.25 = 5.25	–9.75

FIGURE 9-4

Covered Call Writing Sell the April 70 Call and Own XYZ Stock

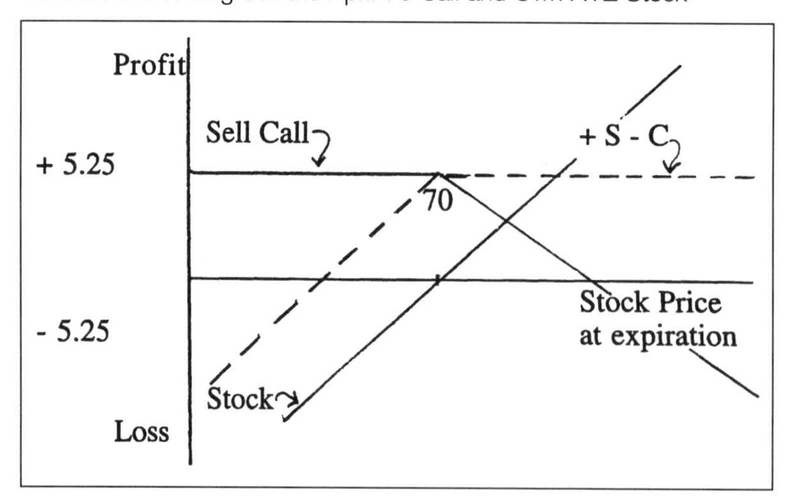

payoff for the stock is a 45 degree line through the current stock price of $70, indicating that the buyer of the stocks gains or loses dollar for dollar as the stock price changes. The payoff at expiration from selling the call is shown by the light solid line. If the stock is $70 or below at expiration, the call seller keeps the entire premium of $5.25. This represents the maximum profit which can be earned from the call. However, if the stock appreciates, the call seller loses dollar for dollar with the price increase. This is the asymmetric payoff provided by an option.

The combined long stock plus short call position produces a payoff indicated by the dashed line. The most the covered writer will earn from selling an at-the-money call is the call premium. If the stock declines, the covered writer will participate in the loss, with some cushion provided by the call premium. The covered call writing strategy appears to be most appropriate in a sideways market. It will produce real losses in a downmarket, and limited gains in a rising market. This is why the strategy underperformed during the bull market of '93 through '97.

Portfolio Return Distributions for Covered Call Writing

Consider a portfolio manager who consistently writes calls against stock in the portfolio. Because the sale of a call caps the portfolio's return at a value equal to the sum of the exercise price and the amount of the call premium received, as shown in Figure 9-4, the majority of returns for the portfolio will be small gains and losses around the mean.

During periods of low volatility, the covered call strategy may produce higher returns than the underlying portfolio. However, when the market appreciates, the covered call writer will not share in large gains because the stock will be called away at the exercise price. Conversely, when the stock declines significantly in price, the portfolio will show a large negative return. Consistently following a covered call strategy through time produces a portfolio return distribution characterized by few large positive returns, many small positive and negative returns near the mean, and a number of larger, negative returns incurred when the market declines. The mean or expected return will be lower than that of the underlying stock portfolio, standard deviation will be less, and skewness will be negative. The distribution will look very similar to the one shown in Figure 9-3B.

To illustrate how the simple strategy of writing covered calls can be used to mold the expected return distribution, consider Figure 9-5. It shows (A) stylized return distributions for a stock portfolio; (B) a covered call portfolio using at-the-money calls; and (C) using 10 percent

FIGURE 9-5

Expected Return Distributions for Various Covered Call Portfolios

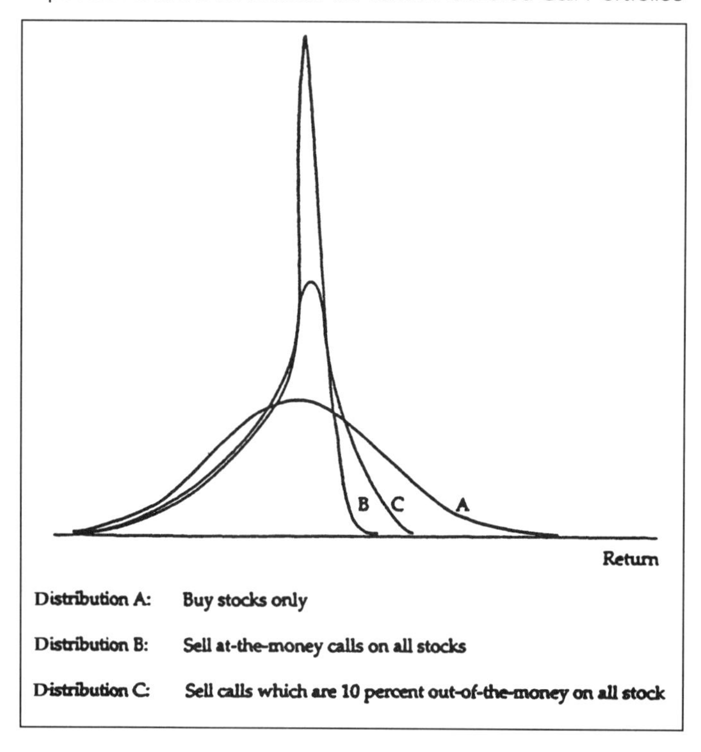

out-of-the-money calls. Note that the degree of undesirable skewness can be altered by varying either the exercise price of the written calls or the value of the portfolio which has options written against it.

Risk Management with Protective Puts—Portfolio Insurance

An alternative strategy to covered call writing is the purchase of protective puts on stocks held in the portfolio. Table 9-5 shows the profit/loss calculations for the protective put strategy (own stock and buy a put), and owning the stock itself. The profit/loss lines are graphed in Figure 9-6. Assume that XYZ stock is at $70 and the 70 put, which expires in 70 days, can be bought for $4.50. As shown earlier, the payoff for the stock is a 45 degree line through the current stock price of $70, indicating that the buyer of the stocks gains or loses dollar for dollar as the stock price changes.

TABLE 9-5

Profit/Loss Table for Protective Put Portfolio at Option Expiration Stock Bought at $70, Put Strike at $70, Put Bought for $4.50

Stock Price at Expiration	Profit/Loss of Stock	Profit/Loss of Long Put	Profit/Loss of Portfolio Long Stock + Long Put
85	+15	0 – 4.50 = –4.50	11.50
80	+10	0 – 4.50 = –4.50	5.50
75	+5	0 – 4.50 = –4.50	–.50
70	0	(70 – 70) – 4.50 = –4.50	–4.50
65	–5	(70 – 65) – 4.50 = .50	–4.50
60	–10	(70 – 60) – 4.50 = 5.50	–4.50
55	–15	(70 – 55) – 4.50 = 10.50	–4.50

FIGURE 9-6

The Protective Put Portfolio Buy the 70 Put and Own XYZ Stock

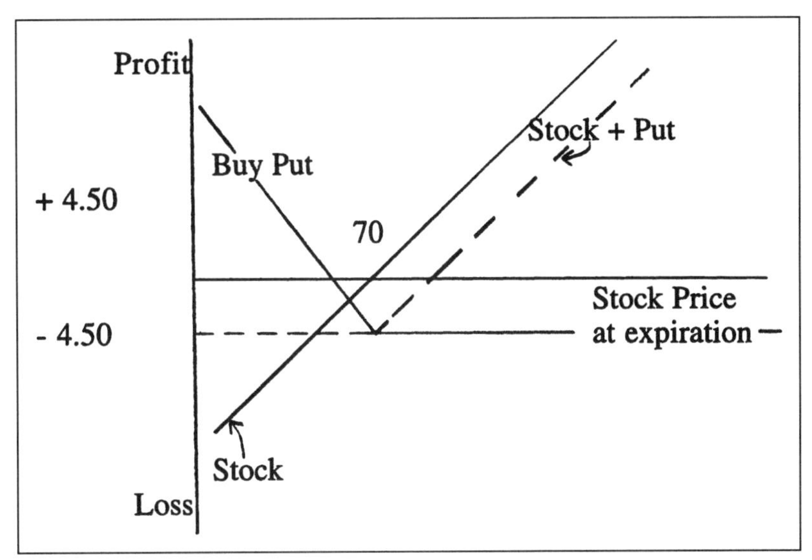

The payoff at expiration from buying the put is shown by the light solid line. If the stock is $70 or above at option expiration, the put seller keeps the entire premium of $4.50, which represents the most the put buyer will lose if the stock goes up. However, if the stock falls in price, the put becomes more valuable, with the payoff limited only by the stock price going to zero. Combining the long put with the long stock position, produces the dashed line shown in the Figure 9-6. When the stock rises in price, the combined position produces a profit equal to the rise in the stock's price less the put premium. If the stock falls, the "stock plus put" portfolio will lose at most the $4.50 premium. For this reason, the purchase of a put against stock owned is called a *protective put* strategy, or *portfolio insurance.*

Portfolio Return Distributions for a Protective Put Strategy

Figure 9-6 can be used to evaluate the impact on portfolio returns from purchasing protective puts. Contrary to covered calls, the protective put strategy provides no cap on the gains which can occur, but truncates the amount of the loss at the difference between the stock's price when the put is bought and the option's exercise price, less the cost of the put. Following this strategy over time, using at-the-money puts will produce a return distribution characterized by a maximum possible loss equal to the average put premium, a large number of small losses and gains caused by put premiums reducing profits in flat markets, and a limited number of large gains earned when stock prices appreciate. This strategy also is termed static portfolio insurance because no change in stock or option positions is required until the options expire. Alternatively, dynamic portfolio insurance is a strategy based on the use of futures contracts, which require frequent trading to replicate the "stock plus put" portfolio payoffs. Explanation of the dynamic insurance strategy is beyond the scope of this case.

Like the covered call strategy, the shape of the protective put return distribution can be controlled by varying the amount or exercise price of the purchased puts. For example, buying puts further out-of-the-money is less costly but reduces the downside protection. Alternatively, protecting only half the portfolio's value by purchasing a reduced amount of puts has a similar effect. Figure 9-7 shows (A) stylized distributions for the stock portfolio only; (B) buying at-the-money puts to cover the entire portfolio; and (C) buying 10 percent out-of-the-money puts on the entire portfolio.

Because the volatility of the portfolio is reduced in a protective put portfolio, the expected return from such a strategy followed continuously through time typically lies below the stock-only portfolio. The lower return results from paying a premium for the insurance protection. The risk as measured by the standard deviation of returns also will be less for the protective put portfolios. Unlike covered call writing, the

FIGURE 9-7

Expected Return Distributions for Various Protective Put Buying Portfolios

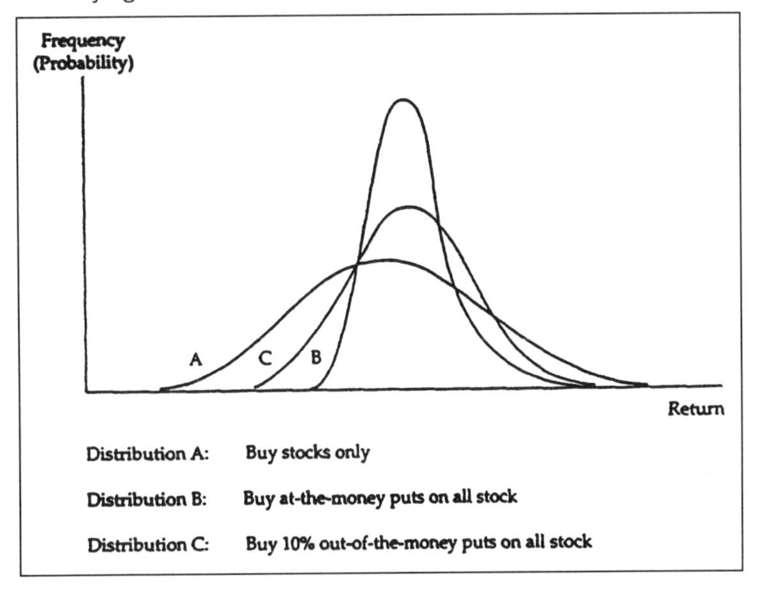

protective put strategy has a return distribution with positive skewness because of the limited losses coupled with potentially unlimited gains. Whether or not individual portfolio managers prefers return distribution A, B, or C in Figure 9-7 depends on their personal trade-offs of defined maximum losses in exchange for potential appreciation.

A protective put portfolio appears most attractive if held when the market suffers a severe decline. For example, if the market index suffers a 15 to 20 percent decline, the put-protected portfolio should decline less than 5 percent. Note in Table 9-1, the returns for the last quarter of 1974 for the S&P 500 Index and the "Insured (Index + Put) Portfolio." The market declined over 20 percent, while the "insured portfolio" lost less than 5 percent. This has a two-fold advantage: (1) The current loss is much less, and (2) when the market rebounds the insured portfolio begins at a much higher level. You don't have to recoup your loss before you can start showing improvement. In sideways or appreciating markets, the strategy will under-perform the market because the put premiums add cost with no tangible benefit.

Strategies using covered calls or protective puts are just two of a broad array of option-based strategies that can be used in portfolio risk management programs. They were chosen as illustrations because of their familiarity to institutional investors, and because of the opposite results which they produce on the skewness of portfolio returns. The main purpose of this part of the presentation is to demonstrate that options are tools for managing risk in portfolios of financial assets. Additional strategies will be developed below, but first, put–call parity is used to explain why equivalent strategies can be created using different kinds of options.

Put–Call Parity and Portfolio Risk Management

To better understand how options can be used in portfolio risk management, it is useful to learn the put–call parity relationship. It is helpful for two reasons: (1) to understand the pricing relationship which must exist between puts and call, and (2) to identify equivalent option strategies using puts or calls. Put–call parity holds precisely for European options on stocks which pay no dividends before option expiration. (European options can only be exercised at option expiration, American options can be exercised at any time.) The put–call parity concept still is appropriate for American options; however, the option prices can no longer be defined as exact values, but within upper and lower limits.

Consider the covered call payoff diagram shown in Figure 9-4. A riskless position can be created by adding a long position in a European put with the identical expiration and strike as a European call. Assuming the stock pays no dividends prior to option expiration, the payoff of the stock-call+put portfolio will be the horizontal dashed line shown in Figure 9-8, at $.75. It is the $.75 difference between the put purchased for $4.50 and the call sold for $5.25. By economic theory, a riskless position should return the riskless rate of return, usually equated to the treasury bill return.

Figure 9-8 shows that the combination of a long stock, short call, and long put, produces a riskless position. Regardless of the stock's price at expiration, the portfolio will be worth the exercise price of the options. This is demonstrated in Figure 9-8. If the stock price rises, the stock's profit will be exactly offset by the call's loss. If the stock price falls, the stock's loss will be offset by appreciation in the put.

A second important point illustrated from Figure 9-8, is that the prices of the put and call are related. If the price of the put and call are "correct," then the investor who creates a riskless position as shown in

FIGURE 9-8

Put–Call Parity Relationship Sell the 70 Call and Buy the 70 Put with Seventy Days to Expiration, Long XYZ Stock

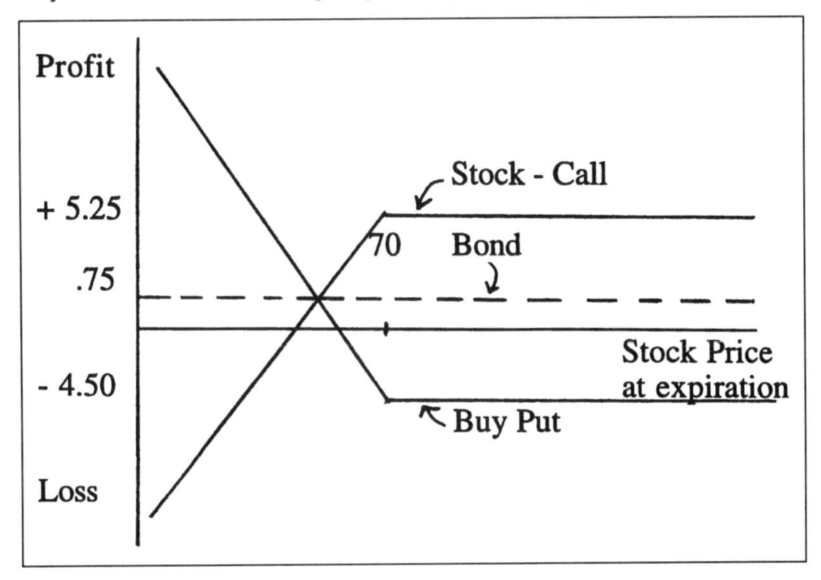

TABLE 9–6

Payoff Diagram for Put–Call Parity European Options, No Dividends[1]

	T_0 Value	Payoff at Option Expiration S < \$70	Payoff at Option Expiration S > \$70
Buy one share XYZ stock	$-S_0$	+S	+S
Buy 1 Put	$-P_0$	(E – S)	0
Sell 1 Call	$+C_0$	0	–(S – E)
Expiration Value		+E	+E

[1]The relationship does not hold exactly if American options are used, or if a dividend payment occurs before expiration. The underlying concept however, remains valid, but the put and call prices can be defined only in terms of upper and lower limits rather than a single value.

Table 9-6, should earn the riskless rate of return. It can be determined if the 75 cent difference between the put and call is appropriate by using the put–call parity equation as shown. There are 70 days until option expiration and the yearly riskless rate is 6 percent. Let E represent the exercise price of the options and e^{-rT} is the symbol for continuous discounting over period T at rate r, thus Ee^{-rT} is the present value of the exercise price. The put–call parity relationship indicates:

$$Ee^{-rT} = \text{S} - \text{C} + \text{P}$$
$$\$70[e^{-(.06/365\text{days})(70\text{ days})}] = \$70 - \$5.25 + \$4.50,$$
$$\$69.20 \cong \$69.25,$$

with the difference attributable to the ⅛ trading increments in option prices. If the call was underpriced, an arbitrageur could create a riskless position by buying the call, selling the put, and shorting the stock. The riskless payoff would provide the exercise price of the option, but the arbitrageur would earn a return greater than the riskless rate. Existence of put–call parity helps keep option prices near their appropriate values.

Developing Equivalent Option Strategies Using Put–Call Parity

By rearranging the put–call parity equation, equivalent strategies using puts or calls can be identified. It will be shown below that (1) buying calls and investing in treasury bills should produce the same payoffs as the protective-put strategy, (2) covered writing is equivalent to selling puts and investing in treasury bills, and (3) a treasury bill can be replicated by holding the stock, selling a call, and buying a put. Using this concept, if the exercise prices are adjusted above and below the current stock price, a *collar* can be created around the range of potential payoffs.

The Call Options and Treasury Bill Strategy

If the put–call parity equation is rearranged to solve for the protective put strategy on the right side, the left side will be equal to a long call plus investment in the riskless bond:

$$Ee^{-rT} + \text{Call} = \text{Stock} + \text{Put}$$

For example, using the data from the case on XYZ stock, the same payoff diagram as shown in Figure 9-9 can be created for the call plus bond portfolio. If the investor buys the 70-day call with an exercise price of 70 for $5.25 and invests $69.25 in treasury bills, the payoff will be the same as for the protective put strategy. If the stock falls below $70, the call is worthless, but the maximum loss incurred is $4.50, the cost of the call, less the interest received from the treasury bill ($-\$5.25 + \$.75 = -\$4.50$) over the holding period of 70 days. If the stock appreciates above $70, the strategy will return the difference between $70 and the stock price (S − $70), plus the $.75 interest on the treasury bill.

FIGURE 9-9

Profit/Loss Diagram for Calls Plus Treasury Bills Strategy
Buy the Seventy-Day 70 Call for \$5.25 Invest \$69.20 in Treasury Bills

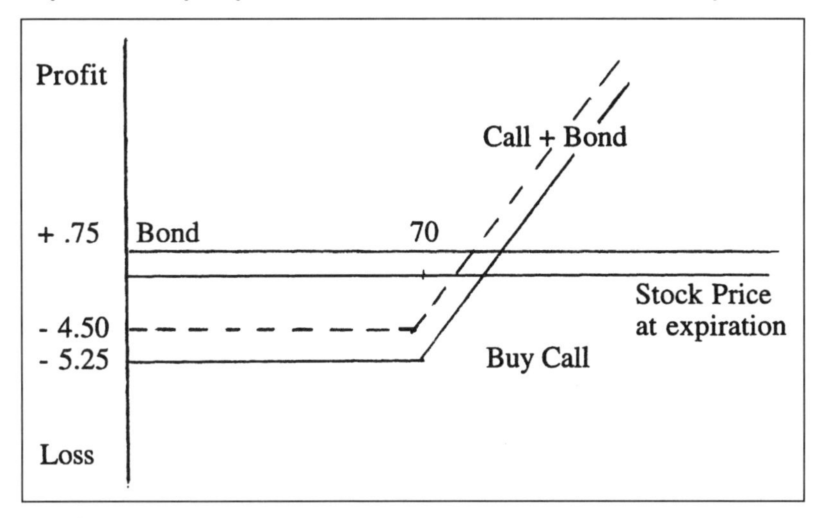

Similar to the protective put strategy, the call option/treasury bill portfolio provides limited downside losses combined with unlimited upside gains. The return distribution will have positive skewness, and should look identical to the return distributions for protective puts shown in Figure 9-7.

Writing Puts and Buying Treasury Bills

If the put–call parity equation is rearranged to express covered call writing (S − C) on the left side of the relationship, it is shown that covered call writing is equivalent to selling puts and investing the discounted exercise price in treasury bills ($Ee^{-rT} - \text{P}$):

$$\text{Stock} - \text{Call} = \text{Put} + Ee^{-rT}$$

Again using the XYZ stock example from the case, the investor sells the 70 put for \$4.50 and combines the proceeds plus other funds equal to \$69.20 in treasury bills. The return distribution from this strategy in a ideal world will exactly match the one shown in Figure 9-5 for the covered call position. Even though the payoff is identical to covered call writing, many institutional investors view the short put strategy differently because when the market declines, it causes the fund to buy stock at prices above then-current market values.

In addition to the equivalence of strategies, the put–call parity relationship is useful to demonstrate that it is possible to combine options with other securities and produce a variety of payoffs. Understanding the put–call parity relationship is essential to grasping the potential which options provide for portfolio risk management.

Alternative Option Strategies

The advantage of options in portfolio risk management is that they enable the manager to mold the portfolio's expected returns to fit almost any market view. Building upon the analysis of the four strategies described above, and strategies described earlier in Chapters 4 and 6, the following option-based plans could be considered by Paige Hanson and the American National Pension Trust as they develop their risk management program.

Short-Term Bearish

If Paige Hanson strongly believes that a short-term correction is at hand, she could buy short-term, at-the-money puts, with a nominal value equal to the underlying portfolio, thus completely insuring the portfolio's value. This protective put strategy would produce the truncated return distribution labeled B in Figure 9-7. If Paige is correct and the market declines, the gain in the value of the puts will offset the losses in the stock portfolio. The greatest loss she can incur is the cost of the puts. If Paige is wrong and the market appreciates, the equity portfolio will participate in the price rise, less the cost of the puts.

The relative "moneyness" and amount of options to buy will depend on Paige's degree of pessimism about the market. Figure 9-10 shows the impact on expected returns achieved by varying the relative exercise price of the puts. If she is extremely bearish, she could consider buying a portfolio multiple of puts, either at-the-money or out-of-the-money, in which case she would profit from the market's decline. Buying a portfolio multiple means buying puts with a total contract value that is greater than the market value of the equity portfolio which is being insured.

It also could be argued that Paige should consider the use of covered call strategies under this market scenario. It should be recommended that American National Pension Trust choose at-the-money calls, with relatively short maturities to get the greatest benefit from the written calls. The two disadvantages of writing calls compared to buying protective puts are that, first, the call premium only cushions the losses if the market falls, thus the portfolio still participates in the falling market. This effect can be reduced if the Trust sells a portfolio multiple of calls to produce more premium income. Second, if Paige is wrong and

FIGURE 9-10

Expected Return Distributions for Protective Put Strategy When Different Proportions of the Stock Is Covered[1]

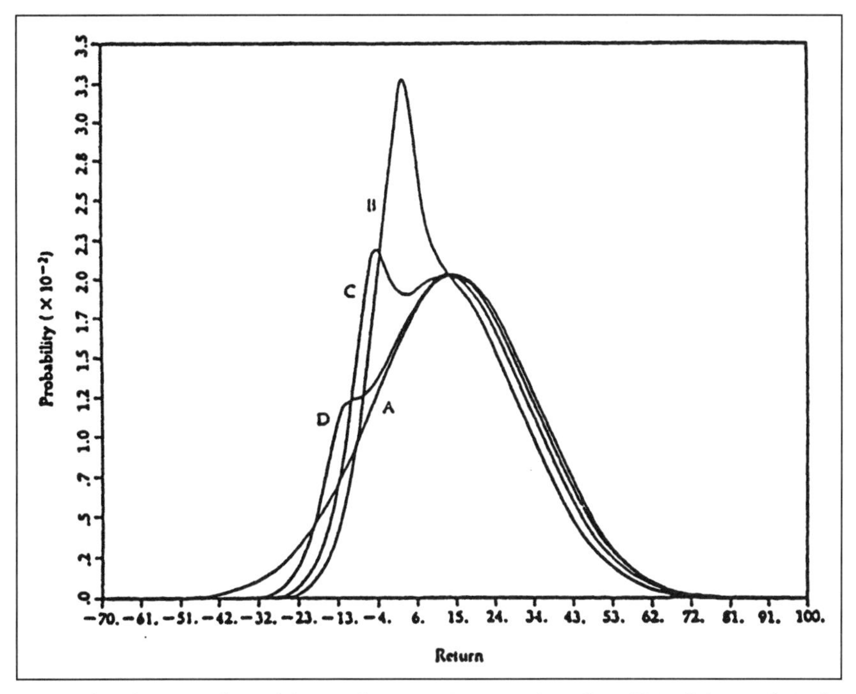

Return distributions of portfolios with put options purchased on 50% of the stock in the portfolio. Distribution (A) is the stock portfolio only, (B) uses puts 10% in-the-money, (C) uses at-the-money puts, and (D) uses 10% out-of-the-money puts.

[1] Adapted from Richard Bookstaber, *Option Strategies for Institutional Investment Management*. Reading, MA: Addison-Wesley, 1983.

the market goes up, the calls will have to be repurchased at a loss. Figure 9-11 shows hypothetical return distributions using different degrees of "moneyness" on the calls.

A contrast between the use of options in risk management and the use of index futures products is appropriate at this point. Under a bearish market view, the appropriate futures strategy is to sell index futures, reducing the equity portfolio's exposure to market changes. The advantage of selling futures rather than the stock itself, is that (1) transaction costs are much lower and (2) no tax consequences are incurred as would happen from selling the stock. If the market falls, the short futures position appreciates, offsetting the loss in the equity portfolio, and the short futures position can be covered whenever the portfolio manager turns bullish.

However, if the portfolio manager is wrong and the market rises, the short futures become a liability, incurring a dollar for dollar loss with the rise in the market. For example, if Paige sells S&P 500 index futures equal in nominal value to 100 percent of the portfolio, she has effectively converted the portfolio to cash. As described earlier, futures do not provide an asymmetric payoff like options, and they will produce different effects on the portfolio's expected returns.

Long-Term Bearish

Under this market scenario, it should be recommended that Paige again choose a protective put strategy and extend the maturity on the purchased puts, possibly choosing puts with exercise prices out-of-the-money. The LEAPS products listed in the case are appropriate vehicles

FIGURE 9-11

Expected Return Distributions for Protective Put Strategy When Different Proportions of the Stock Is Covered[1]

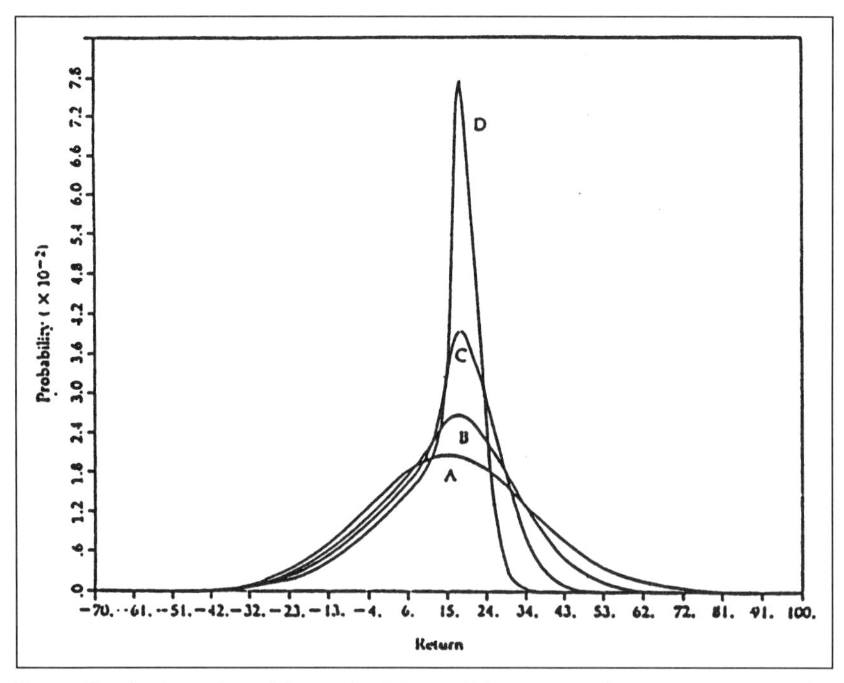

Return distributions of portfolios with 10% out-of-the-money call options written on: (A) 0% of the portfolio (stock only), (B) 25%, (C) 50%, and (D) 75% of the stock portfolio.

[1] Adapted from Richard Bookstaber, *Option Strategies for Institutional Investment Management*. Reading, MA: Addison-Wesley, 1983.

to hedge this market view, because they can be purchased with maturities of up to three years.

Neutral, but Uncertain

An uncertain but negative view about the market calls for strategies that provide some downside protection for the portfolio, but at a reasonable cost. One appropriate strategy is the purchase of protective puts, but choosing those which are out-of-the-money to reduce the costs. Alternatively, the fund might consider the sale of out-of-the-money calls, covered call writing, which will generate current income to offset losses in a market decline.

Another strategy mentioned in the case is the combination of a put and call to produce a *collar* or *fence* around the expected returns. To manage risk under this market expectation, the Trust would buy a short term, out-of-the-money protective put and sell an equivalent call. The call premium received would offset most or all of the put's cost, enabling the strategy to be effected with little up-front cost. Figure 9-12 shows the payoff diagram for a collar strategy using XYZ stock from the previous example. It is assumed that XYZ stock is bought at $70, the 75 call is sold for $3.25 and the 65 put is bought for $2.625. The maximum

FIGURE 9-12

Fence or Collar Strategy Sell the 75 Call and Buy the 65 Put, with the Same Expiration and Hold Long XYZ Stock

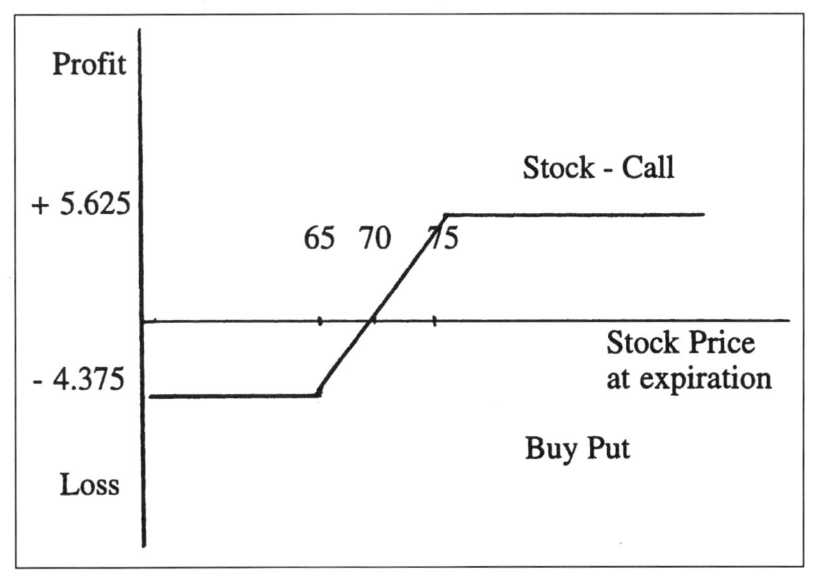

gain for the position is capped at $5.625 [*i.e.*, ($75 − $70) + ($3.25 − $2.625)], which occurs if the stock appreciates above $75. The maximum loss is $4.375, which will happen if the stock falls below $65. The expected return from this portfolio will be truncated between the exercise prices of the put and call.

Institutional investors using this strategy often trade the put and call position with investment banks who take the opposition side of the transactions. The position is priced on a notional value of 100, representing the current value of the index. To create a "costless collar" the strike prices are set so that the premiums of the put and call are equal. Recall that put–call parity shows that the premium of the call will be greater than the equivalent put, thus the call's strike price can be further out-of-the-money than the put. If a 5 percent out-of-the-money put is desired, a typical contract might be structured as a "95 put with a call at 108."

The advantage of entering into a direct contract with the investment bank is that the options can be structured to exactly match the portfolio being hedged. If the portfolio consists of twenty-three stocks, a put and call on the portfolio can be created that mirrors precisely the client's portfolio. The disadvantages of this over-the-counter trade are two: (1) The portfolio manager is basically locked into the position and can only unwind it prior to expiration at a price offered by the banker; and (2) counterparty risk, which means the uncertainty that the bank will fulfill its obligation at expiration. Obviously, these risks are avoided when listed options are used, because the prices are market-determined, and the Option Clearing Corporation (OCC) is the entity behind each contract.

Finally, it should be noted that the collar strategy contains a long put and short call option position, like those used in put–call parity. The difference is that the strike prices are out-of-the-money rather than exactly at-the-money. With this in mind, it is possible to infer the shape of the expected return distribution from the collar. The written call brings in the positive tail of the returns toward the center of the distribution, and the purchased puts truncates the negative tail of returns. Figure 9-13 shows stylized distributions of returns when differing amounts of the stock portfolio are covered with the puts and calls. The limiting case for this strategy, when the strike prices of both options selected are equal to the stock's price, is a vertical line at the riskless rate of return, because the payoff is certain (see Figure 9-8).

Neutral Market View

Option strategies appropriate for the Pension Trust when they are entirely neutral about the market include those which generate current income, but may produce losses if the market moves strongly in either direction.

FIGURE 9-13

Expected Return Distributions from a Fence or Collar When Different Proportions of the Stock Is Covered[1]

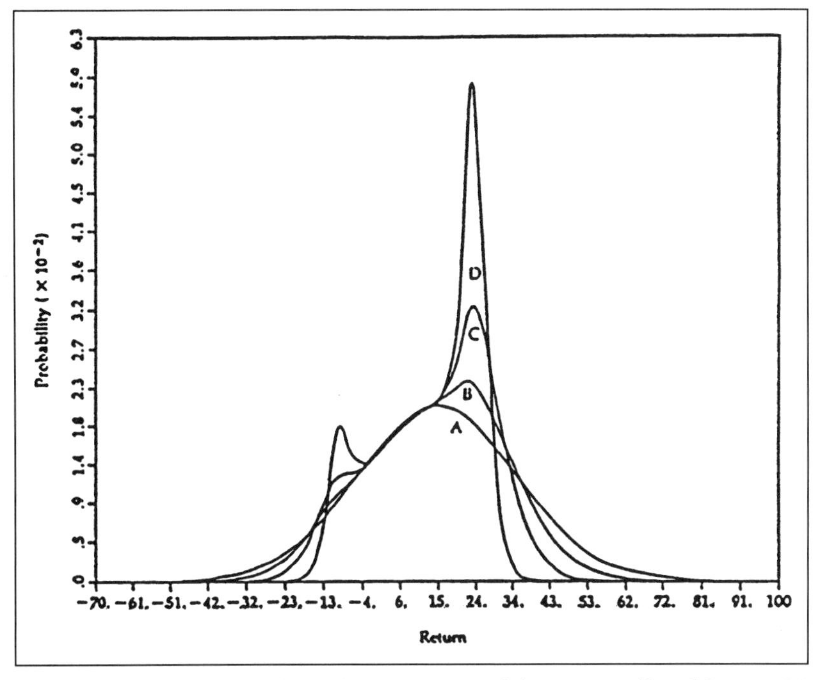

Return distributions of portfolios short a 20% out-of-the-money call, and long a 10% out-of-the-money put. The options cover (A) 0%, (B) 25%, (C) 50%, and (D) 75% of the stock portfolio's value.

[1] Adapted from Richard Bookstaber, *Option Strategies for Institutional Investment Management*. Reading, MA: Addison-Wesley, 1983.

The sale of short-term straddles, *i.e.*, the sale of a put and call with the same strike and maturity, produces this type of outcome. If the market moves strongly in either direction the short put or call becomes a liability and reduces the profits from the strategy. If the market stays flat, most of the proceeds from the option premium are retained by the Pension Trust. The risk of this strategy is that large losses can occur if the market moves a large amount in either direction. The return distribution for this strategy would be relatively peaked, with long tails in either direction.

Another strategy for a neutral market view is writing puts combined with the purchase of treasury bills (*writing escrowed puts*). Because it is equivalent to writing covered calls, the expected return distribution for this strategy can be inferred from portfolio B in Figure 9-5. If the market stays flat or rises, the put premium is retained as current income; however, if the market falls, the fund must be in a position to purchase the stock at the exercise price. This can be done by liquidating sufficient treasury bills. Because writing escrowed puts produces the same payoff as covered call writing, it can be surmised that writing calls against the current equity portfolio is another appropriate strategy for a neutral market view.

Short-Term Bullish

Once the market outlook turns from bearish to bullish, it becomes appropriate to consider option strategies using long index calls. For example, Paige Hanson could increase her exposure to equities by using the call options and treasury bills strategy, in addition to the exposure provided by the underlying equity portfolio. Paige could use a portion of her cash balances to buy short-term index calls and keep the remainder in cash. The moneyness and amount of calls to buy would be a function of her bullish sentiment. If she is highly optimistic, she could buy out-of-the-money calls equivalent to the amount of equity exposure she wanted, or she could buy a multiple of the desired exposure using at-the-money calls. This strategy should be viewed only as a way to increase Paige's equity exposure because of her bullish view. It does not alter the returns from the underlying equity portfolio which Paige currently incurs.

If Paige is correct and the market appreciates, the index call position will increase in value, similar to a position in the underlying stocks. However, if the market stays neutral or declines, Paige will lose no more than the cost of the calls, less the interest earned on the treasury bills. Because this strategy is equivalent to the purchase of protective puts, the return distributions for this strategy are identical to those shown in Figure 9-7.

Options also allow portfolio managers to gain exposure to different markets, such as small-cap stocks, more efficiently than direct investment in these areas. For example, if Paige was bullish about small-cap or high-tech stocks she could implement the "calls and treasury bills" strategy using Russell 2000 options for small-cap stock exposure, or the NASDAQ-100 for high tech exposure. In either case, the downside risk is limited to the cost of the calls less the treasury bill return, while the upside gain is unrestricted.

Long-Term Bullish

An extended-term bullish position can be implemented with the "calls plus treasury bills" strategy using LEAPS calls listed in Table 9-3. The portfolio has exposure to the market from the position in the long calls and intermediate term losses are dampened because of the asymmetric payoff of the calls.

Another strategy which pension fund managers are using in this scenario is the writing of at-the-money or slightly out-of-the-money long-term puts while holding the underlying equity portfolio and cash balances. Those using this strategy rationalize their position because the long-term bias of the market is upward. Put premiums received today add to the current income of the portfolio. If the market retreats, cash balances are used to cover the puts when they expire, and new puts are written at lower strike prices. This is equivalent to buying stock at prices below those currently in the market.

An Evaluation of Listed Option Contracts

The case study includes a list of index option contracts traded on the CBOE. When considered along with the index option contracts traded on other exchanges in the United States and in other markets around the world, it is apparent that listed option products probably can be found to meet the risk management needs of most institutional investors. Some of the main considerations when selecting option contracts for risk management strategies include the option style (American or European), maturity, and index composition. In addition, the portfolio manager should be aware of the differences between individual equity and index options.

American versus European Options

According to information given in the case, only the S&P 100 options are American-style, all others are European. When insuring a portfolio, which style option to use depends primarily on the fund manager's objectives. A European option will be cheaper than an American option on the same underlying security with identical attributes (*e.g.* time to expiration and strike). This happens because there is a cost associated with the privilege of early exercise afforded by the American option. If the objective is to insure a portfolio over some defined time horizon, say one year, and the ability to exercise the option has no value to the fund manager, then it is more cost effective to purchase European rather than American index options. Thus the S&P 500, DJIA, Russell 2000, or NASDAQ-100 options could be considered for "protective put," or "long calls and treasury bills" strategies which involve the purchase of options. If the fund manager is using strategies involving the sale of options, such as escrowed puts or covered calls, and they will not be unhappy if the option is exercised early, then the higher premium income from American options, like the S&P 100, should be considered.

Option Maturity

Standard options such as the OEX, SPX, and DJX are available with maturities up to nine months. To implement option strategies of relatively short duration these contracts should be considered. However, many option strategists select an option with expiration beyond the period they anticipate holding the contract, and trade out of the position at least one month before option expiration. They believe that the market can become volatile for options as expiration approaches, and they wish to avoid trading in a volatile market.

For many risk management strategies, institutional investors often prefer options with maturities from one to three years. Consider a pension fund manager who wants to insure the value of an equity portfolio two years hence. The most effective way to do this is to use puts with a maturity at least as long as the insurance period desired. Once the stock and option positions are established, the manager knows that no further action will be required until the end of the period. Realize that this strategy will, in virtually all situations, produce a different terminal portfolio value than rolling over three-month options every quarter during the next two years. This occurs because realized profits or losses on the options every quarter will not match the profit or loss which the two-year option produces at the end of the insurance period. This phenomenon is called *path dependence*. It is impossible to know at the beginning of the period which strategy will be most costly; all that can be said is that the cost from rolling over three-month options over two years will differ from the cost of using an option with two years to expiration.

Index Composition

Besides the consideration of style and maturity, fund managers should analyze how their particular equity portfolio tracks the portfolio underlying the index option. If the objective is to hedge the portfolio using index puts, the options selected should be those on the underlying index which has returns most closely correlated to the manager's portfolio.

The indexes underlying the index options shown in the case provide a range of alternatives for different types of equity portfolios. The broadly based S&P 500 should produce returns similar to many large equity portfolios held by institutional investors. Also, S&P 500 futures contracts are traded, which facilitates more sophisticated hedging strategies using options, futures, and the underlying equity portfolio. The S&P 100 is a subset of the S&P 500 selected from a broad range of industries. The Russell 2000 is a widely recognized index of small-cap stocks that can be used for risk management of small-cap equity positions. The NASDAQ-100 provides exposure to a different equity market than provided by the other indexes. The DJX provides exposure to thirty of the primary companies in America. Futures contracts also are traded on the DJIA , which facilitates more complicated hedging strategies.

Individual versus Index Options

Individual equity options differ in several ways from index options. Although most institutional investors favor the use of index options, it is useful to be aware of the differences in these products.

Differences in Payoffs. Consider a portfolio of thirty stocks. Strategy one is to buy index puts with a nominal value equal to the value of the portfolio. Strategy two is to buy individual puts on each of the stocks in the portfolio. Payoffs at option maturity will differ between the two strategies depending upon the end-of-period prices of the individual securities in the portfolio. *A priori*, it is not possible to say one strategy is better than the other. However, most portfolio managers use index puts because the prices paid for the index puts is proportionally less than the sum of the individual puts because of the differences in risk as described below.

Differences in Premiums Caused by Differences in Risk. The costs of index puts or calls will be lower than buying an equivalent portfolio of individual options. To understand why, recall that the case indicated total risk (price volatility) can be segmented into two sources. One source is general market factors that affect the prices of all stocks; this is called market risk. The other component is volatility caused by factors unique to each firm, called firm-specific risk. Between 40 and 60 percent of the price changes in the typical security is systematic, related to changes in the market; the remainder can be attributed to unsystematic risk.

The combination of systematic and unsystematic risk represents total risk of the security. As securities are combined into portfolios the firm-specific risk of each security tends to offset the firm-specific volatility in other stocks, and total risk in the portfolio declines as portfolio size increases. This is just a complex way to explain the common sense rule that investors should diversify their holdings.

The act of diversification quickly reduces the firm-specific risk in an equity portfolio toward the limiting value of market risk. Because the market portfolio is fully diversified, it contains only market risk. Considering the two risk components, the differential in pricing between individual and index options can be explained. Recall from Chapters 2 and 3 that the price of an option is directly affected by the volatility of its underlying security. The higher the volatility, the higher the option's price. The volatility estimate used to price options on individual securities reflects the security's total risk, including the market and firm-specific components.

The volatility estimate used to price options on an index, say the S&P 500, represents the total risk of the index, but it contains only market risk because the firm-specific risk in the market portfolio has been removed by diversification. The lower relative volatility estimate means that it will cost less to purchase one S&P 500 index put than individual puts on 500 securities. If the objective is to insure the portfolio, not each security, the purchase of index puts is the most cost effective strategy. Buying individual equity options "overinsures" the portfolio because the total risk of each stock is protected.

Differences in Settlement. The settlement procedure differs between individual and index options and may influence which options the portfolio manager selects. Index options are settled in cash while individual options are exercised by delivery of the underlying stock. Obviously either index or individual options can be traded before expiration on the floor of the options exchange. Also, portfolio insurance strategies involving the purchase of puts or calls are not subject to unexpected exercise because the purchaser determines when the option is exercised. However, strategies involving the sale of options, such as covered writing or escrowed puts, expose the writer to potentially unexpected exercise of the call and sale of the underlying security with accompanying costs and tax consequences.

Trading, Settlement, and Expiration. An additional consideration when using index options is the difference between the last trading day, the settlement day, and option expiration. As the case indicates, most index options cease trading on the day before settlement value is determined. For example, assuming no holidays when markets are closed, trading will cease on Thursday of expiration week. The settlement price to be used in the cash-settlement calculation is determined based on opening prices of stocks in the index on Friday morning. The options will expire on Saturday and the settlement price of the index will be calculated using Friday morning opening prices of the underlying index's stocks. Some risk exists because the opening prices may differ from Thursday's closing values. For options very near-the-money, it is possible to be out-of-the-money Thursday when trading ceases and in-the-money Friday at settlement time.

Evaluating the Data in Tables 9-1 and 9-2

Tables 9-1 and 9-2 provide insight from actual data about the value of portfolio risk management strategies. It should be emphasized that the put and call premiums were determined by the Black-Scholes Option Pricing Model, and do not necessarily reflect the option premiums which would be realized in actual trading. These results are influenced by the option pricing model, the input variables used to calculate the option premiums, the data period, and the assumption that transactions would have occurred at model prices. For example, if volatility estimates used to price the options were systematically lower than those used in actual trading, severe option underpricing would occur, and the insured strategy will appear better than it really could have been. With these caveats in mind, it is useful to compare the returns over time for the two portfolios.

Using Table 9-1, it is possible to compare the quarterly performance of the unhedged S&P 500 Index, and the S&P 500 Index insured with index puts. Note that the insured portfolio shows a smaller gain when the market rises, and a much smaller loss when the market declines dramatically. For example, in the fourth quarter of 1987, the insured S&P 500 Index portfolio suffered a slight loss of −2.82 percent, while the unhedged S&P 500 Index declined by −22.52 percent. For the year of 1987, the insured portfolio earned 20.63 percent compared to a gain of 5.22 percent for the Index. For typical quarters when the Index moves slightly in either direction, the insured portfolio will underperform the index portfolio because of the cost of the put.

The value of using the protective put to avoid a loss early in a year can be illustrated using 1982 data. Note that the Index declined −7.32 percent in the first quarter but the insured index declined only −.65 percent. For the year, the Index earned 22.50 percent, due primarily to the large gain in the fourth quarter. However, the insured Index earned 23.56 percent for the year. Because the yearly returns represent the compound returns over four quarters, by avoiding the large loss in the first quarter, the insured Index outperformed the unhedged Index over the year.

The data in Table 9-1 can be used to produce a histogram of quarterly returns for the S&P 500 Index and the insured Index over the 1971 to 1998 period shown in Figure 9-14. The return on the axis represents the upper value of a "5 percent" interval (*e.g.*, −4.99 percent to 0 percent is shown in the "0.00" return interval). Note that over 40 percent of the quarterly returns for the insured portfolio lie within the −4.99 percent to 0 percent interval, while 23 percent of the unhedged index returns are in this interval. Note also that the insured portfolio incurred no loss greater than −5 percent, and only one gain in the 25–30 percent category. Both distributions have positive skewness, indicated by the larger tail of positive returns. Figure 9-14 illustrates the essence of portfolio risk management, the molding of the index returns to a new shape which may be preferred by management.

Table 9-2 can be used to compare the performance of the Index and the insured portfolio Index over any combination of years between 1971 and 1998. For example, considering the total data period, 1971 to July 1998, the S&P 500 Index earned a cumulative return of more than 3,322 percent. The hedged Index earned only 1,884 percent. The difference reflects the cost of insurance over the total period. Referring back to Table 9-1 and summary statistics shown for the entire data period, it is seen that average yearly return for the Index over the entire data period is 13.71 percent while the hedged portfolio's average return is 11.48 percent. However, risk as measured by standard deviation for the insured portfolio is much less.

Which strategy the portfolio manager prefers is determined by his or her degree of risk aversion. A risk-averse portfolio manager may prefer the insured portfolio compared to the Index itself. This is an especially interesting comparison, when the insured portfolio is compared to an equity/fixed income portfolio allocation which might have equivalent risk. Alternatively, a more volatile portfolio of stocks might be considered because of the insurance provided by the protective puts.

Another interesting comparison of the data in Table 9-2 is to examine the "1971" column, comparing the cumulative year-by-year returns of the Index to the insured portfolio. Except for the years 1971 and 1972, note that the insured portfolio's cumulative return is higher than the

FIGURE 9-14

Histogram of Quarterly Returns: S&P 500 Index and Insured Portfolio

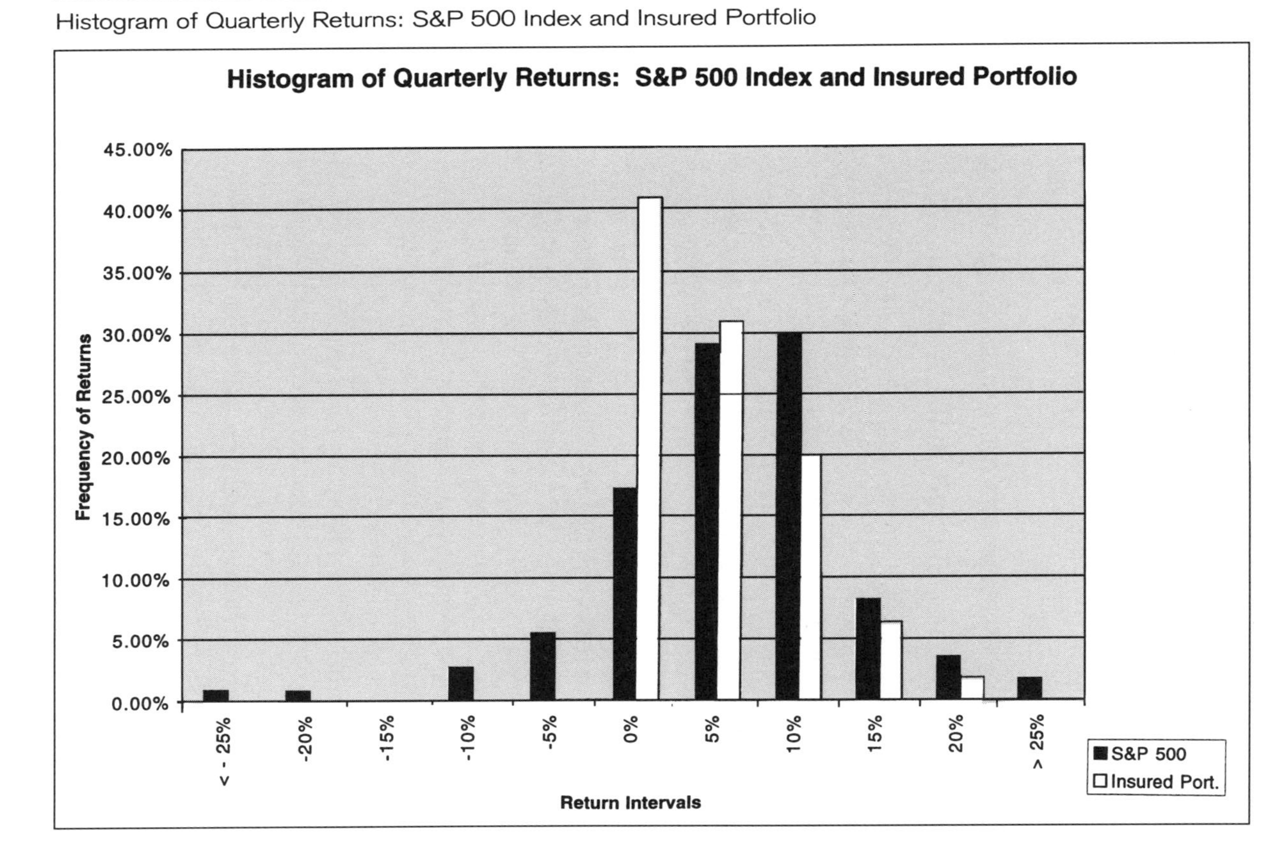

Index's until 1992 (1,044 percent vs. 1,060 percent). An S&P 500 Index portfolio manager who followed a strategy of buying protective puts starting in 1971 would have earned a higher cumulative return until 1992 and incurred less risk. Why? The main reason is that the hedged strategy avoided the large loss in 1974. The benefit of avoiding a large decline in the portfolio was described earlier. Note that if the comparison is begun at any year after 1974, the Index consistently earns a higher cumulative return than the hedged portfolio.

CURRENT RISK MANAGEMENT STRATEGIES FOR PAIGE HANSON

Factors in the Analysis

Paige Hanson is interested in developing a risk management strategy to fit her view of the market and which is appropriate for the current status of the Pension Trust. Factors which should be considered include:

1. Paige is concerned about a market correction from recent highs, but wants to participate in any market advance.
2. The Pension Trust is overfunded and management would like to prevent losses in their portfolio that would increase the cost of providing retirement benefits to their employees.
3. Interest rates seem to be holding steady, and the attitude is that the Fed will not raise rates in the near-term because of the problems in Asia. However, there is increasing evidence of inflation in certain sectors of the U.S. economy, and the balance of trade deficit has been growing. Any surprises regarding interest rates probably would be a rise in rates. This would be bearish for stocks.
4. The Pension Trust is sitting on a relatively large cash balance. Many other institutional investors also have a large amount of cash, which they need to invest. Sitting on cash reduces the portfolio's returns, especially if the market appreciates. If other institutions finally become convinced that the market is not going to fall, their rush to buy could cause a quick rise in the market.
5. There is interest on the part of some members of the strategy committee to gain exposure to small-cap stocks.
6. Evaluation of worldwide economic information suggests a heightened degree of concern. Could the "Asian Contagion" create a panic in financial markets? Will Japan solve the problems

in its banking system or will they continue on the downward spiral begun in the early 1990s? Will China or another major country in Asia devalue its currency and cause instability in foreign exchanges, which spreads to financial markets?

Recommendations

An appropriate strategy for the equity portfolio that the Pension Trust currently holds and which conforms to Paige Hanson's market view of "neutral to negative but uncertain," is to purchase out-of-the-money S&P 500 Index puts with a maturity of six months to a year. The out-of-the-money puts would be cheaper than at-the-money ones, yet provide protection if the market suffers a major pull-back. She also could consider creating a collar by selling out-of-the-money calls on the S&P 500 to help pay for the puts, but this would reduce some of the participation in a market advance. Attention also should be paid to the volatility impounded in option premiums. If volatility is low, options premiums will be cheap, which will favor the purchase of puts but reduce the benefit of selling calls. Paige might want to forgo the collar strategy in favor of just buying puts. If volatility is high, the collar might be worthwhile, because the puts and calls will be expensive, and the call premium will significantly reduce the cost of the insurance.

Data in Table 9-3 can be used to identify the premiums of specific puts or calls and the payoff diagrams using these options can be calculated. Paige can evaluate the differences in cost between using short-term or longer maturity contracts.

With respect to current cash balances in the Trust, the difficulty of timing the market has been documented by many academic studies and Paige's own experience. Thus she is motivated to invest the cash balances to enable the fund to participant in any market advance. Three option-based strategies would enable her to achieve this objective:

- American National Pension Trust could buy selected small-cap stocks which fit portfolio criteria, and control risk exposure by purchasing LEAPS on the Russell 2000 Index. If small stocks enter a decline, the put LEAPS will truncate the losses at the put exercise price.
- If option premium volatility is relatively low, the Pension Trust could allocate 5 to 10 percent of the cash to the purchase of at-the-money, six- to twelve-month, S&P 500 Index or Dow Jones

Average calls. The remainder should be retained cash. This provides a similar return to direct investment in a diversified portfolio of larger capitalization stocks.

- American National Pension Trust could buy Russell 2000 Index or NASDAQ-100 Index calls with 5 to 10 percent of the available cash, and again hold the remainder in treasury bills. The calls provide exposure to small-cap stocks, but the risk is limited to the amount of the call premium.

The recommended strategies illustrate that portfolio risk management using options greatly enriches the portfolio manager's ability to design investment programs to fit individual needs and market perspectives. No other security possesses the asymmetric payoff pattern of options, which makes it easy to mold expected returns to the needs of institutional fund managers. The purpose of this case study was to review the principles of options explained earlier in the book, and to help you apply your knowledge to an actual investment situation.

Case Two, the Norwalk Investment Fund, requires you to focus on another dimension of derivative products in pension fund management. For many funds, the two biggest problems are first, gaining an understanding of options and how they can be used as risk management tools, and second, obtaining approval from their trustees to use derivative products as part of their investment strategy. The Norwalk Fund case provides a setting for your analysis of the issues involved with gaining board approval for investment in derivatives.

CASE TWO: NORWALK PENSION FUND

Because the market correction of 1987, Ann Sawyer and her investment management team have been studying the feasibility of using options and futures in portfolio management strategies. They recently reached the decision that their pension plan, Norwalk Pension Fund, should be using derivative products to manage risk and add value to their portfolios. Sawyer has ten years experience in the money management business and has seen significant advancement in the tools available for investment management during her career. She, and several members of her internal management team, believe that derivative products could add value and enable them to manage risk in ways not possible with other financial products. The problem she faces is persuading the Trustee Committee for the Pension Plan that the plan documents should be amended to allow investment in derivative products.

Sawyer has scheduled a meeting for Thursday morning (tomorrow) to discuss a board presentation about derivatives with two trustees, Tom Jackson and Brenda Woods, who were very interested in this topic. Tom is 66 years old, a recently retired executive of the company who had spent his 38-year career with Norwalk in plant operations. He has no formal training in finance or financial markets, but is respected by the Trustees for his "common sense" approach to investment policy. Tom has the reputation of being very conservative in his views about investment strategies, to the extent that he often questions why the Fund allocates so much to investment in "those risky common stocks." Brenda Woods represents a much different viewpoint. She is 40 years old, has an MBA from the University of Chicago, and is an Executive Vice President for a large manufacturing firm headquartered in New York City. Brenda is the youngest board member, and is considered by the other board members as "very, very bright." Other board members often turn to her for explanations about the technical details of presentations that they receive. Ann thinks the meeting with Tom Jackson and Brenda Woods will be helpful to her as she develops her proposal to the full Committee, which meets in one week.

The Norwalk Electronics Company

The Norwalk Pension Fund is a $12.5 billion pension fund established for employees of the Norwalk Electronics Company headquartered in St. Louis. Norwalk Electronics has been in business since 1922, and is considered one of the premier worldwide manufacturers of commercial electrical and electronic products. Sales are split between commercial and industrial products, which accounts for 58 percent of sales and 52 percent of pre-tax income, and the appliance and construction division, which provides 42 percent of sales and 48 percent of income. Primary products include process control systems, industrial motors and drives, fractional horsepower motors, and appliance components.

Since the 1960s, Norwalk has followed a deliberate strategy of worldwide expansion, both in sales and manufacturing. In 1991, foreign sales revenues amounted to 40 percent of total sales, a figure which has been gradually increasing every year. Most of its low technology manufacturing operations have been moved to low-cost facilities in China and India, enabling Norwalk to keep costs down and prices competitive in the global market. Company prospects are excellent, and the company's pension benefits have traditionally been among the most generous in the industry.

The Norwalk Pension Plan

Norwalk's Pension Plan includes both defined benefit and defined contribution programs, with the larger portion being the defined benefit program containing three-fourths of the fund's value, $9.1 billion. The defined benefit fund is the older plan, and it is funded with common stocks and long-term fixed-income investments. Most of the bonds and Guaranteed Insurance Contracts (GICs) in this plan were purchased in the 1980s when interest rates were much higher than today. Sawyer's plan is to use option and futures strategies for the defined benefit component of the plan.

The fund's assets are divided between nine internal managers and eight external management firms. Ann's internal staff manages $7.1 billion of the fund's assets. The internal management team includes three equity managers, who invest primarily in large-capitalization U.S. stocks; one manager who invests in foreign equities; two domestic fixed-income managers; one manager of foreign fixed-income securities; one real estate manager; two managers of private investments; and one manager of direct placements. The managers' experience ranges from twenty-five months to four years, and they all consider themselves "students of the market," who are interested in new ideas and ways to add value to their portfolios.

The other $5.4 billion of pension fund assets are managed by eight external managers, who follow a variety of strategies. About one-third of the assets are passively managed as an index fund, having the objective of replicating returns from the S&P 500 each quarter. Last year, they requested permission from Sawyer to "equitize" cash balances by using S&P 500 futures, thus remaining fully invested at all times, but she has been unable to authorize this strategy because of plan restrictions. The rest of the assets are placed with smaller managers who specialize in what can best be called active management styles. Most use a bottom-up approach and attempt to identify stocks or sectors that they believe are undervalued.

Ann Sawyer reports to the fund's general manager, Bill Cochran, who is responsible for administration of the entire Pension Plan. She views her role as Director of Investment Planning as one of setting overall investment strategy for the Plan, and of being closely involved in the day-to-day investment decisions. Ann follows a strategy of tactical asset allocation, under a base guideline of 60 percent equity/40 percent fixed-income investments. Fund documents as shown below, provide guidelines for different asset classes allowed by the Plan. In addition to the tactical asset allocation decision, Ann determines the allocation of fund assets between internal and external managers for equity and fixed income investments. Ann is familiar with the academic studies about market efficiency, and the apparent inability of many institutional investors to consistently outperform market indexes like the S&P 500. Her attitude is that, while it is difficult to beat the market, the market is not strictly efficient. She believes that good investment decisions can be made which will produce risk-adjusted returns above the long-term market averages.

Sawyer's Study of Derivatives

The market correction in 1987 probably was the catalyst which motivated Norwalk's managers to examine portfolio risk management. If losses, such as those that occurred during the market crash of '87 were to be avoided, it appeared that derivative products like puts, calls, and futures, offered the best means to do it. In 1998, internal managers at the Norwalk Plan began studying how options and futures could be used to manage portfolio risk. They had considered the technique of dynamic portfolio insurance, but the bad press it received relative to the '87 market crash, made them wary of the technique. Ann then explored strategies using listed puts and calls. Particularly appealing were the strategies of covered call writing, protective put buying, the purchase or sale of straddles, and the sale of escrowed puts.

Like most pension fund managers, Ann Sawyer receives numerous investment ideas from money managers. One manager designed a system that purchases futures as a *cash-overlay* strategy—buying futures to gain market exposure for cash held in the fund's cash management accounts. This is an extension of the cash equitization idea noted above. Two other investment managers proposed strategies based on options. One suggested a technique called *cash equivalent alternatives* and the other suggested a risk control strategy called a *fence* or *collar*.

Cash Equivalent Alternatives

A fund manager in California, Hunter Green Investments, specializes in arbitrage using options. The concept is that a riskless position can be created using the underlying asset and its options, or even just the options themselves. Brief periods of mispricing in options may occur around periods of market movements, and these provide opportunities to create a riskless position that will yield a return above the riskless rate. Hunter Green provided two examples, *forward conversions* and *box spreads*, to illustrate cash equivalent strategies.

TABLE 9-7

Forward Conversion Example ABC Equity Options with 32 Days to Expiration

Bought ABD stock	–$282.500
Bought ABD 280 put (T = 32 days)	– 5.875
Sold ABD 280 call (T = 32 days)	10.750
Net Investment	–$277.625
Receive at expiration	+$280.000
Net Return	$ 2.375 for 32 days
Annualized return	
(2.375/277.625) * (365 days/32 days)	11.24%
versus Treasury bill rate of	6.49%

The Forward Conversion. Shown in Table 9-7 is an example of a forward conversion which uses a long position in ABD stock combined with a short call, and long put. He explained that "it is based on the put–call parity equation" (described in the previous case).

The Box Spread. A box spread is another riskless arbitrage strategy which combines a call (bull) spread with a put (bear) spread. The position value at expiration is totally independent of the price of the underlying instrument. By creating the position for less than the expiring value, a profit will be realized at expiration, as shown in Table 9-8.

The riskless arbitrage idea is attractive to Ann because there is virtually no probability of loss and the return would be higher that treasury securities. It is a way to add value to the portfolio for no increase in risk. If the Plan permits, she would like to place $50 million with Hunter Green on an experimental basis. However, approval must be given by the Trustees before this can happen.

The Fence or Collar. Bob Strong of Newark Trust Company, the master custodian for Norwalk Pension Fund, recently suggested to Ann a risk control strategy using options. The strategy is called a *fence* or *collar* and is created with a put and call on the S&P 500 Index. It can be implemented using listed options traded on exchanges such as the CBOE, or using customized option contracts traded in the over-the-counter market. Table 9-9 shows the details of the Newark Trust proposal.

TABLE 9-8

Box Spread Example S&P 500 Index Options with 210 Days to Expiration

Bull Call Spread Portion		
Purchased:	Aug 1150 S&P 500 Call	–$48.250
Sold:	Aug 1160 S&P 500 Call	43.500
	Cost	–$4.750
Bear Put Spread Portion		
Sold:	Aug 1150 S&P 500 Put	$48.500
Purchased:	Aug 1160 S&P 500 Put	– 53.000
	Cost	–$ 4.500
Total Cost		$ 9.250
Value at expiration (1160–1150)		10.000
Net Profit		$ .750

Annualized Return

($.75/$9.25) * (365 days/210 days) = 14.09%
versus Treasury bill yield of 5.10%

TABLE 9-9

Example of a Fence or Collar Using S&P 500 Index Options

Option Seller:	Norwalk Pension Fund
Option Buyer:	Newark Trust
Notional Amount	$10 million to $500 million
Underlying Index	S&P 500 Index
Option Maturity	1 Year
Option Type	European (OTC)
Settlement	Cash

Strike and Premia

	Strike (% of Spot)	Premium (% of Notional)
Put	90	2.05
Put	80	.53
Call	110	1.30
Call	120	.35

FIGURE 9-15

Fence or Collar Strategy Sell the 1 Year 110 Call, Buy the 1 Year 90 Put, Long S&P 500 Index

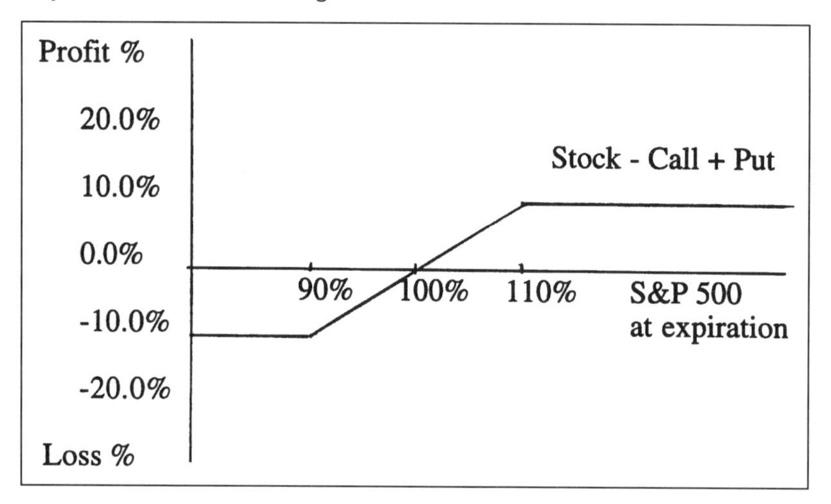

Bob explained that the fence should be considered a risk control position against the underlying equity portfolio. For example, if Norwalk bought the 90 put and sold the 110 call, the fund's cost is only .65 percent of the notional amount. If the S&P 500 is between 90 and 110 percent of its current value at the end of the year, both options expire worthless. However, the long 90 put will protect the fund against losses if the Index falls below 90 percent of its present value. Alternatively, if the S&P 500 appreciates, the 110 call will cap-off or collar the gain at 10 percent above the current level. The profit-loss diagram for this strategy, assuming that the fund holds a portfolio which behaves exactly like the S&P 500, is shown in Figure 9-15.

Ann likes the idea of limiting losses, but is not sure that it is beneficial to cap the gain at 10 percent. She also has questions about doing the trade directly with Newark instead of using listed options to effect the same strategy. Ann told Bob Strong that she would consider his suggestion and present the idea to the board members later in the month.

The Plan Document

The original Pension Plan was written in the late 1940s and has been amended over time to adapt to changes in the investment environment.

As mentioned above, the original pension plan was a defined benefit program, which was popular at the time. The original intent of offering a pension plan was to attract the best workers in the tight labor market after WWII. The defined contribution program was added in the late 1970s and is the only plan available to new employees. Under the defined contribution plan, employees contribute 7.5 percent of salary to the fund, and Norwalk Electronics contributes 8.5 percent. Vesting occurs in two years.

The Plan documents provide specific guidelines relating to investment policy and the type of assets that may be purchased. It states that the fund's objective is to *maximize expected returns subject to the following constraints* designed to control risk:

1. Investment in corporate bonds is limited to those rated AA (S&P) or A^+ (Moody's). No junk, or high-yield, bonds may be purchased. Corporate bonds must be traded on the NYSE Bond Market or widely traded by market makers (more than four).
2. Investment in foreign corporate and government bonds is permitted. These bonds must meet equivalent quality standards as specified for U.S. corporate bonds.
3. Investment in U.S. equities is limited to securities traded on the NYSE, AMEX, or the top 500 in capitalization, traded in the over-the-counter market.
4. Investment in foreign equities is limited to those traded on the country's primary stock exchange, and must be among the top 250 companies in capitalization in the country. (The Plan was amended in 1985 to allow for the purchase of foreign securities.)
5. The Fund may not invest more than 5 percent of its assets in any individual asset.
6. The Fund may invest in equity private investments or new ventures, but such investment is limited to 5 percent of the Fund's assets.
7. The Fund may invest in real estate, but real estate holdings cannot exceed 10 percent of the fund's assets.
8. The Fund will diversify its investments according to the following guidelines:

	Target percent	Minimum percent	Maximum percent
Domestic Equities	55.0	35.0	80.0
Foreign Equities	5.0	0.0	10.0
Domestic Bonds	25.0	10.0	40.0
Foreign Bonds	7.0	0.0	10.0
Real Estate	3.0	1.0	5.0
Private Investments	2.0	2.0	8.0
Cash	3.0	1.0	10.0

"Fund management should invest toward the target percentages, with discretion allowed for their expectations about performance of the different market sectors. Each asset class will be rebalanced at the end of every quarter. Under no instance will the amount invested in any category rise above or fall below the minimum or maximum amounts.

"The performance of both internal and external managers will be evaluated each quarter using rates of return and risk measures (volatility and beta) calculated by an external consultant. These performance measures will be compared to customized benchmarks agreed to by negotiation between the specific fund manager and the Plan director. If performance over a three-year period averages more than 200 basis points below the agreed-upon benchmark, the fund manager will be carefully evaluated before reappointment."

It is specified in the Plan document that any changes to the Plan must be approved by a majority vote of the Fund's trustees. The Plan is silent about options and futures because they were not available in the 1940s when the original Plan document was written. No one has taken the initiative to change the plan document to include or prohibit investment in options or futures.

The Trustees

The Fund's Trustee Committee consists of eight people who meet four times a year, in the month following each calendar quarter. Their main business at each meeting is reviewing quarterly performance of different fund managers, analyzing current market conditions, and evaluating the internal director's expectations about the market over the next 3 to 12 months. Each meeting, Ann makes a presentation about the current asset allocations and the strategy she expects to use over the next quarter.

Ann believes that the Trustees understand their role as one of setting overall investment policy and insuring that the fund is being managed to the greatest possible benefit of the stakeholders. However, she has expressed concern from time to time that the Trustees are overly conservative in their investment policies and the strategies they allow. Concern with their fiduciary responsibilities under the Employee Retirement Income Security Act of 1974 (ERISA) seems to make them resistant to change. She is fortunate that they have not tried to "micromanage" the fund, and have avoided involvement in the day-to-day investment management process.

The eight members have served on the board for at least six years, and one person has been a member for twelve years. The level of sophistication about investments is varied, and most board members are not in financially related occupations. The most knowledgeable member probably is Jack Miller, the 50-year-old CFO of Norwalk. Jack has a Ph.D. in finance from New York University and is quite familiar with portfolio theory and the academic literature about portfolio management and options. The other five board members are executives or managers of companies. Ann frequently uses time during the quarterly meetings to educate them about the latest developments in the field and important issues on the horizon. Over the past two years she has made three or four presentations about options and futures, covering basic terminology, definitions, and strategies. The Board has some curiosity in the opportunities provided by these instruments.

Ann Sawyer's Meeting with Bill Cochran, Tom Jackson, and Brenda Woods

Sawyer: Tom and Brenda, I appreciate your willingness to meet with me and Bill to discuss our interest in trading in options and futures. I'm aware that these instruments are relatively new to some board members and that they will be concerned about how these securities will be used to meet overall objectives of the Fund. I also would like your evaluation of the proposals I sent you about cash equivalent alternatives and the use of fences and collars.

Jackson: I'm glad you asked me to come. I know you are working hard to make money for the fund in a difficult market environment, but I'm concerned about using these securities. They are really risky, why do you want to use them?

Cochran: Tom, we've talked about this several times before. Considered alone, the leverage available in options and futures makes their prices more volatile than the underlying stocks what you refer to as more risky, but we are not going to speculate on the market with these securities. We are talking about using them to control market risk by combining them with positions in other options and the underlying asset.

Jackson: It seems to me that we can do the same thing by changing our tactical asset allocations. If Ann thinks that stocks are going to decline, she can sell stock and go to cash. Why would I want to fool with options if the market is going down?

Sawyer: Tom, I understand your concerns. I'll develop for the Trustees an explanation of portfolio risk management and provide some illustrations about the ways options and futures are used in risk management strategies.

Woods: Ann, I agree that would be useful. I think most members feel pretty comfortable with the concept of portfolio risk management, but they could use some clarification as to why using options differs from the strategy of tactical asset allocation.

Jackson: Another thing, I'm concerned about the legal issues involved. Does ERISA restrict us from trading options or futures? Have you received any opinion from Legal on making these type of investments?

Cochran: Tom, we've also had this discussion before. To my knowledge, ERISA doesn't prohibit investment in options or futures, but it views investment decisions in a portfolio context. It's not the investment tool that's right or wrong, but how the security fits into the entire portfolio.

Sawyer: Before we went very far with our study of these products, we asked our Legal department for an opinion. I'm sure that every pension program is unique with respect to its plan and legal documents, so we cannot make a statement about anyone else, but for us, Legal came back with no substantive issues. The only requirement is that the board must give its approval for us to trade derivative instruments on a routine basis. Our lawyers could find nothing in ERISA which prohibits the use of options. It is not the instrument, but the manner in which any investment is used that can be inappropriate. For example, buying puts while holding an underlying stock portfolio is a means of reducing risk and insuring a portfolio's value. This is a perfectly reasonable and prudent investment strategy. However, buying puts merely to speculate on a market decline is a strategy which may be open to criticism.

Jackson: Ann, it seems to me that we are taking enough of a "prudent" risk by investing in stocks. I think we are open to criticism if we take on more risk with these options and futures just to speculate on the market. For example, I think the board will want to know why you would buy a six-month or one-year option and hold it to expiration?

Sawyer: Tom, the markets in listed options and futures are very liquid so we don't have to hold a position until the option expires. In fact, in most cases we probably would close the position before the expiration month. It appears I will need to describe to the Trustees again how these markets work and the participants in them.

Woods: Ann, another thing which will require more explanation is the cash equivalent strategies. Are you absolutely sure that there is no risk in these positions? I'm not sure that everyone on the board will understand why this is true.

Jackson: Brenda, for once we agree on something. Even if it appears that no risk is incurred, is this an appropriate investment strategy for a pension fund to use? We generally are considered long-term investors, not traders. It sure seems that this strategy will lead to a lot of trading costs which we normally would not incur.

Cochran: Tom, I know this is a different type of strategy than we have used in the past, but we are projecting a positive return above treasury bills after commissions and all other costs of trading. This strategy should be viewed as an alternative to holding cash.

Woods: I'd like to talk about the other strategy presented by Newark Trust called a collar or fence trade. What do you see as the advantages and disadvantages of this transaction?

Sawyer: Brenda, as you probably already know, the fence is one of many strategies which can be used to control the Fund's risk exposure. The advantage of the position is that income from the call offsets most of the cost of the put. Thus we can create a position in which losses are defined to the put's striking price, while the giving up gains above the striking price of the call. An alternative strategy is purchasing only the puts. This will cost more, but we retain all the upside potential.

Woods: You know the Newark Trust proposal indicates that they will be the buyer of the put and call we are selling. I believe this is called an over-the-counter trade. How does this differ from implementing the strategy using options traded in the listed options market such as the CBOE? Is there any difference in risk or cost?

Cochran: The over-the-counter market in options has developed dramatically over the past year. The ability to tailor the options to our particular needs looks attractive, but I have some concerns about entering a position which involves only our Fund and the Trust Company. It sure seems riskier than using market traded options, but I don't have all the information to answer that question.

Sawyer: I agree, Bill. I'll get more information ready on this question prior to the Trustee meeting next week.

Woods: You know that I support your proposal to implement strategies using options and futures for the Fund. While several Trustees share my view, I think that it will be necessary to demonstrate why these strategies should be used. It also will be necessary to overcome some of the biases about the risk of derivatives which have been created by the popular press.

Jackson: I still think this derivatives business is all hocus-pocus. It seems to me that we should just get out of equities or bonds if we think the market is going to decline. Why should we fool around selling calls or buying puts or trading futures to manage our risk? Going to treasury bills when the market looks weak is what I think we should do. You'll have a difficult time making me change my mind.

Preparing the Case Analysis

1. Outline the steps in a strategic plan for implementing the trading of options in a pension fund. In your analysis, be sure to explain the benefits to the Plan of using options and futures strategies. Are options as risky as Tom Jackson believes? Create examples to show how pension funds can use options and futures to reduce risk in their portfolios.
2. Explain the underlying concepts of the cash equivalent alternatives. How is an investment's terminal value locked-in when the positions are first created? Would you support the index fund manager's suggestion to use S&P 500 futures to equitize cash balances?
3. How does the trading of options listed on organized exchanges differ from trading option contracts over-the-counter? What differences in risks exist between the two?

4. Assume it is the end of the Trustee's meeting and they have found Ann Sawyer's arguments persuasive. Before giving final approval, they would like you to draft an amendment to the Plan Document which would permit the use of derivatives. Please prepare this amendment, which should be no longer than two or three paragraphs.

Suggested Analysis for Norwalk Pension Fund Case

Objectives of the Case

In your case analysis, you should develop a strategy that will result in approval from the Pension Fund's Trustee Committee for implementing strategies using options and futures. It may first be helpful to outline a series of strategic steps which can be implemented over time. It should be expected that the approval process will be incremental, with increasing discretion for option and futures strategies being given as the Committee gains confidence in these instruments. Questions raised during the meeting with Jackson and Woods should be answered in that presentation.

The solution for this case builds on the material presented in Chapter 6, Strategies for Institutional Investors, and the suggested solution to the preceding case, American National Pension Trust. Where appropriate, reference will be made to information presented earlier in this chapter. Keep in mind that the information presented here represents only a suggested solution to the case scenario. Different circumstances and personalities may call for a different approach. Study of this material should better prepare you for the process of implementing derivative strategies in your particular situation.

Devising a Strategic Plan

The two fundamental reasons for using options and futures in pension fund strategies are (1) to control risk in ways not possible with other instruments and (2) to add value to the portfolio. Once a decision-maker in the fund has determined that derivative strategies provide these benefits to the pension fund, the following steps may be helpful to gain approval for using these instruments.

Education Of Internal Managers And Trustees

It is necessary to understand that derivative products can be used to control risk in portfolios in ways that cannot be achieved with any other strategy. Managers must become familiar with specific strategies and their appropriate applications. Trustees must understand the concept of risk management and realize a general understanding of strategies and applications.

In the American National Pension Trust case, risk management was defined as controlling the distribution of expected returns from an investment to achieve the objectives of the portfolio manager. Options, because of their asymmetric payoff, can skew the distribution of expected returns either negatively, as shown earlier in Figure 9-5 for a covered call portfolio, or positively, as shown in Figure 9-7, for a protective put strategy. Numerous strategies using options can be devised which will mold the return distribution to the shape desired by the portfolio manager.

When explaining the idea of portfolio risk management, it is imperative that options be considered as part of a total portfolio. Not placing options in a portfolio context is causing Tom Jackson a great deal of difficulty in understanding that derivatives are not speculative gambles, but represent important tools for implementing risk control procedures. Easy examples to use in explaining this concept are the protective put portfolio, long stock + long put, whose profit and loss diagram is shown in the American National Trust Case, Figure 9-6, and the put–call parity relationship, long stock + short call + long put, shown in Figure 9-8.

Once the asymmetric payoff property of an option is understood, it can show how risk management using options differs from asset allocation between stocks, bonds, and cash instruments. As shown in the American National Trust case, Figure 9-2, changing the allocation between stocks and bonds, or stocks and cash, as suggested by Jackson, does not change the normality of the return distribution, it merely reduces its spread.

As an extreme example, consider moving completely out of stocks into cash when the market is expected to fall. The fund manager has locked-in the risk-free rate of return but if the market goes up, the fund will not participate. However, using an option strategy such as the purchase of protective puts gives the portfolio manager the ability to protect against losses if the market should decline, while still participating in the market if it should rise. This is the great benefit of options compared to other investment techniques.

Trial Trading of Options and Futures Products

Trial trading gives managers confidence in using the instruments and uncovers glitches in internal accounting and trading procedures.

Besides gaining familiarity with trading procedures for options and futures, it enables fund managers to determine if their accounting department can develop procedures to book the trades, and if their custodian has the ability to properly account for short and long positions in options and futures.

It may be possible to obtain trustee permission to execute experimental trades of futures and options in what can be defined as risk management positions. Example trades include: (1) Buying S&P 500 futures to maintain market exposure while transitioning between equity managers; (2) buying S&P 500 Index puts to hedge an expected market decline; and (3) selling individual call options on stock which the Plan desires to sell.

The experimental trades can help identify a variety of problems that need to be solved before derivative strategies can be implemented. For example, the accounting department may have no idea of what options or futures are, or why they are being traded. Consequently, they may not have the necessary perspective to relate long stock positions with a hedge created by long positions in index puts. A loss on the puts is viewed as a dumb decision, instead of a premium paid to insure the portfolio from a loss. Short positions in puts and calls also may be baffling to the Fund's accounting department. The accounting system may be set up only for purchases of financial assets, not anticipating that a pension fund would want to short sell anything.

Trial trading can be extremely useful because the problems it identifies usually are not insurmountable. They can be solved with proper education, patience, and modifications in the accounting procedures.

Place Emphasis on Strategies Rather Than Tools

By putting the emphasis on the strategy, two things are accomplished: (1) the focus is kept on the portfolio context of options and futures rather than considering them as individual positions, and (2) it will not be necessary to go to the Trustees for approval when a new tool is brought to market. Trustees seem more comfortable in approving strategies rather than giving a broad brush approval to the use of derivative instruments.

Dynamic changes in the listed options market over the past 20 years underscore the wisdom of emphasizing strategy approval. When listed options were introduced by the CBOE in 1973, only calls were available, and only on sixteen stocks. Trading of individual equity puts started in 1978, and index puts and calls began trading in 1983. LEAPS, options with expirations up to three years, on both individual equities and indexes were introduced in late 1990. If the approval had been granted only for the tool, each time a new product is introduced the fund manager would have to request the trustee's approval to use it. However, if the Trustees are requested to approve a strategy, then the fund manager can use the tool for the strategy that is most appropriate at the time. New tools may make strategy implementation easier.

External Considerations

These will include the trustees' concerns about ERISA, the impact on fund beneficiaries, and the impact on the sponsor's income. Trustees may become overly conservative because of their obligations under the prudent-man rule of ERISA. However, court rulings have upheld the notion that it is diversification, and the portfolio of assets, which is critical to demonstrate prudent investment behavior.

Realize That the Process Will Be Lengthy

It may take more than two or three years before the committee fully endorses all aspects of derivative trading. As noted above, it often is better to request approval of strategies, rather than individual tools for trading. As the trustees become more comfortable with basic strategies, such as portfolio insurance using listed puts, more complex strategies can be suggested and implemented.

The Concept Underlying Cash Equivalent Alternatives

The cash equivalent alternative, which may be called *interest arbitrage*, is often an appropriate way to introduce option trading in a fund. Risk is minimal, but not zero, and the average return should exceed treasury bills. The trustees must understand that funds invested in interest arbitrage represent an alternative to investment in cash assets.

The concept underlying cash equivalent investments is that a pricing relationship exists between different options on the same underlying asset. For a put and call, it can be expressed as the put–call parity equation introduced earlier. In the American National Pension Trust Case, Figure 9-8 shows that the put, call, and underlying stock must be priced correctly, or an arbitrage position can be created which will earn a return greater than the riskless rate. In equation form, put–call parity is given as:

$$\text{Discounted Bond} = \text{Stock} + \text{Put} - \text{Call}$$

If the put and call are properly priced, the right hand side of the equation is equivalent to an investment in a discounted bond paying interest rate r over period T. At expiration the Stock + Put − Call portfolio will be worth the exercise price of the options.

In a forward conversion as shown in Table 9-7, the ABD options both have a strike of 280 and 32 days to expiration. Either the call is overpriced or the put is underpriced, (put–call parity doesn't indicate which). Regardless, the combined Stock − Call + Put portfolio will pay 280 at option expiration. This is a return of 11.24 percent annualized from the 32-day holding period, and much higher than the 5.10 percent available from treasury bills.

An arbitrage profit can be earned from a box spread as shown in Table 9-8, because of mispricing between pairs of puts and calls with different striking prices. To prove that the value of portfolio at maturity is the difference in the two option striking prices consider the following. First the call positions are considered. If the S&P 500 is above 1,160 at expiration, say 1,200, the short 1,160 call costs 40 to cover, but the long 1,150 call is worth 50. A net profit of $10 is the most the position will pay for any Index price above 1,160. If the S&P 500 is between 1,150 and 1,160, the 1,150 call has a value of (Index value − 1,150), and the 1,160 call is worthless. At an Index value below 1,150, both calls are worthless.

Now add the payoff from the put positions. If the S&P 500 is above 1,160, both the 1,150 and 1,160 puts are worthless. In this case, the value from the calls is $10 and the entire portfolio is worth $10. If the Index is between 1,150 and 1,160, the 1,160 put has a value of (1,160 − Index). Adding the call portion gives a value for the entire box spread position of (1,160 − Index) + (Index − 1,150), which equals $10. Finally, if the index is below 1,150, the long 1,160 put is worth (1,160 − Index), and the short 1,150 put costs [− (1,150 − Index)]. Again, this value is (1,160 − Index) − (1,150 − Index), which equals $10. If the box spread can be created with a cost below $\$10e^{-rT}$, then a riskless arbitrage profit can be earned. As shown in Table 9-8, the box spread position can be put on with a cost of $9.25 and will be worth $10 at option expiration. This is a rate of return of 14.09 percent compared to 5.10 percent available on treasury bills.

Describing the interest arbitrage examples to the trustees accomplishes two things. First, it emphasizes the portfolio dimension of options. To evaluate the risk of a position it is necessary to consider the entire investment portfolio rather than an individual component. Second, interest arbitrage is a means to add value to the pension fund, at virtually no risk.

Jackson asks the question, "Is this investment activity appropriate for the Pension fund to pursue?" It seems quite different from their long-term investment philosophy. The answer to his question is that interest arbitrage is important to make the marketplace work. Fund managers who engage in this strategy will bid down the prices of expensive options and bid up the price of cheap ones, until both converge to their appropriate economic value.

The idea of equitizing cash balances using futures suggested by one of Norwalk's external managers is a popular technique used by many funds. It enables the fund to maintain a fully invested equity position without having to make decisions on individual securities every day. As noted above, it is difficult to time the market. To meet equity performance objectives, it is necessary to participate in market advances. Maintaining a fully invested position by the use of futures is one way to insure participation in the market at all times.

It should be noted that options can also be used to maintain a fully exposed position to market appreciation without exposure to market declines. These strategies were discussed in the American National Pension Trust case. For example, the strategy of buying calls and treasury bills appreciates in value if the market rises, but does not lose value below the exercise price of the calls if the market declines. This may be a more appealing strategy than the purchase of futures.

Trading Listed Options on the Exchanges versus Over-the-Counter Option Contracts

The over-the-counter options market is controlled by a few large Wall Street firms who offer customized options to large institutional investors. The primary advantage of the market is that the option's characteristics can be tailored precisely to meet the requirements of the institution regarding strike price, maturity, exercise price, underlying asset, and face amount. Rather complex positions can be constructed to meet whatever conditions are required. Table 9-9 describes a straightforward over-the-counter options arrangement.

The primary disadvantages of using the over-the-counter market are cost, lack of liquidity, and counterparty risk. Remember that prices for exchange-traded options are determined on the exchange floor through interaction between buyers and sellers. Academic studies of these markets indicate that option prices closely track the economic value of the options. In an over-the-counter trade, as proposed in Table

9-9, the prices of the put and call must be negotiated between Norwalk and Newark Trust. It takes a sophisticated investor to determine the correct price of each option. To complicate the issue, institutional investors frequently are offered packages of derivatives rather than a single position. For example, Newark offered a fence strategy to Ann Sawyer, instead of just a protective put position. The package provides more negotiating room in the trade, because the price of both the put and call must be determined. The institution must determine if the price it must pay for the customized product is worth the potentially extra cost.

With regard to liquidity, there is no secondary market for over-the-counter options. If the institution wants to sell the option position before expiration, the Wall Street firm usually takes the other side. However, the institution is at a severe disadvantage in determining the price, because no other buyers would be available.

The third disadvantage of the over-the-counter market is counterparty credit risk. The institution buying the customized options is relying on the creditworthiness of the firm that sells it. If the seller goes bankrupt, the option's protection becomes worthless.

In contrast, creditworthiness for exchange-traded options is provided by the Option Clearing Corporation (OCC). The OCC, as described in Chapter 1, is an indispensable participant in all listed option trading because it provides liquidity and guarantee of performance for all options. The OCC becomes the buyer for every option seller and the seller for every option buyer. This arrangement effectively "delinks" individual buyers and sellers, making possible the secondary trading of contracts. If an option buyer chooses to sell the option before expiration, the position is closed by selling the contract to another buyer on the floor of the exchange. The original buyer's name is replaced on OCC records by the person who bought the option. The original option seller still holds the position and is not involved with the second transaction because the contract is with the OCC rather than an individual. The OCC is only concerned that an equal number of buyers and sellers exists for each contract.

The OCC also performs the function of guaranteeing performance on each contract. As noted above, if the option is exercised, the buyer's brokerage firm delivers the exercise notice to the OCC. The OCC then randomly selects an option seller, beginning with the oldest short positions, to receive the exercise notice. In the rare instance when the seller does not comply, the OCC steps in to honor the contract to the buyer, in effect guaranteeing performance of the contract. The two primary problems of the OTC options market are thus avoided—illiquidity and uncertainty about the seller complying with an exercise notice.

Because of the need for customized option products, the CBOE launched FLEX options in 1993. Institutions with special requirements that cannot be met with available listed options can tailor the contract with respect to underlying stock index, strike price, expiration date, and American or European style. The CBOE's objective is to provide the specialized products more cheaply and with greater liquidity than available in the over-the-counter market. Greater information about FLEX options on the CBOE can be obtained from the CBOE web site: www.cboe.com

Suggested Amendment to the Plan Documents

Any amendment to a pension trust document must be tailored to fit the objectives of the particular pension plan. The question regarding a draft to change the plan document encourages you to consider how to develop a broad statement about using derivatives in a pension fund. We offer an example of one way Norwalk may consider changing their plan document to permit investment in derivatives. Below is a condensed version of the minutes of the trustee's meeting in which a resolution was passed to permit the use of derivative products, and a proposed amendment to the pension trust.

Minutes of Trustee's Meeting to Approve Derivative Investments

The Chairman called upon the Norwalk Pension Plan Administrator to discuss the Trust's use of derivative financial instruments. The Plan Administrator states that derivatives are financial instruments whose value is based upon, or derived from, another security or asset. She further stated that puts and calls on common stock and common stock indexes, and futures on stock indexes, all of which are traded on exchanges, are popular examples of derivatives. However, the advent of the computer has enabled the investment industry to abstract individual features from stock or a debt instrument, such as an income stream, and sell that feature as a separate financial instrument. She then noted that there are times and situations when it would be advantageous for the Trust to purchase only certain features of a security, rather than the security as a whole.

After discussion, on motion duly made and seconded, it was unanimously:

RESOLVED, That the Investment Committee has the authority, and may permit Investment Managers of the Norwalk Pension Trust, to invest on behalf of the Trust in derivative or abstracted features or indices of securities, financial futures, obligations or properties (real or personal), referred to herein collectively as "derivatives," including, but not limited to, financial and commodity index funds, financial swaps, streams of income, or gain or loss abstracted from an asset or whose value is based upon or derived from a feature of another asset, and

FURTHER RESOLVED, That the Norwalk Pension Plan Administrator may take all such actions as are necessary to effect any Investment Committee decision to invest, or permit an Investment Manager to invest in derivatives, his action in such respect to be conclusive evidence of the Investment Committee's approval.

Proposed Amendment to Pension Trust: INVESTMENT OF FUNDS

The first sentence of Article VI is proposed to be amended as follows (italics are additions to the original document):

The Trustees shall have the power to invest and reinvest all funds received by them in such securities, *financial options and futures,* obligations or properties (real or personal), or participations or interests therein, or *derivatives or abstracted features or indices thereof,* referred to herein collectively as "assets" including, without limitation, shares of stock (whether common or preferred), trustees' or receivers' obligations, equipment trust certificates, conditional sale agreements, *financial and commodity index funds, financial swaps, streams of income or gain or loss abstracted from assets or whose value is based upon or derived from a feature of other assets,* lease and purchase agreements, insurance company group annuity investment contracts and agreements, as they may deem advisable in their discretion as though they were the beneficial owners thereof, excluding securities issued or to be issued by any of the Companies, *and futures and options on tangible commodities other than currencies;* provided, however, that investments shall be so diversified as to minimize the risk of large losses unless under the circumstances it is clearly prudent not to do so.

SUMMARY

Development of the listed options market has provided pension fund managers with powerful new tools to manage risk in stock and bond portfolios. The suggested issues and solutions to the two case studies presented in this chapter should provide you with a greater understanding of option strategies that are appropriate for use by institutional investors. However, many pension fund managers who realize how useful options can be to implement their investment strategies face the challenge of obtaining approval from their trustees to implement option strategies within the pension fund.

This case has explored typical questions raised by trustees who are being asked to change their plan document to permit the use of derivative strategies by the fund managers. Although no strategy for gaining approval will be best for every situation, we have observed that most successful plans include the following steps:

- Educate trustees on the reasons for using options in the portfolio: to control risk and to add value.
- Emphasize that options must be considered in the context of the entire pension fund portfolio, and from this perspective, they control, rather than add, risk to the fund.
- Obtain approval for a period of trial trading in derivatives. Your investment managers will become more comfortable with the products, and problems can be uncovered and solved in the accounting and custodial operations.
- Emphasize strategies rather than tools. This will preclude the necessity of obtaining approval for each new option product introduced.
- Be sensitive to the trustee's concerns about ERISA, the fund sponsor, and the fund beneficiaries.
- Finally, recognize that the process will be a lengthy one, but is well worth the effort.

The purpose of these case studies was twofold. First, to make you consider how to address the approval process in your organization, and second, to further your understanding of the appropriate use of options by institutional investors. If you can articulate why and how options should be used by pension fund managers, you can feel comfortable about your knowledge of options and the concepts presented in this book.

CHAPTER 10

THE PREDICTIVE POWER OF OPTIONS

Lawrence G. McMillan

Editor's Note: The ideas and opinions presented in this chapter are those of the author. The Options Institute and the Chicago Board Options Exchange do not endorse and make no representation, expressed or implied, concerning the validity or merits of any ideas or opinions expressed herein. The information in this chapter is presented purely as a format for the explanation of ideas and for the education of the reader. Please feel free to contact the author for a more detailed explanation of the ideas and opinions expressed.

It is possible to use options as an indicator of the forthcoming price movement of the underlying instrument. Using options this way is a form of technical analysis (in fact, they are *sentiment indicators* in most of these situations). Either option prices or option volume can be important as a predictor. In one specific case, option trading activity can be used as a *direct indicator, i.e.,* the option market indicators are pointing in the same direction that the underlying instrument can be expected to move. However, in almost all other cases, measures of options trading activity can be a *contrary indicator*—where you want to take a position in the underlying that is opposite to that of the option-trading public. This chapter describes in detail all of these concepts.

The term "volatility" is important in many of these discussions, so let's lay out the definition one more time: *historic* volatility (also called *actual* volatility) is how fast an underlying instrument (stock, index, or futures contract) has been moving around in the past. It is normally measured in periods of varying length (10 days, 20 days, 50 days, 100 days, etc.—but all volatilities are expressed in annualized terms so they can be compared. Hence, you might say that the 20-day historic volatility of XYZ stock is 30 percent. It is not important, for purposes of this discussion, to know what 30 percent actually means, because we will only be using the volatilities as comparisons. So, if you also knew that the 20-day historic volatility of XYZ stock is generally less than 30 percent, then you would have to surmise that XYZ had been fairly volatile in the last 20 days, or that its stock price had probably been jumping around a bit more than usual.

Implied volatility, on the other hand, is volatility implied by the current price of the options on a particular underlying instrument. Of the factors that go into the makeup of an option's price—stock price, strike price, time remaining, interest rates, dividend, and volatility—all are fixed and known quantities at any one point in time except volatility. So, when you observe the price of a particular option (or all the options on a given stock, say), that price is largely influenced by the marketplace's estimate of volatility. That estimate is really what implied volatility is—and it is often described as the option market's estimate of how volatile the underlying security will be during the life of the option.

OPTIONS AS A DIRECT INDICATOR

In the author's opinion, there is really only one situation where option traders might directly predict the way that the underlying instrument will move. This occurs sometimes in stock options, and it has to do with the fact that, when rumors of fundamental information spread in the marketplace, those who believe the rumors will sometimes trade options because of the leverage available. These situations are sometimes predictive of corporate news events—mostly either takeover bids or earnings surprises. It is illegal for someone who knowingly possesses material non-public information to trade upon it, but it is perfectly legal for observers of the option market to try to discern what other traders are doing and to follow their "tracks." In this case, those "tracks" would usually be in the form of increased option volume, although a sharp increase in option prices (*i.e.*, in implied volatility) may be a good clue as well.

Option Volume as a Direct Indicator

One possible clue that corporate news might be released soon is a sharp increase in the option volume of a particular stock. Typically, any time that the total option volume across all series jumps to unusually high levels, then you might want to do some further investigation to see if there is a reason to believe that the company is about to make an announcement that might materially affect the price of the stock.

Not all situations where option volume dramatically increases should be assumed to be potential takeover candidates or earnings surprises; there is a lot of "noise" that could cause increased volume in the normal course of option business. For example, an institution might have executed a large covered call write, or an arbitrageur might have established a conversion arbitrage, or some large trader or market maker may have executed a hedged—more or less neutral—spread. In any of these situations, option volume jumps dramatically, but it does not portend any type of corporate news announcement. One must investigate before actually executing any option trades. If, after such investigation, it seems that the option volume might be predicting something for the stock, then you would want to buy calls (or buy the underlying stock) if call volume is heavy. On the other hand, if put volume is heavy, then perhaps something negative may be forthcoming, and you would then want to buy puts or sell the underlying stock short.

Before getting into some case histories where option volume was a good predictor of forthcoming stock movement, let's first look at some general situations. A trader might be looking at newspaper listings of the previous day's option trading, or using the CBOE's web site (www.cboe.com), to observe the option trading in a particular stock's individual options. The following examples show how to weed out the "noise" and attempt to zero in on the potentially more profitable situations.

Suppose you are looking at stock XYZ sometime early in the month of April, and you know that the average option volume is about 200 contracts per day. Yet, you notice that it traded over 1,500 contracts yesterday. On the surface, this is a situation that might be very interesting—option volume more than seven times its normal levels. However, upon closer examination, you see the pattern presented in Table 10-1.

In Table 10-1, most of the option volume is in the July 55 call, which is out-of-the-money and has an expiration date three months hence. This is not the type of option that someone with short-term fundamental information would buy—such a person would most likely concentrate on short-term options. Hence, you can surmise that this heavy trading in the July 55 calls is probably the result of a covered call write

TABLE 10-1

Example 1 of "Noise" in Option Volume

XYZ Price: 50	April 50 call volume: 50 May 55 call volume: 20 July 55 call volume: 1500

by an institution. Therefore, this stock is discarded from any list of potential takeover candidates.

Another fairly common "noise" pattern often arises as a result of arbitrage. This time, suppose stock ABC options are noticed to be very active early in the month of April. The option volume in Table 10-2 is well in excess of ABC's normal option activity.

The biggest clue in Table 10-2 that this volume is a "noise" pattern and not a pattern indicative of any sort of trading based on fundamental information is that there is an (approximately) equal number of puts and calls traded with the same terms—same expiration month and strike price. What most likely has occurred is a forward conversion or a reverse conversion: an arbitrage strategy in which an arbitrageur establishes positions in the options and offsets the risk by trading stock. The arbitrageur does this to make a small amount of money from interest rates, or perhaps a market maker helps to facilitate an institutional or customer order; it mostly likely is not activity that indicates a pending fundamental event. Therefore, this one also is discarded as a potential takeover situation.

One more example should help to identify "noise" when option volume has increased heavily. Suppose that stock XXX has heavier than normal option activity, but, when examined, that volume is shown to be of the form presented in Table 10-3.

TABLE 10-2

Example 2 of "Noise" in Option Volume

ABC Price: 60	May 50 call volume: 1000 May 50 put volume: 1000 April 55 call volume: 30

TABLE 10-3

Example 3 of "Noise" in Option Volume

XXX Price: 25	April 20 call volume: 800
	April 25 call volume: 800

In this case, there is approximately equal option volume in the same expiration month, but at two different strikes—and not much other option volume at all. This might well be the result of a vertical option spread, either a bull spread or a bear spread. Once again, put yourself in the mindset of someone who has information about a fundamental development at a company. Would you buy a spread? Most likely not; you would just buy calls or puts and look to profit from any move in the underlying stock when the news came out.

So, what pattern is the one to look for? Table 10-4—an illustration of volume figures you might see in a newspaper listing of option trading—is more typical of the situation where a strong rumor might be influencing option trading. Assume that the table represents the trading in XYZ's call options on some day early in the month of April, and that on a normal day, only about 300 options trade on XYZ. Furthermore, assume that put volume was sparse and not great in volume.

Table 10-4 shows total option volume on this day is 3,095. That is not only more than ten times the average volume, but the pattern is a good one: most of the activity is concentrated in the near-term at-the-money April 50 calls, or the near-term April 55 calls that are one strike

TABLE 10-4

Option Volume That Might Indicate Rumors

XYZ Price: 50

Month:	April	May	June	July
Strike:				
45	300	100	50	70
50	800	300	100	75
55	600	350	50	
60	200	100		

out-of-the-money. Traders believe these will give them the most leverage if a corporate news announcement is released. A lot of the rest of the volume may come from market makers attempting to hedge their positions. The market makers are most likely those who have sold the April 50 and 55 calls to the aggressive buyers. Typically, market makers do not want to have the short-market exposure that a short call position represents, so they hedge their position. One commonly employed hedging strategy that market makers use is to buy other options offered when they sell some options. This hedging by market makers accounts for some of the volume in other series (calls or puts with the same underlying but with different strike prices or expiration dates). Market makers may also buy the underlying stock to hedge the risk of their option position. On-floor market makers might even contact some off-floor traders—perhaps institutions who own XYZ common stock—and notify them that aggressive buying has surfaced in XYZ stock, and that the implied volatility of the options has risen. They might suggest that the off-floor traders write some covered calls against a portion of the stock that they own. The upstairs traders know that rumors are frequently false, and thus they might sell some calls, their motivation being that, even if a positive announcement occurs, they still have received a good price for the covered write. And they still own some stock that does not have any calls written against it. So these off-floor traders might sell longer-term calls, and the on-floor market makers might buy them for the purpose of hedging some of their risk from having sold the shorter-term calls, the April 50 and April 55 calls in Table 10-4. The net result is a pattern of option trading in XYZ that resembles that shown in Table 10-4. It is the author's opinion that this is the pattern of option volume that most often precedes a true corporate news event.

To get a little more specific, you might use the following criterion, although there are other volume constraints that could be profitably used as well. *If the total option trading volume in a particular stock on a given day is at least double the average option volume, then you should take notice.* "Average" option volume is typically defined as the 20-day moving average of total option volume in a particular stock's options. This type of option analysis does not work for index options or futures options, because it is only in individual stocks where rumors sometimes spread in advance of an actual corporate news announcement.

Figure 10-1 and Figure 10-2 show some examples of how option volume increased just before a takeover of the underlying stock. Figure 10-1 is American Cyanamid (ACY) during the summer of 1994. Total

FIGURE 10-1

American Cyanimid (ACY) Option Trading Deficit, February–July 1994

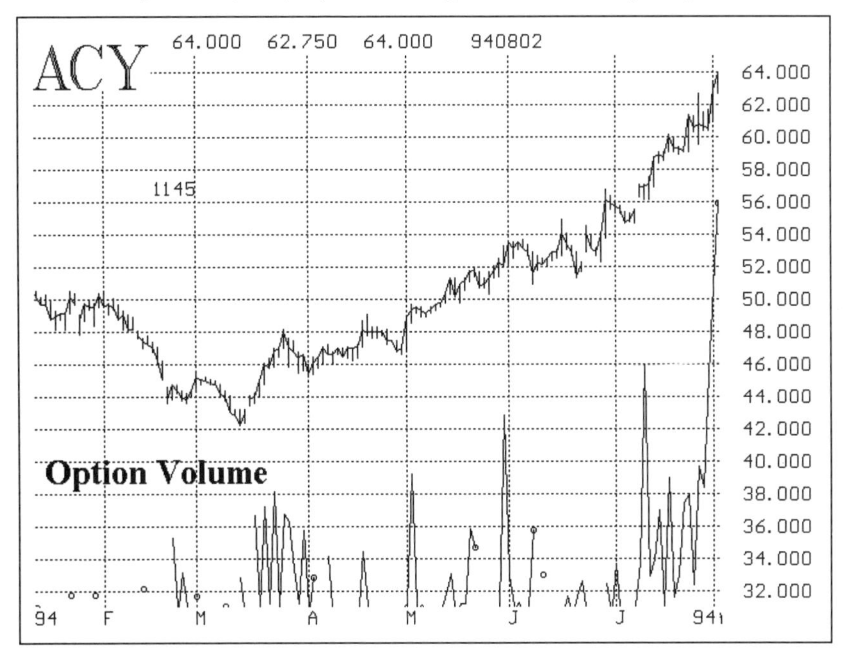

FIGURE 10-2

Duracell Option Volume, June–Sept. 1996

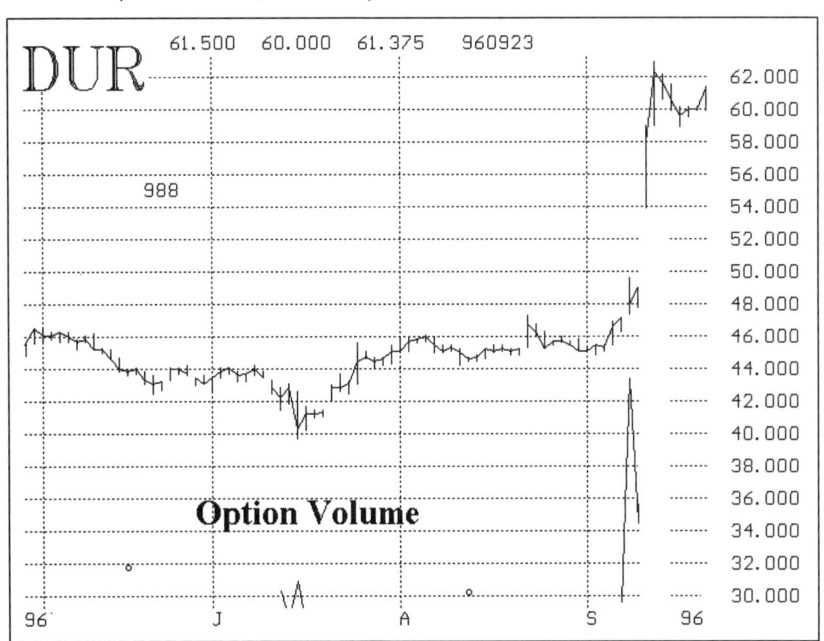

daily option volume is shown on the bottom of the chart, below the bar chart of the daily stock prices. The stock ended a declining period in March 1994, and traded from about 42 up to 64 by early August. Along the way, there were some spurts of increasing option volume, but it was most noticeable in July when option volume became heavy on nearly every trading day. Eventually the volume rocketed to record levels at the end of July and the beginning of August. Although this graph does not show it, this option volume was mostly call volume and had the distinct look of the "takeover" pattern described above. On the next day, after the above chart ends, ACY received a takeover bid at $91 per share.

Here is another example of a "takeover" pattern. Figure 10-2 is the chart of Duracell (DUR) stock price overlaid on top of total option volume. There was virtually no option volume on most days during June, July, and August of 1996. Then, in early September, DUR broke out over old highs at 47 and traded up to 48. At the same time, option volume ballooned to record levels—literally out of nowhere. It is often a favorable pattern to see the underlying stock break out above a previous resistance area while being accompanied by such an increase in option volume. The next day, the stock moved up slightly, and on the following day, a takeover bid was received at a price in excess of $60 per share.

If you are going to trade these "rumor" stocks—and these situations pose a great amount of risk—it might be better to trade either the underlying stock or in-the-money, short-term options. An in-the-money, short-term option has a very high delta and relatively little time value premium, so it behaves more or less like the stock itself. If you are holding these contracts and enough time passes so that the expiration date is near, but nothing has happened to the stock price, then you may want to roll your contracts forward to the next month if you still feel that the stock is a possible takeover candidate.

As an exit strategy, use a predetermined stop price to close the option position if the stock goes down instead of up. On the positive side, when you take a position based on heavy option activity, you should be looking to exit when and if the anticipated news is announced. Once the news is out, then there is no advantage to reading the option volume. This is even true during the "rumor" stage: if the rumor is widely publicized then that publicity itself accounts for the option volume. It is the author's opinion that, if you chase every rumor that is printed in the paper or mentioned on TV, you will generally be pursuing a losing strategy. Thus, an additional thing to look for when scanning for heavy option volume situations is a lack of news. The less news the better, for that means that any unusually heavy option activity that you observe stands a better chance of being a truly attractive situation (assuming that the activity is not just "noise" as defined above).

Option Prices as a Direct Indicator

The above discussion centered on using option volume as an indicator. However, sometimes option volume does not increase, even though rumors are spreading in the marketplace. In those cases, however, option prices may increase. And that increase in prices can be a good clue as to a forthcoming corporate news announcement.

For example, consider this scenario: an off-floor trader who is eager to act on what he or she considers to be a valid rumor checks the market in the calls of a very thinly traded stock, *i.e.*, one with low option volume. After asking for a market, the off-floor trader is told that the last trade in at-the-money option was at 4 and the current market is bid at 4 and offered at 4¼ and that 10 contracts are offered. The trader buys the ten at 4¼ and asks for a new bid and ask. The response is 4¼ bid and offered at 4½ with 10 contracts offered. Again, the off-floor trader, being eager to initiate a position, buys the 10 calls that are offered at 4½, and requests a new bid and ask. Because there is typically little activity in these options, the market makers are wary of establishing too big of an unhedged position in these options, so the new market is something like 4⅜ bid and 4⅝ offered, with 10 contracts offered. At this point, the off-floor trader decides that 4⅝ is too expensive and refuses to pay that price.

What has happened? Only 20 contracts traded, so that is certainly not going to show up on any high volume screens that volume-watchers would observe. However, the options have gotten significantly more expensive: the price has risen from 4 to 4½ without a stock price change. That information shows up as an increase in implied volatility of the options on this stock. Therefore, it is often an important clue to find cases where implied volatility spikes higher, often in the absence of news or option volume. Such cases may quite often portend corporate news announcements, just as heavy option volume might.

Here is a classic example. Gerber (GEB) was a stock that had very thinly traded options during the early part of 1994. There are two accompanying charts: Figure 10-3 shows Gerber's stock price overlaid above option volume, while Figure 10-4 shows Gerber's stock price overlaid with option implied volatility.

FIGURE 10-3

Gerber Stock Price Overlaid Above Option Volume, 1993–1994

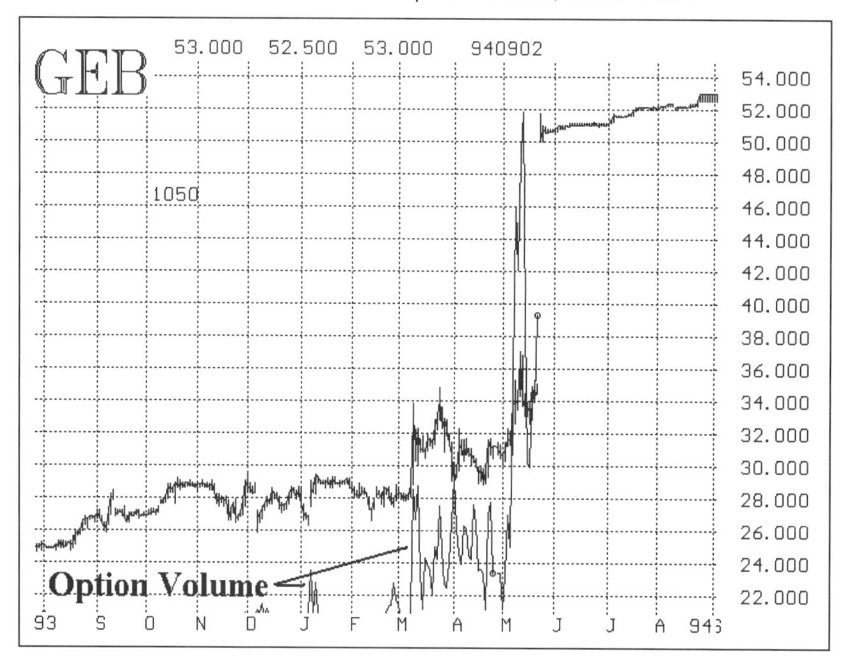

FIGURE 10-4

Gerber Stock Price Overlaid Above Option Implied Volatility

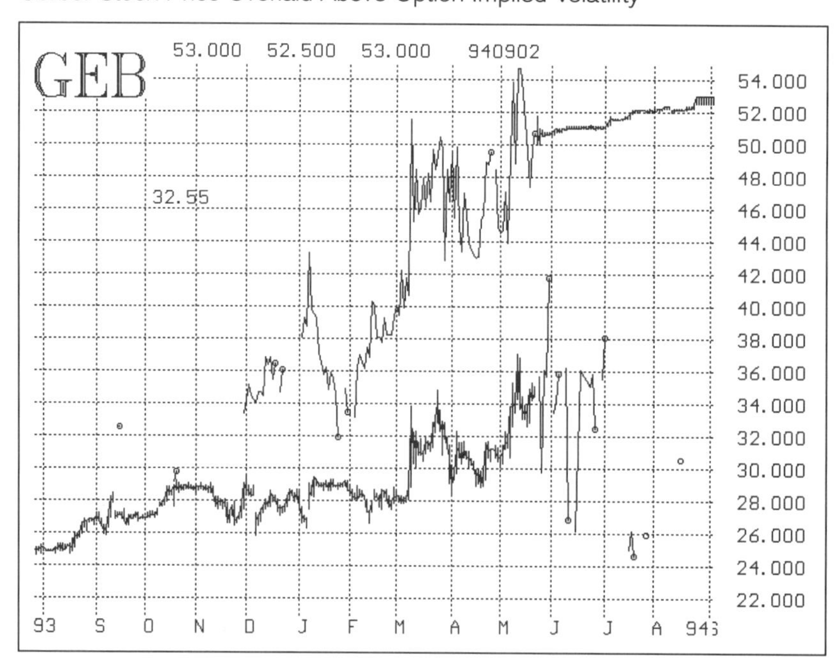

On the volume chart, you can see that Gerber did not have much option volume all the while the stock was trading in the $27–29 range up through February of 1994. Then, when the stock broke out to the upside, option volume increased heavily. However, that first breakout was followed by a pullback all the way to $29. Eventually the stock moved higher in May and was eventually taken over in mid-May. Thus, if you were only watching option volume, you would have probably gotten long near the $32–34 area on that breakout in March 1994. You might have stopped yourself out on the pullback to $29, depending on how tight of a stop you use.

Traders who also watch option implied volatility, however, had a much easier time of it. Notice that implied volatility first began to increase in January 1994, while the stock was still in its $27–29 trading range. If you had bought stock or short-term in-the-money options at that time, you would have been fully positioned for the breakout when it occurred (assuming you continued to roll your options forward as they neared expiration). Even when the pullback to $29 occurred, you would have still been in a winning trade and probably would not have stopped yourself out.

Thus, implied volatility was the first, and perhaps best, clue to the takeover that eventually occurred in Gerber stock that year. Thus, it is also important to watch for situations where implied volatility increases, even in the absence of option volume. In those situations, you may want to buy stock rather than options because, by definition, the options are going to be somewhat expensive if implied volatility is increasing dramatically. However, the purchase of a short-term in-the-money option may still be an attractive alternative to a common stock purchase.

You may also find situations where both implied volatility and option volume increase heavily. This usually occurs after a rumor becomes public, but it is generally a positive combination. Often, traders who act on early rumors buy at nearly any price if they believe the news is imminent. That can cause both volume and implied volatility to skyrocket.

Other Useful Indications from Option Prices

There are other situations where it behooves you to pay attention to increases in implied volatility. These generally have to do with some known news item that is pending. To the extent that implied volatility increases because of pending corporate news, these situations resemble the takeover or earnings surprises that were the focus of the previous examples. However, these are different in that the reason for the news is generally publicly known—it is just that no one knows what the news will be or how it will affect the stock.

For example, a small biotech company may have its best product up for review by the FDA. Everyone knows when the FDA ruling is going to be handed down, but no one—believe me, no one—knows in advance how the FDA is going to rule in any situation. Whichever way the FDA rules will generally cause this stock to experience a rather large gap opening: upwards if the FDA rules in favor of approval for the drug, or downwards if the FDA denies approval. Thus, before the ruling, the options become very expensive—high implied volatility—because the option market knows that a gap has an increased chance of occurring and the option prices have such a potential move factored in.

Similar events occur when an important lawsuit is pending: no one knows how a judge or jury will rule, but they more or less know when that ruling will take place, so option prices inflate, *i.e.*, the implied volatility rises in advance of the verdict. These could be patent infringement lawsuits or perhaps lawsuits filed by the FTC to block a merger or takeover.

Figure 10-5 illustrates a fairly typical example. Liposome Technology (LIPO) is a biotech company and had an important review date before the FDA in June of 1997. As the date approached, the stock rose during April and May from a price of $18 all the way up to $28. Implied volatility began to increase fairly heavily during May and into June. The stock price rise might have fooled you into thinking that the news was going to be positive—and perhaps that was the general consensus among traders, but general consensus opinions are often wrong. The increase in the implied volatility of the options, however, was a warning that the stock could gap in <u>either</u> direction. In fact, when the FDA made its ruling in late June, it was negative; the stock suffered a tremendous fall from $25 to 9!

The option markets are fairly efficient at warnings of this type, so even stock owners who do not trade options should pay attention to the implied volatility of options, for it may be foretelling a serious upcoming price gap for the underlying stock. If you already own the stock in these cases, it is probably a wise idea to lighten up some—either sell out some of your stock or sell some expensive (*i.e.*, high implied volatility) covered calls.

FIGURE 10-5

Dhiposome Technology Trading Before and After an FDA Ruling on Its Product

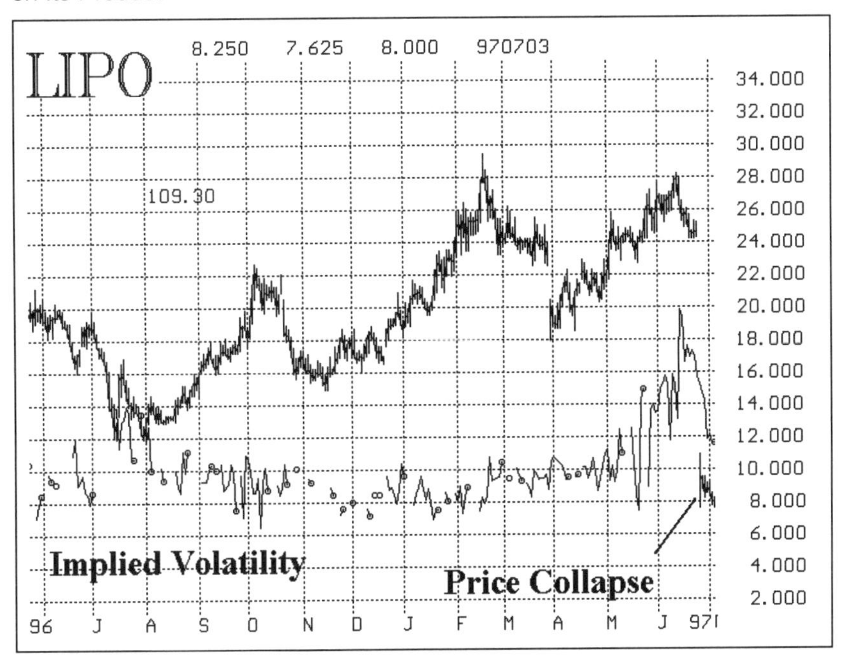

If you do not already own stock, there is a fairly high-risk strategy that can sometimes prove profitable: wait until the day before the event is to take place (the FDA ruling, for example) and buy the near-term at-the-money straddle. Sell the straddle as soon as the stock re-opens after the ruling has been made (or the lawsuit verdict is announced). This is called *event-driven straddle buying*, and it relies upon the fact that the gap in the stock price after the news is announced is often greater than the marketplace had been anticipating. This is especially true for FDA rulings and lawsuit verdicts. In some cases, the strategy might be tried with earnings surprises, in cases where option implied volatility increases dramatically in advance of the earnings announcement itself.

The trading gaps for FDA rulings and lawsuits are generally more dramatic and more of a surprise than are the trading gaps associated with earnings surprises. Again, the increased risk of paying two option premiums, especially when implied volatility is high, places this strategy in the high-risk category.

OPTIONS AS A CONTRARY INDICATOR

The preceding discussion describes the use of options as direct indicators. Option market contrarians, however, believe that the option trading activity can be used as a contrary indicator. Consequently, a contrarian wants to do the opposite of what the majority of option traders appear to be doing. If too many option traders are bullish, the contrarian wants to be bearish; and if too many are bearish, the contrarian wants to be bullish.

Of course, the key is in defining and measuring how many is "too many." On a day-to-day basis, not much can be interpreted from option volume or option pricing patterns. However, when they reach an extreme, that is when it is time to sit up and take notice. It is often said that "the public is right all during a move, but wrong at the extremes." According to this theory, the public buys during a bull market while it is rising, but then they go in head over heels right at the top. The same sort of thing is believed to happen in option trading.

The option indicators that are applicable to contrarian investing are option volume, in the form of put–call ratios, and option pricing (implied volatility). For put–call ratios, you watch for signs that too many calls or too many puts are being bought. For implied volatility, you watch to see when it is "too high" or "too low." These measures will be defined and further described in the following sections.

Using Implied Volatility—Option Prices—As Contrary Indicators

One of the most common phenomenon found in option trading is that the options on a particular underlying instrument often trade at a very high implied volatility, if the underlying declines sharply. This can be true of the broad market, of individual sectors, of individual stocks, and sometimes even of futures contracts.

Using the VIX Index

Perhaps the most widely followed application of this theory occurs while using the CBOE's *Volatility Index (VIX)*. VIX measures implied volatility of the most liquid OEX options, and OEX options in turn are some of the most liquid vehicles on the broad market. Thus, OEX is a good place to look for contrary sentiment. When the stock market begins to fall rapidly, VIX will often rise sharply (beginning with the sharp market decline in October 1987, most corrections and bear markets in the broad market have taken place in a very short period of time, so a rapid decline is a relatively frequent event). When VIX spikes upward and then forms a peak, the theory indicates that it is the time of utmost panic by put buyers. Their panic buying inflates the prices of both puts and calls, and that shows up as a sharp increase in the VIX. When this happens, it is usually a good time to buy the market. Figures 10-6 and 10-7 show VIX and OEX over comparable time periods from 1996 to 1998. (This theory is believed to hold true for all time periods, not just this one.)

In Figure 10-6, note that VIX is not constantly at one level. It moves around quite a bit. Typically the VIX remains in a range, although the parameters of the range change for different market environments. For example, back in 1993 and 1994 (not shown), the whole stock market was in a fairly non-volatile period, and VIX ranged between about 9 and 13 in those days. By the time period shown in Figure 10-6, the market had become much more volatile, and VIX thus traded at significantly higher levels.

FIGURE 10-6

Volatility Index (VIX) Chart for 1996–1998

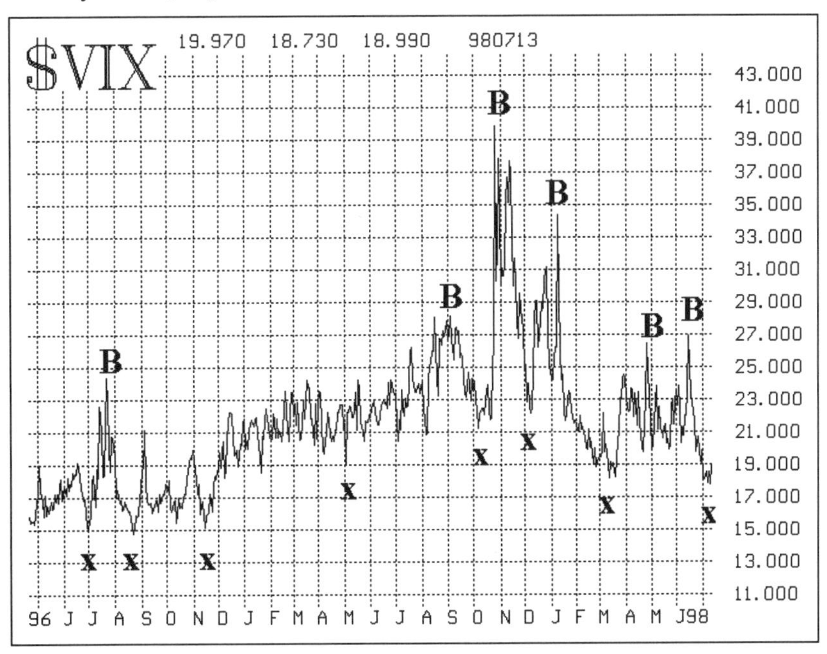

FIGURE 10-7

OEX Options Trading, 1996–1998

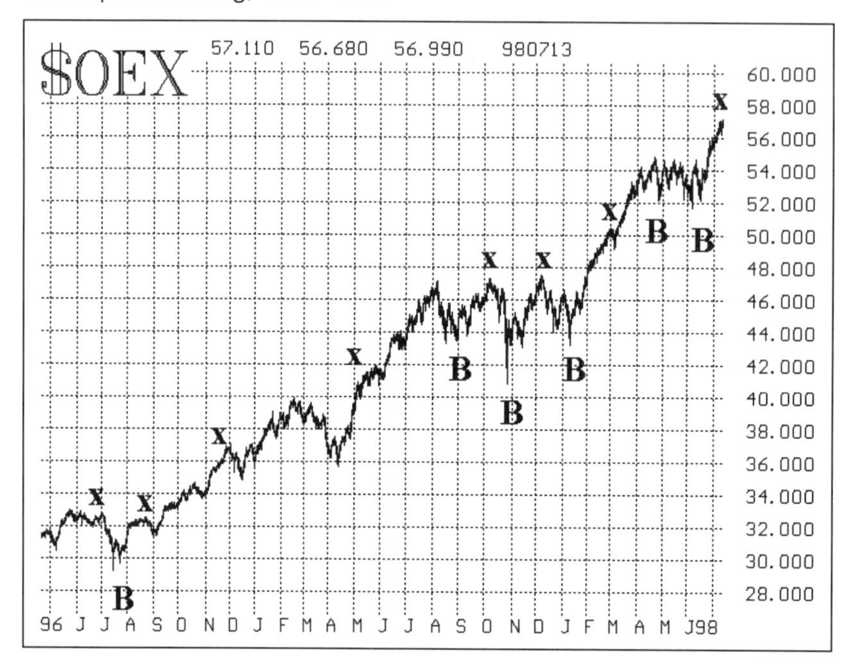

On the VIX chart (Figure 10-6), spike tops have been marked as buy signals ("B"). Figure 10-7, the chart of the OEX itself, shows where those buy signals occurred in terms of OEX prices. You can see that the Bs are mostly at or very near significant bottoms in the market. Consequently, according to this approach to market timing, a spike top in VIX, which appears when the market has been declining, can often be used as a signal to take bullish positions on the broad market. This is a form of contrary opinion trading, because the contrary trader wants to buy when everyone else is panicking and is paying top dollar to buy puts (which shows up as a sharp increase in VIX).

Note that there is no one level at which VIX gives this buy signal. Rather, it occurs when VIX spikes upward out of its then-current trading range and subsequently forms a peak. You do not want to buy while VIX is still rising; you want to wait until it has peaked before stepping in. Two dramatic downside moves in recent years, the 508-point decline in October 1987 (when VIX rose above 100 during the trading day), and the 554-point decline in October 1997, for example, resulted in the highest VIX readings on record. The aftermath of both of these declines were excellent buy points after VIX had peaked.

The observation of VIX for the purpose of generating a buy signal has become fairly widespread. Some technicians have even more sophisticated methods using *moving averages, Bollinger bands, oscillators,* and *stochastics* on the VIX itself. I am not sure if it is necessary to get that sophisticated in interpreting VIX. All you really need to be certain of is that the VIX has reached a peak before you buy. In any case, this is a good indicator to add to your arsenal.

You may also have noticed the x's on the VIX chart (Figure 10-6). Those are placed at periods of low volatility. Again, there is not a single level where one can say VIX is "too low." You must look at the chart and mark relative lows. For example, on the extreme right-hand side of Figure 10-6, VIX—which had been trading in a range of 19 to 27, with a few forays out of the range on the high side—suddenly slumped to 18. This is marked with the "x" on the rightmost part of the chart. At the time (July 1998), it was impossible to tell if VIX was going to go lower or higher from there, but it was certainly the lowest that it had been in quite some time. The other x's on Figure 10-6 mark similar periods of low implied volatility.

What causes VIX to get "too low?" And how should it be interpreted? These are the two important questions whose answers describe how this part of the VIX trading theory works. VIX gets too low because there is too much complacency in the market. Option buyers lower or cancel their bids, according to the theory, because they have been losing money in a stable market environment and/or sellers become more aggressive because they have been making money from option time decay. In any case, practitioners of contrarian opinion know that if there's too much complacency, then you might expect a market "explosion." Interpret this condition to predict a violent market move when VIX is trading at relatively low extremes.

You do not know if such a violent market move is going to occur on the upside or the downside, only that it is a good bet that a large move will occur in one direction or the other. In the past, conditions of low implied volatility have led to both types of moves (see Figure 10-7 of the OEX where the x's have been marked in the same spots). Most of the moves have been on the downside, although there have been some on the upside as well, most notably the move beginning in early 1995 (not shown on chart), when VIX was at its lowest levels ever. This preceded a long and steep rise by the broad market.

The strategy for this case is to buy straddles, *i.e.*, buy a put and a call with the same expiration date and same strike price. This allows you to profit when—and if—the market "explosion" (either up or down) comes. In fact, if the market explosion occurs on the downside, the straddle holder most likely will receive the additional benefit of seeing implied volatility rise higher. Again, it is important to understand the risk of buying not one, but two option premiums that might erode to zero if the expected move in the underlying stock does not occur before the option's expiration.

In summary, it behooves a contrary-opinion trader to pay attention to VIX. When VIX gets "too expensive" by spiking quickly higher during a sharply declining market, then that can be used as a buy signal. On the other hand, if VIX is "too low," there is too much complacency in the marketplace, and you can consider buying straddles in preparation for a market explosion in one direction or the other.

It Works for Stocks Too

This same philosophy of observing the implied volatility of the options in order to trade the underlying applies quite well to stocks, too. The same sort of thinking as was used with the VIX example above is applicable to stocks: when implied volatility shoots higher during a sharply declining phase in the stock's price, then a good buy signal may be at hand. Alternatively, if the implied volatility of the options on that stock are at or near historic lows, then you could expect the stock to be ready for an explosive move in one direction or the other.

Some readers may recognize this general philosophy of trading as something called *volatility trading*. What many do not realize is that volatility trading is really a form of contrary trading, *i.e.*, you take a position opposite that of the majority, for it is the majority that has caused implied volatility to get "out of line" in the first place.

Figure 10-8 is a chart of AT&T (symbol T). The price of AT&T stock is on the top, and the implied volatility of AT&T options is on the bottom of the chart. The implied volatility of AT&T stock options generally traded in a fairly tight range from late 1994 through June of 1996. At that time, however, the stock began to fall rather sharply, dropping from $47 to $35. As that happened (and such a move can be considered a rather panicky sell-off for a non-volatile stock, such as AT&T was at the time), implied volatility began to rise rapidly. Finally, at point "a" on the chart, implied volatility peaked. That was a good buy point for the underlying stock.

It should also be pointed out that, when implied volatility reaches a peak such as this, there is another strategy, besides merely going long, that is appropriate: you might quite reasonably sell covered calls on

FIGURE 10-8

AT&T Stock Price and Implied Volatility of Options, 1994–1996

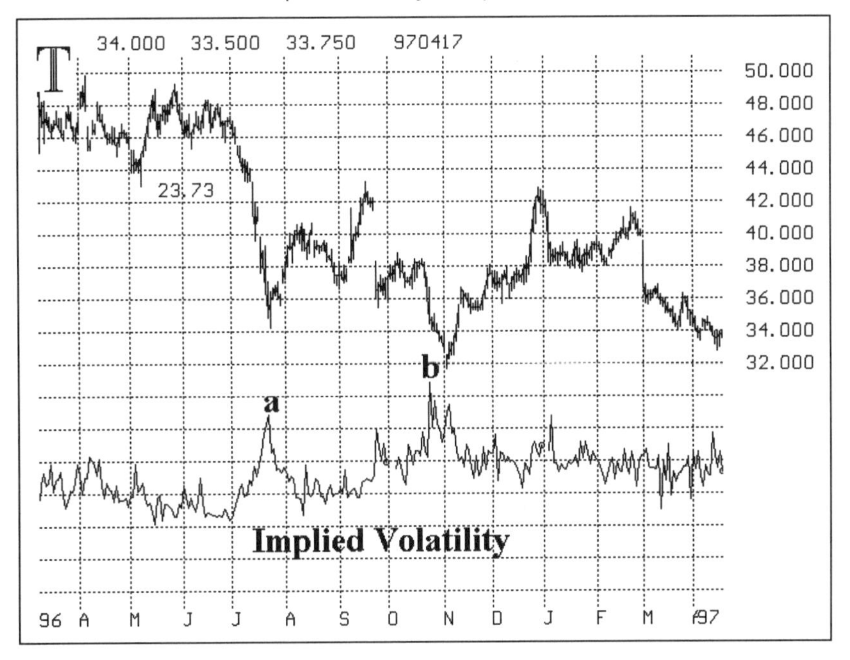

AT&T stock (or sell cash-secured puts, a strategy with a similar risk profile, if you have the approval of your broker to use it). Because the options are, by definition, expensive (implied volatility is at a high peak when this strategy is entered), and because covered call writing and cash-secured put selling are strategies that benefit from sideways or rising stock prices, they benefit from both the expensive options and an upside bias derived from the peak in implied volatility.

Returning to the AT&T example, a few months later, the stock once again began to slide—making new relative lows—and implied volatility rose once again. This time it peaked at the point marked "b" on Figure 10-8. That was actually slightly before the bottom in AT&T's stock price, and it was certainly a good time to buy the stock or to set up covered call writes (or cash-secured put sales).

I'm going to include one example of futures options in this section also, because it highlights this entire concept extremely well. It also encompasses a news event that most people remember quite vividly. In the spring of 1996, "mad cow disease" received quite a bit of attention as a worldwide news item. There was a scare that eating beef could give humans this disease.

Figure 10-9 is a chart of August Live Cattle futures during that time period. You can see that prices began to fall rather dramatically in April of 1996, dropping from a high of 64 to a low of 57 in just a few weeks. That is a large move for cattle futures over such a short period of time. As the move gathered momentum on the downside—due to the scare stories reported nightly on TV news—implied volatility rose. In fact, it rose sharply, eventually reaching a peak that is quite visible on the chart.

That also marked the exact bottom day for the decline in cattle futures B, which went into a two-year-long bull market from that point. Once again, a peak in implied volatility in a rapidly declining market was an excellent buy signal (at least a great point for covered call writes).

Having established that stocks (and sometimes futures) can be bought when implied volatility spikes too high, let's look at the opposite side of the coin. What happens when the implied volatility of a

FIGURE 10-9

Live Cow Futures Trading During "Mad Cow Disease" Scare, 1996

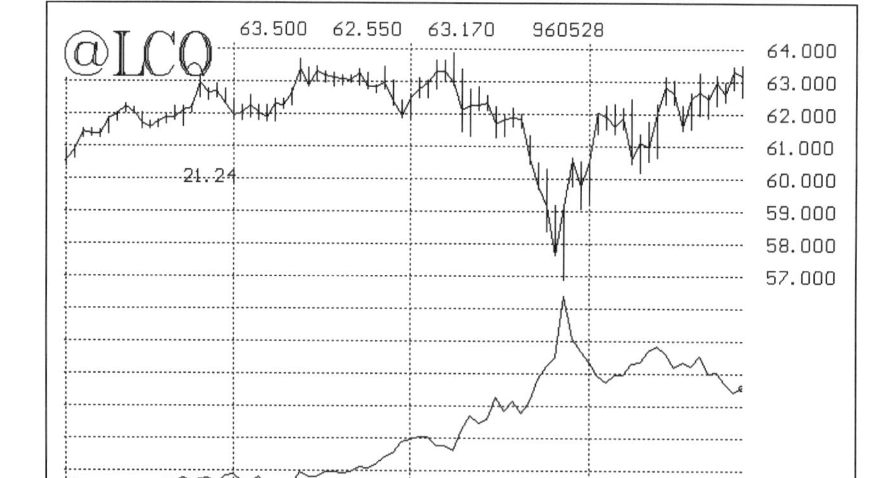

stock's options gets "too low"? As with the VIX above, (the x's in Figure 10-6) you can expect a stock to explode shortly after the implied volatility of its options gets too low. And, as before, you might consider buying straddles—both a put and a call with the same strike and expiration—to take advantage of this. Because one cannot be certain of how soon the stock will make its move, it is best to buy a straddle with at least three months of time remaining. This allows ample time for the stock to have a chance to break out in one direction or the other, while not being overly susceptible to time value decay during the first month that the straddle is held.

Figure 10-10 contains a chart of the stock price of Ascend Communications (ASND) above a chart of its implied volatility. For a year and a half, from May of 1996 through January of 1998, implied volatility traded in a range. But then, inexplicably, in late February 1998 implied volatility dropped to a multi-year low, as marked on Figure 10-10. An arrow shows the same point with regard to Ascend's stock price. The stock was near $35 at the time and a 3-month straddle cost about $7.

FIGURE 10-10

Ascend Communications Stock Prices and Implied Volatility of Options, May 1996–Jaunary 1998, Showing Extreme Volatility

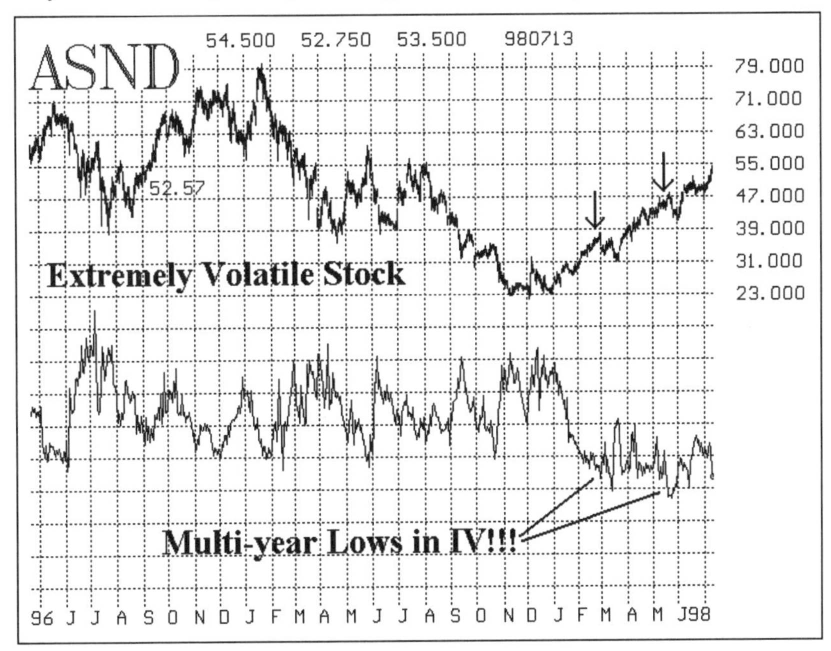

For a month or so, the stock traded within a range of $31 to 38, but then it broke out strongly on the upside and reached $50 before option expiration, and the straddle traded near $15. A second opportunity also arose in this stock in mid-May when options on Ascend traded at an even lower implied volatility (also marked with an arrow). This also proved to be a good straddle buying opportunity.

Why was the implied volatility so low for ASND options near the beginning of March, 1998? No one will ever know for sure. In retrospect, it was "too low" because the stock price easily outdistanced the straddle price over the allotted time. You must remember that implied volatility is nothing more than the option market's best guess as to how volatile the underlying security will be during the life of the option. Sometimes that guess is right. Sometimes it is wrong. In fact, if implied volatility is at a low extreme, history has shown that there is good chance it will be wrong. This explains the contrary-opinion part of the volatility trading theory. Given all the input into option pricing, the option market had priced the ASND options too cheaply. This does not mean that the option market makers were not doing their job, because they merely balance supply and demand. In this case, demand was weak and supply was aggressive, so the options got "too cheap," *i.e.*, the level of implied volatility decreased so low that, with hindsight, the stock price moved outside of the implied range.

Numerous charts could be displayed to illustrate this concept. On almost any day of the week, you are able to find at least a few stocks whose options are "too cheap." Even in a severe down market, where we know implied volatilities tend to rise, there will be some stocks with extremely low implied volatilities. Consequently, the strategy of purchasing straddles on stocks whose options are trading at a very low implied volatility can be used frequently and—if employed properly—can sometimes result in good profits. It must be kept in mind, however, that the purchase of two option premiums, as required by this strategy, involves the risk of losing the entire premium paid if the underlying stock or index fails to move beyond one of the break-even points before option expiration.

Implementing the Purchased Straddle Strategy

The following section outlines the way to implement this strategy which, in the author's opinion, gives you the best chance of earning a profit. There are four steps:

1. Identify cheap options

2. Verify that the current situation has a high probability of success
3. Verify that the underlying has been able to make such moves in the past
4. Check the fundamentals to see if there is any obvious reason for the low implied volatility

For Step 1, identify cheap options, in theory, each of the options on a particular underlying security—IBM for example—should have the same implied volatility at any one point in time. In reality, that is not true, but they are often fairly close to each other or they follow a uniform pattern, called a *volatility skew*. To simplify things for analysis purposes, it is common practice to combine all the individual option's implied volatilities into just one number that tracked over time. It is beyond the scope of this chapter to describe that combining process, but suffice it to say that if one has a "history of implied volatilities" (which is not the same as historic volatility), then the current reading can be compared with past readings for the purpose of estimating whether the current reading is high or low.

Concentrate on situations in which the current reading of implied volatility is in the tenth percentile or lower of past implied volatility readings. Look, for example, at the daily implied volatility levels for the past two years or so—that encompasses over 500 trading days—and compare today's current implied volatility. If today's reading is lower than 90 percent of all those past readings, then it is in the tenth percentile or lower, and the options can be considered cheap. This then satisfies Step 1.

In order to perform this analysis, it is probably necessary to subscribe to a service that tracks volatility, or to buy one of the more sophisticated option software systems and have a data feed so that the system can track it. There is also one free Internet site that has the data.

Step 2, verifying that the current situation has a high probability of success, is a more difficult process, but necessary to ensure that the options are truly cheap. In situations such as these, you should consider buying a straddle with at least three months of time remaining. First retrieve the current prices of the options and the underlying security. With this information, you can determine the break-even points of the proposed position.

As an example, Ascend Communications (ASND) is trading at $36 on March 2. The June 35 calls are offered at 4⅝, and the June 35 puts are offered at 2⅞. Thus, the cost of the straddle is 7½. This straddle has about 3½ months of time remaining, so it satisfies the "time remaining" condition as well. Next, determine the break-even points. A delta-neutral trader would not necessarily buy an equal amount of puts and calls, but, for the purposes of this example, assume that one put and call are being purchased. That makes the upside break-even: $35 + $7½ = $42½, and the downside break-even: $35 − $7½ = $27½.

Next, you need a probability calculator of some sort. There are several available in the marketplace for reasonably cheap prices. Also, if you own the more sophisticated option software systems, a probability calculator is included as well. Some calculators do more than others, but all will calculate the probability of the stock's price exceeding the break-even points at the end of the option's life. A more advanced calculator will calculate the probability of the stock ever trading at the break-even point at any time during the option's life.

To use these calculators, you must make an estimate of the actual volatility that the underlying will experience during the life of the straddle. This is something of a difficult task. It does not suffice to merely assume that the forthcoming actual volatility will equal the current reading of implied volatility. Most likely it will be higher, if you believe in the contrary theory of this approach (the options are cheap because the market forces of supply and demand have forced them to be; but because, according to the theory, the market is often wrong at major turning points, the theory says that it is reasonable to expect a price explosion by the underlying security). So, you need a better estimate of forthcoming volatility. Here is a simple approach that often yields a workable number: use the minimum of the 10-day, 20-day, or 50-day historical volatility. Because these numbers measure how quickly the underlying has recently changed in price, they may prove to be good guesses of short-term forthcoming volatility without overstating it.

So, now we have all the inputs needed for the calculator—a time horizon, a distance (the break-even points), and an estimate of actual volatility. The output of the calculator should show a probability of success of at least 75 percent. If it doesn't, then drop this straddle from consideration and go on to evaluate another one.

Continuing with the ASND example, the following data is available: from March 2, 1998. There are 77 trading days remaining until option expiration on June 19. The current reading of implied volatility is 44 percent, and it is in the first percentile (see Figure 10-10 that shows ASND's implied volatility had traded down to a multi-year low). Looking at the recent historic, or actual, volatilities of ASND, assume they are as follows:

10-day historic volatility: 50%
20-day historic volatility: 52%
50-day historic volatility: 60%

The minimum of these three numbers is 50 percent, so that is the one to use in the probability calculator. Given the current stock price of $36, you want to determine the probability of either trading above $42½ or below 27½ in 77 trading days. Assume that the calculator returns a favorable reading in excess of 80 percent (in fact, the actual probability was higher than that at the time).

Thus, the second step in the analysis process is satisfied. A straddle has now been identified with a good probability of success of ownership (Step 2), and the implied volatility of the options in that straddle are quite cheap when compared with past readings of implied volatility (Step 1).

So far, the analysis has been based on mathematics, but it is still theoretical to a certain extent. To perform Step 3, verifying that the underlying has been able to make such moves in the past, we want to actually look at a chart of the underlying security and verify to ourselves that it has really been able to move the required distance in the required time. The trick here is it determine how big a move to look for. The straddle itself costs 7½, but the stock might make moves bigger than that and yet the straddle purchase could be a losing proposition! How could that happen? Consider the case where ASND might rise $5 to a price of $41, then fall $10 to 31, then rise $10 to 41, then fall $10 to 31, and so forth. Even though the stock was making moves bigger than 7½, it never reached the break-even points of $42½ on the upside and 27½ on the downside.

If you look for past stock price moves of twice the cost of the straddle, then it is more likely that you are looking at cases where the straddle purchase would be profitable if the stock can move that far. There is no case where the ASND can make a 15-point move from any price level within the break-even points and not hit one or the other of the break-even points.

Figure 10-11 is the chart of ASND once again, only this time we are only interested in the price action of ASND in the past, so the price scale has been expanded somewhat. There is a box on the upper right of the chart that is 15 points high and 3½ months wide. That box represents the parameters that must be exceeded for this current straddle purchase to be profitable. That is, if ASND can move 15 points in 3½ months or less, then this will be a profitable straddle purchase. Visually, or with some

FIGURE 10-11

Ascend Communications Stock Trading Prices, 1996–1998

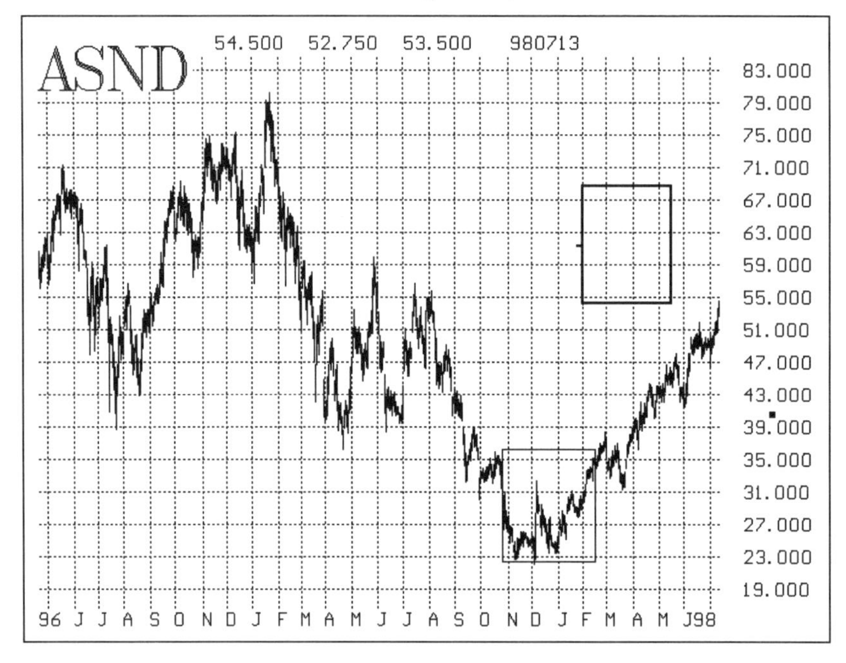

sort of a template, see just how often in the past ASND has been able to be contained within a box like this.

The box must initially be positioned with the little tab on the left-hand side equal to the stock price on the day you are using to begin the test. For example, a box has been drawn over the chart from October 1997 through February 1998 that just barely contains the stock price. That box was positioned with the tab equal to the ASND closing price on the day that represents the left-hand side of the box—October 27, 1997, in this case.

However, you will find it is hard to position the box in many places on the chart that ASND would not have broken out of it. Most of the time, ASND had no trouble either rising 7½ points or falling 7½ points from its starting price during any 3½ month time period. Thus, this looks like an attractive straddle purchase from this "visual test" as well. This process can be quantified using historic price data and some simple programming. In fact, it should be if you are looking at straddles on a stock that has changed its general price level greatly over the two-year period. In that case, you might want to make the height of the box a percentage move for the underlying, rather than just an absolute number as has been done in this example.

The final test, Step 4, checking the fundamentals to see if there is any obvious reason for the low volatility, is just to ensure that there isn't some unforeseen reason for the straddle to be so cheap. Check to see if a takeover offer has been made and accepted; typically, once that happens, the options lose almost all of their volatility because, if the takeover is consummated as forecast, there is very little room for the stock to move up or down. The computer and the concomitant mathematics would not know that, of course, so the straddle would show up as a very attractive buy. In reality, though, it would not be a good buy because of the fundamental news. It behooves you to check through the past news stories just to be sure that the stock will be a freely trading vehicle during the time the straddle is in place.

When all four of these criteria are satisfied, you can feel confident that you have found a straddle purchase with a high probability of success. Of course, that does not guarantee that a profit will be made, but it certainly gets you off on the right foot. Note that this strategy can be used equally well with index options and with futures options. As long as one has access to the implied and historic volatility levels going back over time, the same analyses can be made in those markets as well.

The preceding discussions have not really addressed much about how to handle a position once it has been established, but the following comments may be helpful for straddle buyers. Once the stock (or index or futures contract) makes a move large enough to exceed the break-even points, some profits should be taken. At that time, most of the risk is in the long calls if the stock has moved up, or in the long puts if the stock has moved down. The position is not delta neutral any more.

If that move also includes a breakout of a technical level, then it is best to try to ride the trend. Sometimes very large moves ensue after a straddle is purchased, and you would like to be there on the trend for as long as possible. Therefore, I do not recommended trading against the straddle. Rather, hold onto the profitable side for as long as the underlying does not violate a trailing moving average (say, the 20-day moving average. If the underlying security makes a really large and fairly uniform move, you will make excellent profits as you ride along with the trend.

As for taking losses, a general rule of thumb might be to risk about 60 percent of the initial straddle purchase price. Even if no stock movement occurs right away, this criterion allows you to hold the straddle for at least a couple of months (assuming the straddle had at least three months of life remaining, to begin with). If nothing has happened by then, it might be best to sell this straddle and to look for another one to buy. By the very nature of the strategy, though, the risk is limited to 100 percent of the purchase price, and you should hope for more winners than losers if the initial analysis was done properly (perhaps even some large winners if the stock breaks out and begins a long trend). Therefore, some practitioners of this strategy prefer to hold the straddles all the way to the end. They figure that the stock might make that big move during the last month of the life of the position, and if they exit with a month to go, they will miss it.

In this regard, operating the strategy can become something more of an art than a science. Overall, when operated properly within the guidelines set forth above, this strategy is one of the easiest to monitor. Straddle buyers sleep well at night, because they know that no big, dramatic move in the market or in an individual stock or index can cause them harm; their risk is limited, and any such big move might even be a welcome sight.

Using Option Trading Volume as a Contrary Indicator

So far, we have concentrated on option prices as a contrary indicator. Experienced option traders know that volatility is extremely important—and that is what we have been discussing. But traders should also keep an eye on volume. At the beginning of this chapter, it was shown that volume in stock options can be a harbinger of takeovers or earnings surprises, but that such an approach does not really work for index or futures options. Now, we are going to look at option volume as a contrary indicator for indices and futures. This approach may sometimes work for stock options, but I believe it is less reliable there.

For any given group of options, you can quite easily calculate what is called the *put–call ratio*. It is simply the volume of puts traded on a particular day divided by the volume of calls traded on that day. You might simplistically do this for a small group of options—say, just the OEX Index. Or you might do it for a far broader group such as all equity options, for example. But you should not mix options that do not have any relationship to each other. For example, it is valid to calculate the put–call ratio for options on gold, and it is valid to calculate the put–call ratio for options on corn futures. But there would be no purpose in calculating the put–call ratio for a combination of the two, because gold and corn have little or nothing to do with each other.

The general theory is that if the put–call ratio gets too high, then the majority of investors are buying too many puts—that is, they are too bearish. Consequently, a contrary investor would want to consider a bullish strategy. Conversely, if the put–call ratio gets too low, then, according to this theory, too many people are buying calls and, hence, they are too bullish. In this case, the contrary investor would think about bearish strategies such as selling stock short or buying puts.

It is useful to keep track of the put–call ratio on any sector options that are at least reasonably active. This might include the Banking Index ($BKX) or the Mexico Index ($MEX) or the NASDAQ-100 Index ($NDX) and a lot of others. All the major futures options can be calculated as well. A separate put–call ratio would be calculated for corn, wheat, soybeans, crude oil, sugar, and so forth. When calculating the ratio for futures options, one must sum the puts and calls traded over all the trading months, as far out in time as they go.

Once the daily put–call ratios are calculated and recorded for all groups of options desired, they can be expressed as moving averages going back over time to smooth out any rough daily fluctuations. A 21-day moving average of the put–call ratio has been a good indicator to use in the past.

Figure 10-12 is the 21-day moving average of the CBOE's equity-only put-call ratio. On a daily basis, this ratio includes all the puts and calls traded on stocks (no indices or any other sorts of option) on the CBOE. The daily figure is available from the CBOE's web site at www.cboe.com, and a good deal of historic data is available there as well. Note that the left-hand scale in Figure 10-12 indicates that the equity-only put–call ratio usually ranges between about 30 and 55 on a daily basis. If it gets outside of that range, it is considered to be a fairly extreme daily reading.

Buys and sells have been marked on the charts at local peaks and valleys. A peak indicates "too much" put buying, so it is a buy signal for the market, according to contrary trading theory. A valley indicates "too much" call buying, so it is a sell signal. Figure 10-13 contains OEX price history, and the buys and sells from Figure 10-12 are marked on Figure 10-13 as well. The put–call ratio has given some good signals, going back many years.

Note that no attempt is made to absolutely define the level of where the buys and sells take place. That is, you cannot say that the

FIGURE 10-12

CBOE's Equity-Only Put–Call Ratio, 21-Day Monthly Average

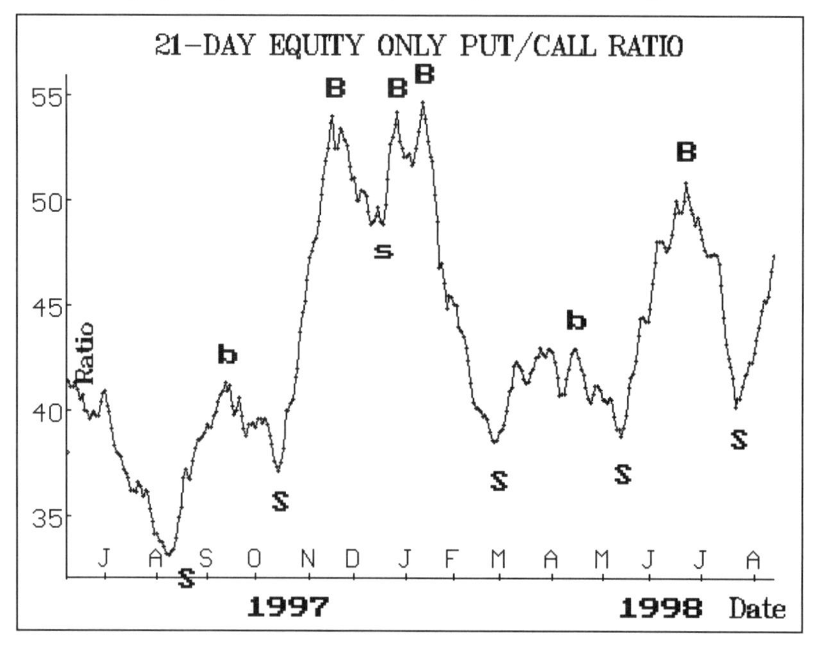

FIGURE 10-13

OEX Price History Chart with Put–Call Buy and Sell Traded

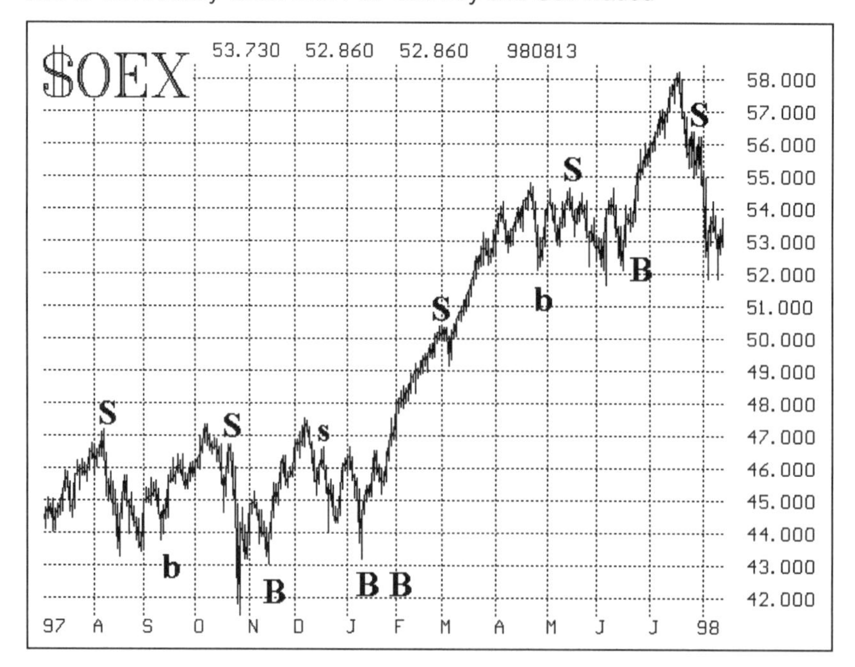

put–call ratio is "too high if it climbs to 50." A signal occurs at local peaks and valleys—wherever they may appear. Understandably, the buy signals take place near the top of the chart, while sell signals occur near the bottom. This dynamic interpretation is important for two reasons: 1) you are not forced to buy into a declining market—which you might have to if you use static numbers as buy levels; and 2) the concept can be extended to any other market without having to define static numbers to constitute a buy or a sell in that other market.

Contrary opinion theory works best when the actions of speculators are observed. Other activities such as arbitrage or institutional covered writes or spread trading add volume to the mix, but these factors do not constitute the type of trading that is useful for forming contrary opinions, because it affects the put–call ratio of the OEX index. Both SPX and OEX options are used heavily by institutions seeking to buy protection for their vast stock holdings. That activity increases the put–call ratio, but it is not useful as a contrary indicator. As a result, the OEX put–call ratio has a spotty track record in the mid-1990s in terms of being a contrary indicator. To understand why, consider this question: If an institutional money manager buys stock and buys puts, is that manager bullish because stock is being purchased or is that manager bearish because puts are being purchased? One could debate the answer, but the point is that a manager has a protected bullish position when trading such a strategy. Therefore that manager's actions cannot be interpreted in contrary opinion theory.

The purest broad-based index options to be used for contrary opinion trading seem to be the S&P 500 futures options, because they are least used by large institutions. Hence they seem to reflect a fair amount of speculative activity, which is what contrary opinion theory needs. Of course, in the future, things may change. Institutions might decide to buy equity put options instead of index put options for protection. If that were to happen, then the OEX put–call ratio might once again take on importance for contrary opinion trading, and the equity-only put–call ratio lose some importance. Back in the 1980s, in fact, the OEX ratio was quite useful (institutions had not yet adopted put buying as a form of insurance, at least not en masse) while the equity-only put–call ratio was not as useful—there was a lot of off-floor conversion and reversal arbitrage in equity options back then that distorted the ratio (that arbitrage is virtually nonexistent now). So, it behooves you to keep track of all the broad-based ratios and to monitor their efficiency.

As stated earlier, the put–call ratio can be used for sector index and futures options, too. When attempting to use the put–call ratios for sector indices or for futures contracts, a general rule of thumb is that the more active the contract is, the better its signals will be. If the option volume is too thin in a particular sector, one or two large orders may distort the ratio unnecessarily. As a result, whenever you are about to utilize a signal in a thinly traded sector, you should check how past signals have fared—just to check the reliability. One word of warning: the put–call ratio does not work well in the grain futures options (wheat, corn, and soybeans), probably because of commercial hedging activity, something akin to what has happened in OEX.

In general, the put–call ratio is a useful technical indicator to have in your arsenal. It can show extreme public opinions, and those opinions can be traded against by contrary opinion traders. The main caution in the use of these ratios is that you should be certain that there is sufficient volume in the options and that there is not a preponderance of hedging activity which might distort the speculative activity that you truly wants to measure.

SUMMARY

Option volume and option prices often are used as an aid in trading the underlying instrument. In this chapter, it was shown that increased volume or a spike in volatility in a stock's options might portend an important corporate news item. Moreover, implied volatility levels (*i.e.*, option prices) and volume can be used as a contrary theory indicator in many other cases. Although no indicator is infallible, these do have a good track record, and the serious option trader may benefit from a working knowledge of implied volatility extremes and put–call ratios.

CHAPTER 11

ELECTRONIC RESOURCES

Floyd Fulkerson

The marvelous thing about the Internet is that anything we write about it in this chapter will be out of date by the time it reaches the printer, much less over the years as this book sits on your bookshelf. Why is this marvelous? In a word, change—big change for the individual investor. At the time of this printing, the number of new people globally gaining access to the World Wide Web doubles, by some accounts, every few months. Along with an expanding user base, the number of Web sites is booming as well. Internet stocks are as hot in the marketplace as the hype about the Net itself. However, most people would admit that high expectations about ways the Internet would have impacted their life by now have not been met. In our opinion there is one major exception to this—investing.

In the early 1990s electronic trading of stocks and options via the Internet was almost unheard of. As we go to press there are already on the order of 3,000,000 online trading accounts established in the United States alone. By some estimates, the electronically traded portion of all retail stock transactions made on a yearly basis is something over 15 percent. This number will only swell in the coming years, especially with more and more investors beyond our borders investing in U.S. equities with a few clicks of a mouse. How can this phenomenon be explained? By two things principally: price and information.

Today there are approximately 50 online brokerage Web sites. We will delve into the world of online trading a little later, but what we can consider now is simply the impact this has had on investing. Intense competition has resulted in plummeting transaction costs for making stock and option trades. This has brought many price-sensitive investors into the world of online investing. Beyond ease of use, speed, and costs, how else are online brokerages to compete with one another?

The method of choice so far has been to provide (or even overwhelm) individual investors with a wealth of information—more valuable information than one could possibly use. In the past the type of investment information now available for free was accessible primarily to professional traders and large institutions for a fee. With respect to market timing at least, easy access to information perhaps gave some professionals an edge. Today easy access to information has brought the independent investors to the World Wide Web—those who like to cull through financial data and reports, assimilate what they consider important, and make their own investment decisions. And what is the result of all this? Certainly a new breed of investor—one who has direct access to the tools that were once the professional's.

Although we make no judgment about the advantages or disadvantages of dealing with an account executive at a full-service brokerage firm relative to trading online, consider what the Internet brings to the individual investor, assuming that investor wants it: efficiency, in terms having direct access (in some cases) to and the ability to trade immediately with the funds in one's account; quotes from any exchange for any type of financial product; all types of financial information about a stock or fund, including the filings each might have made with the government; released and/or projected government economic figures; and remote but direct access to an equity or option trading floor where an order may be sent electronically. All this is accessible whenever the investor wants it, without ever having to contact another human being. By any account, all this appeals to a great many people: investors who relish the affordability and independence the Net provides them; investors who wish to take more direct control over investment decisions and their financial futures; and investors who consider themselves as informed and on a more level playing field with the armies of professional traders who can move markets. What a decade ago took time searching publications for financial data, time consulting with brokers and financial advisors, and more time for placing trades and receiving reports by phone or mail, can now be accomplished at one's leisure with a cell phone and a handheld computer from virtually anywhere.

All this new technology has most certainly been a boon to the financial community, in terms of trading volume on equity and options exchanges alike. More and more, U.S. exchanges find customers coming from abroad—foreign investors sending orders to our trading floors through this proliferation of online brokerage Web sites. The same phenomenon works in reverse. U.S. investors find themselves accessing quotes and financial news from an increasing number of offshore marketplaces. This is the global marketplace we all read and hear about, and anyone with a computer has access to it. As economies and marketplaces become more globally involved, investors' accessibility to these markets improves and trading volumes increase. What kind of potential downside to all of this can exist in the future? Only time will tell what impact these benefits and efficiencies might have on future market volatility.

CHICAGO BOARD OPTIONS EXCHANGE (CBOE)

The Chicago Board Options Exchange (CBOE) has always been a pioneer in the options industry with respect to electronic improvements to its trading systems—from the many ways orders can enter and reports can leave the trading floor, as well as the speed with which option transactions can be executed and cleared during the day on the trading floor. By maintaining an edge in trading technology CBOE strives to offer its customers the best service in the industry. Another area in which the CBOE leads the industry is investor education. Electronically, this is achieved with a highly respected, stand-alone piece of software, *The Options Toolbox,* and a very widely used web site, www.cboe.com, shown in Figure 11-1.

The CBOE Web Site: www.cboe.com

At times "surfing" the Web for the right resources can be a frustrating way to spend a few hours, often yielding slim results. Although financial and economic information for the investor abounds and is generally not hard to locate, information specifically about equity and index options is a little harder to come by. The CBOE's web site offers the option investor convenient one-stop shopping. It is a complete online option trading resource from an authoritative source—the Chicago Board Options Exchange, the first listed option exchange in the world. There you will find a complete array of general market information, option and stock quotes, historical data, educational materials, and interactive tools to assist option investors with their trading decisions. Once you've

FIGURE 11-1

CBOE's Home Page (www.cboe.com)

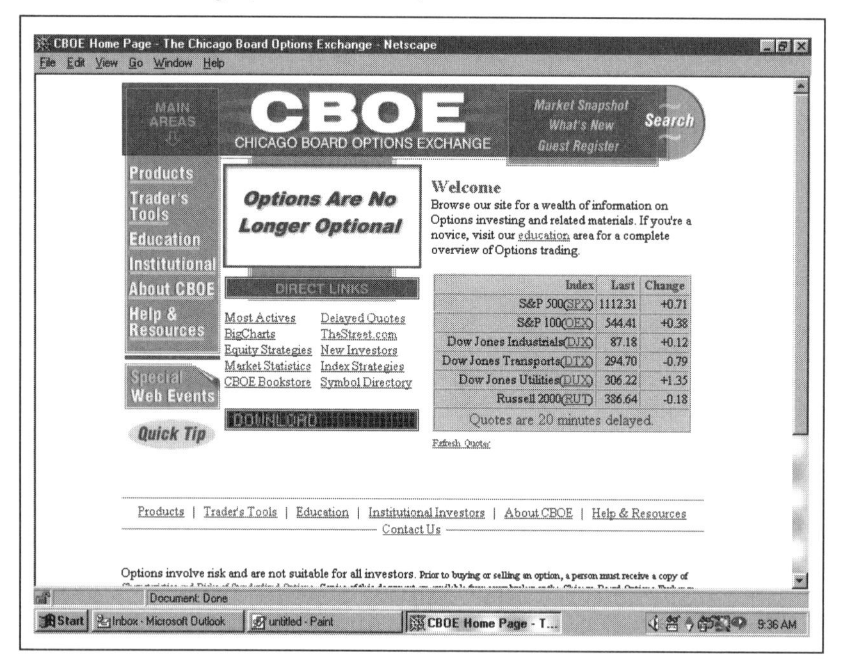

made your decisions, you'll be ready to contact your broker or online brokerage site to make a trade.

At the CBOE Web site, you will find a fully functioning, online version of the CBOE's educational software piece, *The Options Toolbox*. You may order a free copy of program on CD or download the installable files from the site. There is also a functioning option pricing calculator program, apart from the calculating capabilities of the *Toolbox*.

If you love or hate a particular stock, and want to know if it is optionable—*i.e.*, has option contracts listed and traded on it—and if it is, what months and strikes are available, and what their symbols are—the site offers a list of optionable stocks and indexes traded on all U.S. options exchanges. This list is updated monthly and is downloadable, including all new issues for the previous month.

Need to find codes, symbols, or quotes? At cboe.com, for each optionable stock and index listed and traded in the United States on any exchange you will find a full array of option quotes, symbols, and codes.

If you are interested specifically in index options, for each optionable index you will find its component stocks and their weightings in the index (if applicable). In addition, you will find historical settlement values, updated price code changes, and dividend yields.

Need to know what's hot in the options marketplace on any given day? You'll find lists of the Chicago Board Options Exchange's most active option classes (stocks or indexes), and most active calls and puts, updated during the trading day. The site also includes updated historical volatility numbers for all U.S. optionable stocks, stock split and merger announcements, and an interactive stock and index charting facility. Investors looking for financial fundamentals about corporations issuing optionable stock will find research reports including corporate descriptions, past financial performance, and future earnings projections.

The site also has an e-mail service that can be personalized to alert you to changes in much of the timely information posted and updated on the sight, daily or as necessary.

Like most other sites on the World Wide Web, the CBOE's web site offers multiple hyperlinks to many other sites of interest to the option investor. Take advantage of all the free things the Chicago Board Options Exchange is offering the investing public; after a visit to www.cboe.com you'll be ready for a talk with your broker.

The Options Toolbox

Released by the CBOE in 1996, *The Options Toolbox* (Figure 11-2) offers both the option and stock investor a comprehensive, highly interactive course on both equity and index options, as well as LEAPS on both equity issues and equity indexes. Consider *Toolbox* an electronic textbook. The exchange offers it (CD or diskette) free of charge by telephone (800-OPTIONS). Alternatively, there are various ways an investor has access to the *Toolbox* from www.cboe.com.

Section 1 - Basic Definitions

Build a sound foundation by learning more about:

- The difference between a Call option and a Put option
- The benefits of a standardized option contract
- Understanding the rights afforded the option owner
- Understanding the obligations faced by the option writer
- The difference between time value and intrinsic value
- The mechanics of exercise and assignment

FIGURE 11-2

CBOE's Options Toolbox

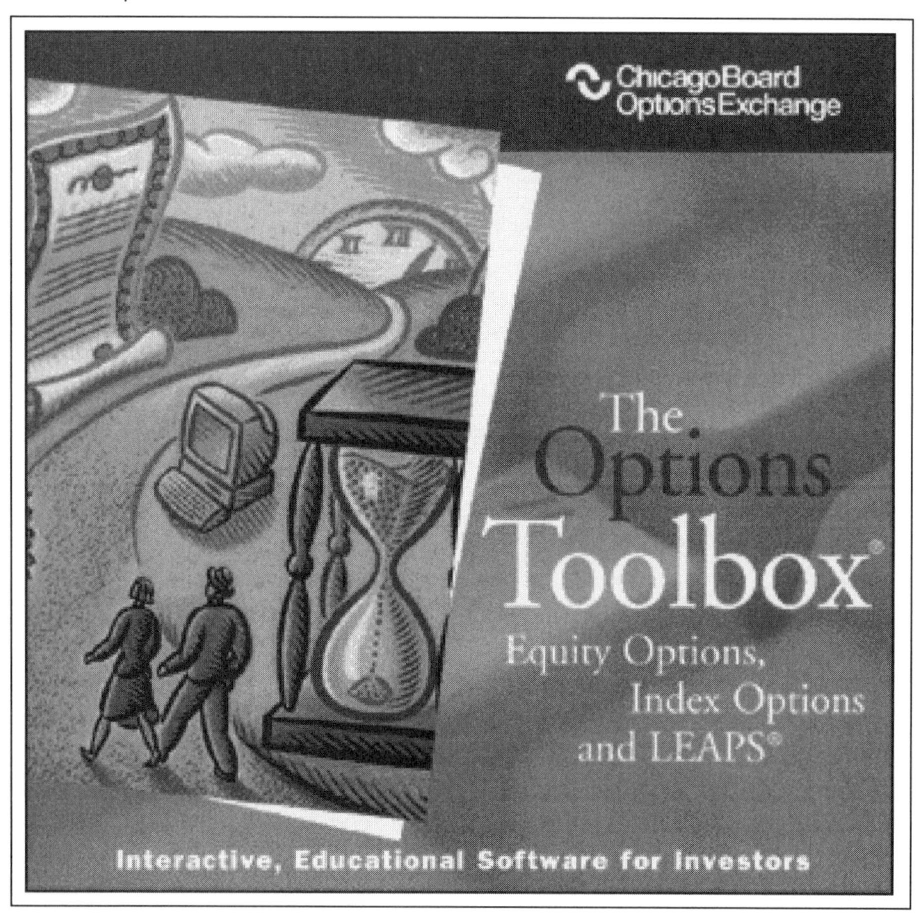

Section 2 - Principal Factors Affecting an Option's Price

Learn how the value of an equity option is impacted by:

- The underlying stock's price and the option's strike price
- Volatility
- Time until expiration
- Dividends and interest rates
- Supply and demand
- Market sentiment and liquidity

Section 3 - Placing an Option Order

Use sample option order tickets to:

- Learn about simple buy and sell orders as well as spreads
- Learn about market orders, limit orders, and order contingencies
- Prepare to talk to a broker
- Practice building buy, sell, and spread orders

Section 4 - Construct and Test a Simulated Option Position

Simulate the marketplace with *The Options Toolbox* to:

- Review and test strategies that are consistent with any given market opinion
- Review and test the many strategies useful to stock investors
- Test a strategy to understand how the passage of time affects an option position
- Test a strategy to understand how changes in volatility impact an option position
- Perform "what-if" analyses on your own simulated strategy, with prices of your choice

LEAPS (Long-Term Equity AnticiPation Securities)

Expand your knowledge of equity options by:

- Understanding the LEAPS contract in contrast to a conventional equity option
- Appreciating the many benefits to the option and/or stock investor of using LEAPS
- Understanding how conventional option pricing factors might be different with LEAPS
- Testing the *Toolbox's* many example LEAPS strategies
- Performing "what-if" analyses on your own simulated equity LEAPS strategies

Index Options

To improve your understanding of index options this section covers:

- The differences between equity and index options
- Types of indexes and how they are calculated
- Special considerations when using index options and cash settlement
- The mechanics of exercise and assignment of European options
- Commonly used index option strategies

If you have an opinion on an underlying stock, the broad market, or a particular sector of the market, but don't have a strategy in mind, a unique feature of the *Toolbox* is the large number of examples you can analyze and test over time. Alternatively, you can create and test a strategy of you own. Because each strategy can have its own risk/reward profile, you can decide which one fits your own financial situation and comfort level before jumping into the marketplace.

PAPER RESOURCES

There are plenty of ways to keep informed about the world of electronic investing without having to fire up your computer. There's certainly nothing wrong with reading a newspaper, a magazine, or even a book now and then—paper can be good! For those of you eager to reap the most current information and enjoy the latest technology the web has to offer, keep an eye on major financial publications for regular columns, feature articles, cover stories, and even special issues on electronic investing.

Investing Online For Dummies, from the publishers of the now-familiar yellow and black "For Dummies" book series, covers dozens of online resources of general interest to the online investor. The book comes with a CD that includes demo or limited-use versions of various investing software pieces, as well as an installable, fully working version of CBOE's *The Options Toolbox*.

Online Investor is a magazine specifically for the cyber-minded, which can be found on major newsstands. It focuses on specific financial web sites of all types, and some specifically for option traders. It will also keep you up-to-date on quote services and the latest in the electronic brokerage business. Their site address is www.onlineinvestor.com.

Futures magazine is a monthly publication geared to today's futures and option investor. Once a year, in late spring, it prints a special

"Guide to Computerized Trading" issue, which is a directory of products and services for the electronic investor. It lists vendors of software products, online/Internet services, and sources for financial quotations and data. If you do not subscribe to this magazine you can purchase the special issue by itself. Check their Web site, www.futuresmag.com, and/or give them a call (toll-free at 888-898-5514).

Stocks & Commodities is a monthly magazine that regularly reviews software for stock and option investors, and archives these reviews in special issues from time to time. For more information about these issues, and for lots of links to sites useful to the option investor, see their Web site, www.traders.com.

The weekly financial newspaper, *Barron's* (www.barrons.com), prints a regular column titled "Electronic Investor." In addition, it occasionally features extensive articles (sometimes cover stories) on top financial Web sites for investors and top online brokerage sites. For those who read *Investor's Business Daily* (www.investors.com), there are columns and forums all over the newspaper where you will frequently find mention of various financial Web sites for investors, as well as discussions on different aspects of electronic investing.

Various popular glossy financial magazines, *e.g., Smart Money* (www.smartmoney.com), *Forbes* (www.forbes.com), and *Business Week* (www.businessweek.com) will from time to time cover, rate, and keep you abreast of the latest in financial Web sites.

THE WORLD WIDE WEB

Let's now hop online and cruise the Web to visit a few sites of interest to the option investor. Before we do, however, there are some issues that you should consider each time you warm up your browser and search engine: reliability, validity, rumors, and manipulation. As you probably know, the Web is a largely unregulated environment. Anyone who wants to post a fact or comment on the Internet is free to do so with little or no accountability. Facts and figures don't have to be correct, and frequently aren't.

Watch out particularly when you go to Internet user groups, chat rooms, discussion groups, and bulletin boards—sites where users can interact live or post material for others to read. Although sites such as these allow reliable information to be shared among credible, honest people, they can also be incubators for rumors being spread by those motivated by greed or by malcontents. Just be aware that stories can get started with the intention of manipulating the market price of any type of traded security. Would you blindly follow a tip you overheard at the

Xerox machine or in a crowded bar full of strangers? Why then would you follow such a tip from an anonymous source on the Web?

Hopping from one web location to another via the hyperlinks conveniently offered on most sites can make surfing easier and more interesting. However, when linking from one place to another you can frequently find yourself in unknown territory. Try to know where you are going, or at least understand where you are when you get there. Have some confidence in the source of the information you are reading, and cross-check facts and figures as much as possible. Make sure the information you are reading is up-to-date. These are things that we should be concerned with when doing any kind of research, but the quantity of information on the Internet can overwhelm those searching for quality.

Online services like Prodigy, America Online, and CompuServe offer their customers pre-packaged personal finance information, frequently conglomerations of services from vendors you've heard of already. If you're using one of these services just know that there is a wealth of information on the World Wide Web beyond those walls. The more popular Internet search engines like Yahoo, AltaVista, Lycos, or Excite offer investors direct access from their interfaces to pages offering market summaries, quotes, financial news, and links to brokerage services. Of course these pages range in degree of usefulness depending on the investor.

Here is a look at a few specific Web sites of different categories to just introduce you to the Web world of investing if you are unfamiliar with it. If you are on the Web investing already, some of these sites might serve as additional resources for you. For every Web site mentioned here there are tens of sites not mentioned, and in the future who knows how many more. No attempt is made to get too specific about the sites, nor to make judgments about quality of interface or content. That task is left up to online and print journalists who can do this on a more timely basis.

SITES OF GENERAL INTEREST TO INVESTORS

What's the market doing? Where was XYZ trading two weeks ago, or two hours ago? What's the Fed doing? How's consumer confidence? Which corporate insiders are selling or buying? These are things you might find yourself wondering about from time to time. Instant answers to these and many similar questions can be found at these sites and others like them. You will likely encounter some sites that seem to contain little beyond a number of hyperlinks. Carefully take advantage of these.

Big Charts (http://www.bigcharts.com). They've got the data, you pick the parameters and indicators. You can create almost any kind of stock or index chart you could want. For the market technician it's a valuable site. Free service. Access provided also through CBOE site.

Briefing.com (http://www.briefing.com). Quick, concise, objective analyses of important news affecting stocks and bonds. Analyses of individual stocks and the factors that affect share (and option) values tomorrow. Some free services, some fee-based.

CBS MarketWatch (http://cbs.marketwatch.com). Includes the usual market quotes, company information, portfolio tracking, stock screening, charting utilities, etc. Its strength lies in its reporting and analysis of breaking new stories, a variety of market commentary, and a primer for the new investor. Weaker on option quotes, but an excellent resource. Currently all free stuff.

CNNfn's Market Briefing (http://www.cnnfn.com/pushthis/briefing.html). E-mailed daily to you just as U.S. financial markets close and Asian markets open, this Market Briefing delivers a roundup of each day's trading activity from all the major stock and commodities exchanges, as well as breaking business news and feature stories.

DailyStocks.com (http://www.dailystocks.com). Option quotes, stock quotes, and more quotes, from the United States and abroad. Market summaries and plenty of market indicators. Corporate earnings and research, SEC filings, charting. Incredible number of links. More info than you could ever need.

FinancialWeb (http://www.financialweb.com). Good site made up of links to other free financial sites, including an option analytic site.

Hoover's Online (http://www.hoovers.com). Comprehensive source for information on corporations and industries. Fee based, but some free info provided.

InfoBeat (http://www.infobeat.com). Receive a daily e-mail message containing closing prices and news for a personalized portfolio of market indices, mutual funds, and securities from the three major U.S. equity exchanges. If you like, receive news alerts during the day if hot news comes across the wire for one of the companies you follow.

Quote.com (http://www.quote.com). Requires subscription, but quotes on over 12,000 stocks, 100,000 options, 300 indexes, 500 commodity futures, 7,500 mutual funds. Some free stuff.

Securities and Exchange Commission (SEC) (http://www.sec.gov). Archive of corporate filings with the SEC including annual reports, prospectuses, corporate insider trading reports, and the like.

TheStreet.com (http://www.thestreet.com). Heavy on market news and insightful, timely commentary, but includes some standard investing tools (tracking, charting, etc.). Market updates sent by e-mail. Limited access provided by CBOE site. Some free, some fee-based stuff.

White House Economic Statistics Briefing Room (http://www.whitehouse.gov/fsbr/esbr.html). Provides easy access to current Federal economic indicators and statistics: *e.g.*, unemployment figures, price indexes, production numbers, interest rates, etc.

Wired News (http://www.wired.com/news). Specializes in news about technology stocks. Receive newsletters, intra-day and end-of-day, about the high tech industry. Free.

Yahoo Finance Home Page (http://quote.yahoo.com). Highly regarded mega site. Quotes, numerous news sources, company profiles, research, earnings forecasts and surprises, chat rooms and forums, charting. Look for a feature called "My Yahoo!" that allows you to filter news reports to reflect your portfolio.

Zack's Investment Research (http://www.zacks.com). Offers quotes, daily e-mail portfolio alerts, comprehensive fundamental analyses of companies, insider trading reports, and detailed analyst recommendations. Some fee-based. Access to free stuff also via CBOE site.

Sites Specifically for Option Investors

As this text goes to press there are not many sites on the World Wide Web designed specifically for the option investor, but more are being planned. Notice that the last site mentioned in this category is peripherally related to listed-option trading. It is a site for employees of corporations who are at least in part compensated with company-issued stock options.

American Stock Exchange (http://www.amex.com). Of particular interest is the area of the site devoted to Amex Derivative Securities.

Chicago Board Options Exchange (http://www.cboe.com). This site covered in detail above.

McMillan Analysis Corporation (http://www.optionstrategist.com). Lawrence G. McMillan's *(The Option Strategist)* short-term stock and options trading site on the latest techniques and strategies for trading a variety of innovative options products. Free weekly commentary, quotes, historical and implied volatility data, and other tools. Lots of free info; some features for a fee.

OptionsAnalysis.com (http://www.optionsanalysis.com). This site offers quotes, graphs of your personal trades, paper trading, historical backtesting of option trades of the past, an option trade scanner, and all sorts of volatility rankings. Fee-based, but some free data.

Options Clearing Corporation (http://www.theocc.com). The OCC's site offers access to the latest bulletins about specific option classes and underlying stocks from the different options exchanges as well as links to their sites.

Options Industry Council (http://www.optionscentral.com). Keep an eye on this site for a list of free seminars offered around the country as well as access to an options symbols guide.

Power Analyzer (http://www.optionstrader.com). Searches the marketplace and rates option trading opportunities (*e.g.*, covered calls, straddles, spreads) according to its parameters. Fee-based.

StockOptions.com (http://www.stock-options.com). Internet resource for employees, officers, and directors of public corporations for calculating, inventorying, and analyzing employee stock-options. Free.

Value Line (http://www.valueline.com). This well known service for investors offers a product specifically for the option trader, *The Value Line Daily Options Survey*. A proprietary rating system searches the marketplace and recommends bullish and bearish strategies. Fee-based with trial versions available.

If you employ an Internet search engine to hunt for new option-related Web sites, using "option" as the key word for your search will not be specific enough. You will most likely get a list of literally millions of sites having nothing to do with listed options. Use terms such as: *option pricing, option analysis, option valuation, option strategies,* or *option trading*. Or, you might try searching for terms related to pricing models: *option pricing model, Black-Scholes, binomial pricing model,* or *option theoretical values*.

Sites as Resources for Electronic Trading and Education

Futures Magazine (http://www.futuresmag.com). Online resource with links to vendors of quotes and data, news services, online analysis, and stand-alone software. Purchase their special yearly issue, "*Guide to Computerized Trading*," from here.

Gomez Advisors (http://www.scorecard.com). Rates online brokerage sites with respect to many parameters: *e.g.*, ease of use, customer confidence, overall cost.

Technical Analysis of Stocks & Commodities Magazine (http://www.traders.com). A resource for archived software reviews from the magazine as well as links to software producers. Offers lots of links to other potentially useful sites for the online investor.

Traders' Library (http://www.traderslibrary.com). This bookstore offers a comprehensive selection of titles for the stock and option investor. Order online, by phone, or request a paper catalogue. There are usually books on sale, with some free stuff (sample newsletters and magazines) offered.

Wall Street Directory (http://www.wsdinc.com). A central resource with links to software producers, quote and data vendors, news services, etc. Includes a calendar of seminars. You can buy books and subscribe to newsletters and magazines.

Online Brokerage Services

As of press time, there are over 50 sites on the World Wide Web for brokerage. Most of these sites execute option trades as well as listed and over-the-counter equity transactions. If your full-service brokerage firm does not have a reduced-fee site, it might soon. Some of the discount sites that full-brokerage firms offer investors have names markedly different from the parent company's, so consult your broker. Financial periodicals, especially newspapers, are full of advertisements for online brokerage services. Competition is intense so keep up with what is going on with respect to changing fees and enhanced services.

There will no doubt be consolidation in this new industry at some point in the future. There will also be advances in technology that improve service. For these reasons we do not attempt to be specific about any particular brokerage site, nor to rate them in any way. The media are constantly doing this for you, as well as a couple of rating services accessible on the Web. We suggest you keep the following issues in mind when choosing an online broker:

- Who are these people? Know who you are dealing with; *i.e.*, who is backing your trades? Decide for yourself if the company is reputable. Do they have insurance, specifically SIPC? There are too many reputable firms doing business on the Net to let a couple of bad ones tarnish the concept.
- Ease of use. How many screens do you need to access to make a trade? In other words, do you have to go to one or more places for a quote, then go to another place to actually fill in the data to make a trade? Ideally this should all happen on one page. Some sites are much more cumbersome than others. How are the trades verified: on your computer screen, via fax, or by telephone? Do you get an instant trade report, or are the trades not "logged" and your account updated until the next day? You have to decide how important this is for you. Does the site have dummy areas for testing the order input process? If you are trading online and under the impression that your trades are being transmitted electronically, make sure that they really are. You might find the occasional site that accepts a trade from the Internet via your computer but then transmits that order to another place by telephone.
- One-stop shopping. Is there an adequate amount of data other than that needed immediately for trading on the site? How about market summaries? What stocks or options are the most active for the day? Are there news headlines? Some sites allow you to customize what you see. You might as well keep abreast

of what's happening in the marketplace on the site where you trade.

- Reliability and/or accessibility. How easy is the site to pull up time and time again? Are there memory-intensive graphics slowing things down, and is the server always accessible? You might test this on busy days as well as when the market isn't doing much. As important, are there alternate means of communication with this brokerage firm, in case there is a problem with a trade or with your computer? Is the customer service department open at all hours, or just during market hours? (A word to the wise: when you receive trade confirmations or notices of cancellation ("outs") always print them for your records.)
- Commissions. This is really a personal issue, *i.e.*, balancing service you find adequate or satisfying with the cost. To reduce surprises understand on what your commissions are based, *e.g.*, on number of contracts, or underlying value of contracts. Is there a minimum charge? Does it cost more to ring up the broker than to make the trade on your computer?
- Quotes. What type of quotes are you seeing at the site? There are *delayed* quotes (commonly by 15–20 minutes) and *real-time* quotes (not delayed , *i.e.*, current when you push the Enter key on your keyboard). If the quotes are real-time, are they *dynamic* (update automatically as the quotes changes in the marketplace without your having to keep pressing your Enter key)? Know how fresh the quote is when you are both making an investment decision or preparing to enter a trade!
- Cash. Is there a minimum (opening) cash balance? Are you paid interest on cash balances in your account?

OPTIONS VALUATION SOFTWARE

Below is a list compiled by The Options Institute of option valuation software offered by different vendors. Some programs are considered "stand-alone" because they do not require a separate feed of option price data (extra expense) from another service; they rely on the user to input some simple figures for calculations. Other, more elaborate programs that require a data feed tend to be much more expensive and are usually oriented more to a serious, active trader rather than an occasional user.

(The Options Institute and the Chicago Board Options Exchange make no representation, expressed or implied, concerning the suitability or merits of any software for option valuation. It simply would be impossible for us to test and evaluate each software package under all the various scenarios; it would be improper for us to recommend a software program without understanding your unique needs as a user. Please feel free to call these vendors, certainly check their Web sites and, if offered, request or download a demonstration version and make your own informed choice.)

The Options Toolbox. Free. Windows. Chicago Board Options Exchange, 400 S. LaSalle Street, Chicago, IL 60605. 800-OPTIONS. www.cboe.com. Everything you ever wanted to know about equity options, index options, and LEAPS in an interactive, tutorial course. Includes in-depth discussions and analyses for many example strategies as well as a strategy of user's own design. Call for a free CD or download from Web site.

Op-Eval-3. $25. Windows. Op-Eval, 2501 N. Lincoln Avenue, #200, Chicago, IL 60614. Very good, basic equity or index option pricing calculator with changeable inputs. Multiple option positions and graphing capability. Demo disk available.

Options Laboratory. $89.95. Windows. Mantic Software Corporation, 5918 Jordan Drive, Loveland, CO 80537. 800-730-2919. www.manticsoft.com. Software with textbook. Good interactive pricing of single or multiple positions with changeable inputs. Graphs for profit and loss, delta, and gamma. Other products available.

Option Wizard. $99.95. Windows and Mac. Sarkett & Associates, Inc., 485 Sunset Road, Winnetka, IL 60093. 847-446-2222. www.option-wizard.com. Microsoft Excel product that prices options. Calculates historical and implied volatilities, in-the-money probabilities, percent-to-double, delta, and premium decay. Other products available.

Option Simulator. $145. Windows. Bay Options, 1235 Walnut Street, Berkeley, CA 94709. 510-845-6425. www.bayoptions.com. Evaluates strategies with options on stock, indexes, or futures. Graphs position, profit and loss, delta, gamma, theta, and vega. Other products available.

Option Master. $159. Windows and Mac. Institute for Options Research, Inc., P.O. Box 6586, Lake Tahoe, NV 89449. 800-407-2422. www.options-inc.com. Stand-alone options analysis piece. Calculates Greeks (delta, gamma, theta, etc.), historical and implied volatility, and probability of profit for single options or strategies. More products available.

OptionTrader. $595. Windows. AustinSoft, Inc., 2 World Trade Center, Suite 1888, New York, NY. 10048. 800-538-3822. www.austin-soft.com. Industrial-strength software for the average consumer. Takes data feed for options analysis, viewing and scanning marketplace and option strategies. Other products available.

OptionOracle. $695 Windows. Deltasoft Financial Technologies, Inc., 3639 Harbor Boulevard., Suite 213, Ventura, CA 93001. 800-250-7866. www.option-oracle.com. Combines options analysis, charting, and portfolio tracking. Takes end-of-day data and scans for trades based on your outlook and assumptions. Other products available.

Option Vue. $995. Windows. Option Vue Systems International, Inc., 1117 S. Milwaukee, Libertyville, IL 60048. 800-733-6610. www.option vue.com. Complete decision support system for serious options trading. Takes live or delayed data. What-if scenarios, strategy evaluation, trade recommendations, historical data. More products / services available.

Option Pro On-Line. $795-$1,295. Windows. Essex Trading Co., 107 N. Hale, Suite 206, Wheaton, IL 60187. 800-726-2140. www.essextrading.com. Complete option valuation, trade searching, volatility ranking, and position tracking analytics for stock, stock index, and futures options. Versions for either live or end-of-day data feed. Other products available.

Option Station. $1,799. Windows. Omega Research, Inc., 8700 W. Flagler Street, Suite 250, Miami, FL 33174. 305-551-9991. www.omegaresearch.com. For serious option traders. Takes live, delayed, or end-of-day data. Searches for strategies based upon user-defined assumptions. What-if scenarios, graphing. Other products available.

QUOTATION AND DATA SERVICES

Beyond quotes received for free or for a fee on the Internet, some investors have the need for better access to quotes, especially option investors needing comprehensive option quotes for stand-alone, option evaluation software. In addition to daily quotes, some investors might want historical option data: time and sale information or perhaps volatility numbers. All these things are available to individual investors. With respect to market quotes, real-time quotes are generally more expensive than delayed or end-of-day quotes. With real-time quotes, dynamic quotes cost more than those that are not updated dynamically. You might have to pay fees to the individual exchanges from which you want quotes. You might not be given bid and ask prices, just last sales. For historical data a general rule is: the older the data the more expensive. There are all sorts of quote and data packages to choose from at widely varying rates. Following is a list of some of the more popular services (each has a Web site):

- BMI Quotes (www.bmiquotes.com)
- DBC Online (www.dbc.com)
- InterQuote (www.interquote.com)
- PC Quote (www.pcquote.com)
- Standard & Poor's ComStock (www.spcomstock.com)
- Trak Data (www.tdc.com)

GLOSSARY

adjusted strike price Strike price of an option, created as the result of a special event such as a stock split or a stock dividend. The adjusted strike price can differ from the regular intervals prescribed for strike prices. See **strike price interval**.

adjusting A dynamic trading process by which a floor trader with a spread position buys or sells options or stock to maintain the delta neutrality of the position. See **delta**.

aggregate exercise price The total dollar value transferred in settlement of an exercised option.

American option An option that can be exercised at any time prior to expiration. See **European option**.

arbitrage A trading technique that involves the simultaneous purchase and sale of identical assets or of equivalent assets in two different markets with the intent of profiting by the price discrepancy.

ask price The price at which a seller is offering to sell an option or stock.

assignment Notification by The Options Clearing Corporation to the writer of an option that a holder of the option has exercised and that the terms of settlement must be met. Assignments are made on a random basis by The Options Clearing Corporation. See **delivery** and **exercise**.

at-the-money A term that describes an option with an exercise price that is equal to the current market price of the underlying stock.

automatic exercise Same as **exercise by exception**.

averaging down Buying more of a stock or an option at a lower price than the original purchase so as to reduce the average cost.

backspread A delta-neutral spread composed of more long options than short options on the same underlying stock. This position generally profits from a large movement in either direction in the underlying stock.

barrier options Various options that cease to exist or are automatically exercised once a stated level is reached by the underlying. Mostly traded over-the-counter. CAPS are exchange-traded barrier options.

bearish An adjective describing the belief that a stock or the market in general will decline in price.

bear spread One of a variety of strategies involving two or more options (or options combined with a position in the underlying stock) that will profit from a fall in price in the underlying stock.

bear spread (call) The simultaneous sale of one call option with a lower strike price and the purchase of another call option with a higher strike price.

bear spread (put) The simultaneous purchase of one put option with a higher strike price and the sale of another put option with a lower strike price.

beta A measure of how closely the movement of an individual stock tracks the movement of the entire stock market.

bid price The price at which a buyer is willing to buy an option or stock.

Black-Scholes model A mathematical formula used to calculate an option's theoretical value from the following inputs: stock price, strike price, interest rates, dividends, time to expiration, and volatility.

book Same as **public order book**.

box spread A four-sided option spread that involves a long call and short put at one strike price as well as a short call and long put at another strike price. In other words, this is a synthetic long stock position at one strike price and a synthetic short stock position at another strike price.

break-even point A stock price at option expiration at which an option strategy results in neither a profit nor a loss.

broker A person acting as an agent for making securities transactions. An "account executive" or a "broker" at a brokerage firm deals with customers. A "floor broker" on the trading floor of an exchange actually executes someone else's trading orders.

bullish An adjective describing the belief that a stock or the market in general will rise in price.

bull spread One of a variety of strategies involving two or more options (or options combined with a stock position) that will profit from a rise in price in the underlying.

bull spread (call) The simultaneous purchase of one call option with a lower strike price and the sale of another call option with a higher strike price.

bull spread (put) The simultaneous sale of one put option with a higher strike price and the purchase of another put option with a lower strike price.

butterfly spread A strategy involving four options and three strike prices that has both limited risk and limited profit potential. A long call butterfly is established by buying one call at the lowest strike price, selling two calls at the middle strike price, and buying one call at the highest strike price. A long put butterfly is established by buying one put at the highest strike price, selling two puts at the middle strike price, and buying one put at the lowest strike price.

buy-write Same as **covered call**.

CBOE The Chicago Board Options Exchange. CBOE opened in April 1973, and is the oldest and largest listed options exchange.

CFTC The Commodity Futures Trading Commission. The CFTC is the agency of the federal government that regulates commodity futures trading.

calendar spread Same as **time spread**.

call option A contract that gives the holder the right (but not the obligation) to purchase the underlying stock at some predetermined price. In the case of American call options, this right can be exercised at any time until the expiration date. In the case of European call options, this right can only be exercised on the expiration date. For the writer (or grantor) of a call option, the contract represents an obligation to sell stock to the holder if the option is exercised.

carrying cost The interest expense on money borrowed to finance a stock or option position.

cash settlement The process by which the terms of an option contract are fulfilled through the payment or receipt in dollars of the amount by which the option is in-the-money as opposed to delivering or receiving the underlying stock.

Christmas tree spread A strategy involving six options and four strike prices that has both limited risk and limited profit potential. For example, a long call Christmas tree spread is established by buying one call at the lowest strike, skipping the second strike, selling three calls at the third strike, and buying two calls at the fourth strike.

class of options A term referring to all options of the same security type—either calls or puts—covering the same underlying security.

closing price The final price at which a transaction was made, but not necessarily the settlement price. See **settlement price**.

closing rotation See **trading rotation**.

closing transaction A reduction or an elimination of an open position by the appropriate offsetting purchase or sale. An existing long option position is closed out by a selling transaction. An existing short option position is closed out by a purchase transaction.

collateral Securities against which loans are made. If the value of the securities (relative to the loan) declines to an unacceptable level, this triggers a *margin call*. As such, the investor is asked to post additional collateral or the securities are sold to repay the loan.

combination spread An option technique involving a long call and a short put, or a short call and a long put. Such strategies do not fall into clearly defined categories, and the term *combination* is often used very loosely. This tactic is also called a *fence strategy*. See **fence**.

commodities See **futures contract**.

condor spread A strategy involving four options and four strike prices that has both limited risk and limited profit potential. A long call condor spread is established by buying one call at the lowest strike, selling one call at the second strike, selling another call at the third strike, and buying one call at the fourth strike. This spread is also referred to as a *flat-top butterfly* or a *top hat spread*.

consecutive expiration cycle See **cycle**.

contingency order An order to conduct one transaction in one security that depends on the price of another instrument. An example might be, "Sell the XYZ Jan 50 call at 2, contingent upon XYZ being at or below $49½."

contract size The amount of the underlying asset covered by the option contract. This is 100 shares for one equity option unless adjusted for a special event, such as a stock split or a stock dividend. For index options, the contract size is the index level times the index multiplier.

conversion An investment strategy in which a long put and a short call with the same strike price and expiration are combined with long stock to lock in a nearly riskless profit. The process of executing these three-sided trades is sometimes called *conversion arbitrage*. See **reverse conversion**.

cover To close out an option position. This term is used most frequently to describe the purchase of an option to close out an existing short position for either a profit or loss.

covered call An option strategy in which a call option is written against long stock on a share-for-share basis.

covered combination Same as **covered strangle**.

covered option An open short option position that is fully collateralized. If the holder of the option exercises, the writer of the option will not have a problem fulfilling the delivery requirements. See **uncovered option**.

covered put An option strategy in which a put option is written against a sufficient amount of cash (or T-bills) to pay for the stock purchase if the short option is assigned.

covered straddle An option strategy in which one call and one put with the same strike price and expiration are written against 100 shares of the underlying stock. In actuality, this is not a "covered" strategy because assignment on the short put would require purchase of stock on margin.

covered strangle A strategy in which one call and one put with the same expiration—but different strike prices—are written against 100 shares of the underlying stock. In actuality, this is not a "covered" strategy because assignment on the short put would require purchase of stock on margin. This method is also known as a *covered combination.*

credit Money received in an account either from a deposit or a transaction that results in increasing the account's cash balance.

credit spread A spread strategy that increases the account's cash balance when it is established. A bull spread with puts and a bear spread with calls are examples of credit spreads.

curvature Same as **gamma**.

cycle The expiration dates applicable to the different series of options. Traditionally, there are three cycles: January/April/July/October, February/May/August/November, and March/June/September/December. Today, equity options expire on a sequential cycle that involves a total of four option series: two near-term months and two far-term months. For example, on January 1, a stock traditionally in the January cycle will be trading options expiring in January, February, April, and July. Index options, however, expire on a consecutive cycle that involves the four near-term months. For example, on January 1, index options will be trading options expiring in January, February, March, and April.

day trade A position that is opened and closed on the same day.

debit Money paid out from an account either from a withdrawal or a transaction that results in decreasing the cash balance.

debit spread A spread strategy that decreases the account's cash balance when it is established. A bull spread with calls and a bear spread with puts are examples of debit spreads.

decay See **time decay**.

delivery The process of meeting the terms of a written option when notification of assignment has been received. In the case of a short call, the writer must deliver stock and in return receives cash for the stock sold. In the case of a short put, the writer pays cash and in return receives the stock purchased.

delta A measure of the rate of change in an option's theoretical value for a one-unit change in the price of the underlying security.

delta-neutral spread A trading strategy, sometimes used by professional market makers, that matches the total long deltas of a position (long stock, long calls, short puts) with the total short deltas (short stock, short calls, long puts).

diagonal spread A strategy involving the simultaneous purchase and sale of two options of the same type that have different strike prices and different expiration dates. Example: Buy 1 May 45 call and sell 1 March 50 call.

discount An adjective used to describe an option that is trading below its **intrinsic value.**

discretion Freedom given to the floor broker by an investor to use his judgment regarding the execution of an order. Discretion can be limited, as in the case of a limit order that gives the floor broker ⅛ or ¼ point from the stated limit price to use his judgment in executing the order. Discretion can also be unlimited, as in the case of a market-not-held order. See **market-not-held order**.

dynamic hedging A short-term trading strategy generally using futures contracts to replicate some of the characteristics of option contracts. The strategy takes into account the replicated option's delta and often requires adjusting.

early exercise A feature of American options that allows the holder to exercise an option at any time prior to the expiration date.

edge (1) The spread between the bid and ask price. This is called the *trader's edge.* (2) The difference between the market price of an option and its theoretical value using an option pricing model. This is called the *theoretical edge.*

equity In a margin account, this is the difference between the securities owned and the margin loans owed. It is the amount the investor would keep if all positions were closed out and all margin loans paid off.

equity option An option on a common stock.

equivalent position Same as **synthetic position**.

European option An option that can be exercised only on the expiration date. See **American option**.

ex-dividend date The day before which an investor must have purchased the stock in order to receive the dividend. On the ex-dividend date, the previous day's closing price is reduced by the amount of the dividend because purchasers of the stock on the ex-dividend date will not receive the dividend payment.

exercise To invoke the rights granted to the holder of an option contract. In the case of a call, the option holder buys the underlying stock from the option writer. In the case of a put, the option holder sells the underlying stock to the option writer.

exercise by exception A procedure used by The Options Clearing Corporation to exercise in-the-money options, unless specifically instructed by the holder of the option not to do so. This procedure protects the holder from losing the intrinsic value of the option because of failure to exercise. Unless instructed not to do so, The Options Clearing Corporation will exercise all equity options of 75 cents or more in-the-money in customer accounts, and 25 cents or more in firm and market-maker accounts. For index options subject to cash settlements, The Options Clearing Corporation, unless instructed not to do so, will exercise all index options 25 cents or more in-the-money in customer accounts, and a penny or more in firm and market-maker accounts.

exercise cycle See **cycle**.

exercise limits The total number of puts and/or calls that a holder is allowed to exercise during any five consecutive trading days.

exercise price The price at which the holder of an option can purchase (call) or sell (put) the underlying stock from or to the option writer.

exotic options Various over-the-counter options whose terms are very specific, and sometimes unique. Examples include *Bermuda options* (somewhere between American and European type, this option can be exercised only on certain dates) and *look-back* options (whose strike price is set at the option's expiration date and varies depending on the level reached by the underlying security).

expiration date The date on which an option and the right to exercise it cease to exist.

extrinsic value Same as **time value**.

fair value A price that favors neither buyer nor seller. In the case of options, this term is often used to describe the theoretical value of an option derived from a mathematical formula.

fence A strategy involving a long call and a short put, or a short call and long put at different strike prices with the same expiration date. When this strategy is established in conjunction with the underlying stock, the three-sided tactic is called a *risk conversion* (long stock) or a *risk reversal* (short stock). This strategy is also called a *combination*. See **conversion** and **reverse conversion**.

FLexible EXchange options (FLEX) Customized equity and equity index options. The user can specify, within certain limits, the terms of the options, such as exercise price, expiration date, exercise type, and settlement calculation. Can only be traded in a minimum size, which makes FLEX an institutional product.

floor broker A trader on an exchange floor who executes trading orders for other people.

floor trader An exchange member on the trading floor who buys and sells for his own account and therefore functions as a market maker.

frontrunning An illegal securities transaction based on prior nonpublic knowledge of a forthcoming transaction that will affect the price of a stock.

fundamental analysis A method of determining stock prices based on the study of earnings, sales, dividends, and accounting information.

fungibility Interchangeability resulting from standardization. Options listed on national exchanges are fungible, while over-the-counter options generally are not.

futures contract A contract calling for the delivery of a specific quantity of physical goods or a financial instrument (or its cash value) at some specific date in the future. There are exchange-traded futures contracts with standardized terms, and there are over-the-counter futures contracts with negotiated terms.

gamma A measure of the rate of change in an option's delta for a one-unit change in the price of the underlying security. See **delta**.

good-'til-cancelled (GTC) order A type of limit trading order that remains in effect until it is either executed (filled) or cancelled, as opposed to a day order, which expires if not executed by the end of the trading day.

grantor Same as **writer**.

guts The purchase (or sale) of both an in-the-money call and an in-the-money put. A box spread can be viewed as the combination of an in-the-money strangle and an out-of-the-money strangle. To differentiate between these two strangles, the term *guts* refers to the in-the-money strangle. See **box spread** and **strangle**.

haircut Similar to margin required of public customers, this term refers to the equity required of floor traders on equity option exchanges. Generally, one of the advantages of being a floor trader is that the haircut is less than margin requirements for public customers.

hedge A position established with the specific intent of protecting an existing position. Example: an owner of common stock buys a put option to hedge against a possible stock price decline.

hedge ratio Same as **delta**.

historical volatility A measure of how volatile a stock has been over a given period of time. Usually defined as the annualized standard deviation of a stock's daily returns.

holder The owner of a long stock or option position.

horizontal spread Same as **time spread**.

implied volatility The volatility percentage that justifies an option's price. Calculated from the options current price, the price of the underlying, the exercise price, time to expiration, interest rate and dividends, and "backing out" the volatility estimate.

index A compilation of several stock prices into a single number. Example: the S&P 100 Index.

index option An option whose underlying entity is an index. Generally, index options are cash-settled.

institution A professional investment management company. Typically, this term is used to describe large money managers such as banks, pension funds, mutual funds, and insurance companies.

intermarket spread A strategy involving opposing positions in securities related to two different underlying entities. Example: long OEX calls and short SPX calls.

in-the-money An adjective used to describe an option with intrinsic value. A call option is in-the-money if the stock price is above the strike price. A put option is in-the-money if the stock price is below the strike price.

intrinsic value The in-the-money portion of an option's price. See **in-the-money**.

iron butterfly An option strategy with limited risk and limited profit potential that involves both a long (or short) straddle and a short (or long) strangle.

jelly roll spread A long call and short put with the same strike price in one month, and a short call and long put with the identical strike in another month. This is the combination of synthetic long and short positions in different months. Generally only floor traders use this spread.

kappa Same as **vega**.

lambda A measure of leverage. The expected percent change in the value of an option for a 1 percent change in the value of the underlying.

last trading day The last business day prior to expiration during which purchases and sales of an option can be made. For equity options, this is generally the third Friday of expiration month. For other types of options, the specification of the last trading day varies.

leg A term describing one side of a position with two or more sides. When a trader *legs into* a spread, he establishes one side first, hoping for a favorable price movement so the other side can be executed at a better price. This is, of course, a risk-oriented method of establishing a spread position.

leverage A term describing the greater percentage of profit or loss potential when a given amount of money controls a security with a much larger face value. For example, a call option enables the holder to assume the up-side potential of 100 shares of stock by investing a much smaller amount than that required to buy the stock. If the stock increases by 10 percent, for example, the option can double in value. Conversely, a 10 percent stock price decline can result in the total loss of the purchase price of the option.

limit order A trading order placed with a broker to buy or sell a security at a specific price.

liquid market A trading environment characterized by high trading volume, a narrow spread between the bid and ask, and the ability to trade larger sized orders without significant price changes.

listed option A put or call traded on a national option exchange with standardized terms. In contrast, over-the-counter options usually have nonstandardized or negotiated terms. See **FLexible EXchange options**.

local A floor trader on a futures exchange who buys and sells for his own account, thus fulfilling the same role as a market maker on an options exchange.

long position A term used to describe either (1) an open position that is expected to benefit from a rise in the price of the underlying stock such as long call, short put, or long stock; or (2) an open position resulting from an opening purchase transaction such as *long call, long put*, or *long stock.*

Long-term Equity AnticiPation Securities (LEAPS) Long-term equity and index options. There are no differences between equity LEAPS and equity options except the longer exercise term of the LEAPS.

margin The minimum equity required to support an investment position. To buy *on margin* refers to borrowing part of the purchase price of a security from a brokerage firm.

market basket A group of common stocks whose price movement is expected to closely correlate with an index.

market maker An exchange member on the trading floor who buys and sells for his own account and who has the responsibility of making bids and offers and maintaining a fair and orderly market.

market-maker system A method of supplying liquidity in options markets by having market makers in competition with one another.

market-not-held order A type of market order that allows the investor to give discretion to the floor broker regarding the time and price at which a trade is executed.

market order A trading instruction from an investor to a broker to immediately buy or sell a security at the best available price.

mark-to-market An accounting process by which the price of securities held in an account are valued each day to reflect the last sale price or market quote if the last sale is outside of the market quote. The result of this process is that the equity in an account is updated daily to properly reflect current security prices.

married put strategy The simultaneous purchase of stock and the corresponding number of put options. This is a limited risk strategy during the life of the puts because the stock can be sold at the strike price of the puts.

mixed spread A term used loosely to describe a trading position that does not fit neatly into a standard spread category.

multiply-listed options Options (most often equity options) that are traded on two or more security exchanges.

naked option Same as **uncovered option**.

net margin requirement The equity required in a margin account to support an option position after deducting the premium received from sold options.

net order Same as **contingency order**.

neutral An adjective describing the belief that a stock or the market in general will neither rise nor decline significantly.

90/10 strategy An option strategy in which an investor buys Treasury bills (or other liquid assets) with 90 percent of his funds, and buys call options with the balance.

nonequity option Any option that does not have common stock as the underlying asset. Nonequity options include options on futures, indexes, interest rate composites, physicals, and so on.

nonsystematic risk The portion of total risk that can be attributed to the particular firm. See **systematic risk**.

not-held order A type of order that allows the investor to release the floor broker from the normal obligations implied by the other terms of the order. For example, a limit order designated as "not-held" allows the floor broker to use his discretion in filling the order when the market trades at the limit price of the order. In this case, the floor broker is not obligated to provide the customer with an execution if the market trades through the limit price on the order. See **discretion; market-not-held order**.

OTC option An over-the-counter option is one that is traded in the over-the-counter market. OTC options are not usually listed on an options exchange and generally do not have standardized terms.

omega Same as **vega**.

opening rotation See **trading rotation**.

opening transaction An addition to or creation of a trading position. An opening purchase transaction adds long options (or long securities) to an investor's total position, and an opening sell transaction adds short options (or short securities).

open interest The total number of existing option contracts.

open outcry The trading method by which competing market makers make bids and offers on the trading floor.

option A contract that gives the buyer the right, but not the obligation, to buy or sell a particular asset (the underlying security) at a fixed price for a specific period of time. The contract also obligates the seller to meet the terms of delivery if the contract right is exercised by the buyer.

optionable stock A stock on which options are traded.

option period The time from when an option contract is created to the expiration date.

option pricing curve A graphical representation of the estimated theoretical value of an option at one point in time, at various prices of the underlying asset.

option pricing model A mathematical formula used to calculate the theoretical value of an option. See **Black-Scholes model**.

Options Clearing Corporation (OCC) A corporation owned by the exchanges that trade listed stock options, OCC is an intermediary between option buyers and sellers. OCC issues and guarantees all option contracts.

option valuation model See **option pricing model**.

option writer The seller of an option contract who is obligated to meet the terms of delivery if the option holder exercises his right.

order book official An exchange employee in charge of keeping the public order book and executing the orders therein.

out-of-the-money An adjective used to describe an option that has no intrinsic value, *i.e.*, all of its value consists of time value. A call option is out-of-the-money if the stock price is below the strike price. A put option is out-of-the-money if the stock price is above the strike price. See **intrinsic value** and **time value**.

over-the-counter option Same as **OTC option**.

overvalued An adjective used to describe an option that is trading at a price higher than its theoretical value. It must be remembered that this is a subjective evaluation, because theoretical value depends on one subjective input—the volatility estimate.

overwrite An option strategy involving the sale of a call option against an existing long stock position. This is different from the covered-write strategy, which involves the simultaneous purchase of stock and sale of a call.

parity An adjective used to describe the difference between the stock price and the strike price of an in-the-money option. When an option is trading at its intrinsic value, it is said to be *trading at parity*.

physical option An option whose underlying entity is a physical good or commodity. For example, currency options traded at the Philadelphia Exchange and many OTC currency options are options on the currency itself, rather than on futures contracts.

pin risk The risk to a floor trader with a conversion or reversal position that the stock price will exactly equal the strike price at option expiration. The trader will not know how many of his long options to exercise because he will not know how many of his short options will be assigned. The risk is that on the following Monday he will have a long or short stock position and thus be subject to the risk of an adverse price move.

pit Same as **trading pit**.

position The combined total of an investor's open option contracts and long or short stock.

position limits The maximum number of open option contracts that an investor can hold in one account or a group of related accounts. Some exchanges express the limit in terms of option contracts on the same side of the market, and others express it in terms of total long or short delta.

position trading An investing strategy in which open positions are held for an extended period of time.

premium (1) Total price of an option: intrinsic value plus time value. (2) Often this word is used to mean the same as **time value**.

primary market (1) For securities that are traded in more than one market, the primary market is usually the exchange where the most volume is traded. (2) The initial sale of securities to public investors. See **secondary market**.

profit graph A graphical presentation of the profit-and-loss possibilities of an investment strategy at one point in time (usually option expiration), at various stock prices.

profit profile Same as **profit graph**.

public order book The limit buy and limit sell orders from public customers that are away from the current market price and are managed by the order book official or specialist. If the market price moves so that an order in the public order book is the best bid or offer, that order has priority and must be the first one filled at that price.

put option A contract that gives the buyer the right (but not the obligation) to sell the underlying stock at some predetermined price. For the writer (or grantor) of a put option, the contract represents an obligation to buy stock from the buyer if the option is assigned.

ratio calendar combination A term used loosely to describe any variation on an investment strategy that involves both puts and calls in unequal quantities and at least two different strike prices and two different expirations.

ratio calendar spread An investment strategy in which more short-term options are sold than longer-term options are purchased.

ratio spread (1) Most commonly used to describe the purchase of near-the-money options and the sale of a greater number of farther out-of-the-money options, with all options having the same expiration date. (2) Generally used to describe any investment strategy in which options are bought and sold in unequal numbers or on a greater than one-for-one basis with the underlying stock.

ratio write An investment strategy in which stock is purchased and call options are sold on a greater than one-for-one basis.

realized gain and losses The net amount received or paid when a closing transaction is made and matched together with an opening transaction.

repair strategy An investment strategy in which an existing long stock position is supplemented by buying one in-the-money call (or one at-the-money call) and selling two out-of-the-money calls, all calls having the same expiration. The effect of this strategy is to lower the break-even point of stock ownership without significantly increasing the risk of the total position.

resistance A term used in technical analysis to describe a price area at which rising price action is expected to stop or meet increased selling activity. This analysis is based on historic price behavior of the stock.

reversal Same as **reverse conversion**.

reverse conversion An investment strategy used by professional option traders in which a short put and long call with the same strike price and expiration are combined with short stock to lock in a nearly riskless profit. The process of executing these three-sided trades is sometimes called *reversal arbitrage*.

rho A measure of the expected change in an option's theoretical value for a 1 percent change in interest rates.

risk arbitrage Commonly used term to describe the purchase of a stock subject to takeover rumors with the hope of selling at a significant profit to a company effecting the takeover. The risk is present because there is never any guarantee that a takeover will materialize.

risk conversion/reversal See **fence**.

rolling A trading action in which the trader simultaneously closes an open option position and creates a new option position at a different strike price, different expiration, or both. Variations of this include roll up, roll down, and roll out.

rotation See **trading rotation**.

SEC The Securities and Exchange Commission. The SEC is the federal government agency that regulates the securities industry.

scalper A trader on the floor of an exchange who hopes to buy on the bid price, sell on the ask price, and profit from moment to moment price movements. Risk is limited by the very short time duration (usually 10 seconds to 3 minutes) of maintaining any one position.

secondary market A market where securities are bought and sold after their initial purchase by public investors.

sector indices Indices that measure the performance of a narrow market segment, such as biotechnology or small capitalization stocks.

sequential expiration cycle See **cycle**.

series of options Option contracts on the same underlying stock having the same strike price and expiration month.

settlement price The official price at the end of a trading session. This price is established by The Options Clearing Corporation and is used to determine changes in account equity, margin requirements, and for other purposes. See **mark-to-market**.

short option position The position of an option writer which represents an obligation to meet the terms of the option if it is assigned.

short position Any open position that is expected to benefit from a decline in the price of the underlying stock such as long put, short call, or short stock.

short stock position A strategy that profits from a stock price decline. It is initiated by borrowing stock from a broker-dealer and selling it in the open market. This strategy is closed out at a later date by buying back the stock.

specialist An exchange member who manages the limit order book and makes bids and offers for his own account in the absence of opposite market side orders. See **market maker**.

speculator A trader with an expectation of a particular market price behavior.

spread A position consisting of two parts, each of which alone would profit from opposite directional price moves. These opposite parts are entered simultaneously in the hope of (1) limiting risk, or (2) benefiting in a change of price relationship between the two.

spread order Trading order to simultaneously make two transactions, each of which would benefit from opposite directional price moves.

standard deviation A statistical measure of variability in a data series. Used as an estimate of volatility.

stock index futures A futures contract that has as its underlying entity a stock market index. Such futures contracts are generally subject to cash settlement.

stop-limit order A type of contingency order placed with a broker that becomes a limit order when the security trades, or is bid or offered at a specific price.

stop order A type of contingency order placed with a broker that becomes a market order when the security trades, or is bid or offered at a specific price.

straddle A trading position involving puts and calls on a one-to-one basis in which the puts and calls have the same strike price, expiration, and underlying entity. A long straddle is when both options are owned and a short straddle is when both options are written.

strangle A trading position involving out-of-the-money puts and calls on a one-to-one basis. The puts and calls have different strike prices, but the same expiration and underlying stock. A long strangle is when both options are owned, and a short strangle is when both options are written.

strap A strategy involving two calls and one put. All options have the same strike price, expiration, and underlying stock.

strike price Same as **exercise price**.

strike price interval The normal price difference between option exercise prices. Equity options generally have $2.50 strike price intervals (if the underlying security price ranges from $5 to $25), $5.00 intervals (from $25 to $200), and $10 intervals above $200. Index options generally have $5 strike price intervals at all price levels. See **adjusted strike price**.

strip A strategy involving two puts and one call. All options have the same strike price, expiration, and underlying stock.

suitability A requirement that any investing strategy fall within the financial means and investment objectives of an investor.

support A term used in technical analysis to describe a price area at which falling price action is expected to stop or meet increased buying activity. This analysis is based on previous price behavior of the stock.

synthetic long call A long stock position combined with a long put.

synthetic long put A short stock position combined with a long call.

synthetic long stock A long call position combined with a short put.

synthetic position A strategy involving two or more instruments that has the same risk–reward profile as a strategy involving only one instrument. The following list summarizes the six primary synthetic positions.

synthetic short call A short stock position combined with a short put.

synthetic short put A long stock position combined with a short call.

synthetic short stock A short call position combined with a long put.

systematic risk The portion of total risk that can be attributed to the overall market. See **nonsystematic risk**.

tau Same as **vega**.

technical analysis A method of predicting future stock price movements based on the study of historical market data such as the prices themselves, trading volume, open interest, the relation of advancing issues to declining issues, and short selling volume.

theoretical value An estimated fair value of an option derived from a mathematical model.

theta A measure of the rate of change in an option's theoretical value for a one-unit change in time to the option's expiration date. See **time decay**.

tick (1) The smallest unit price change allowed in trading a security. For a common stock, this is generally $\frac{1}{16}$th point. For an option under $3 in price, this is generally $\frac{1}{16}$th point. For an option over $3, this is generally $\frac{1}{8}$th point. (2) The net number of stocks upticking or down ticking. For example, if there are 10 stocks total with 7 having traded on upticks and 3 having traded on downticks, then the tick is +4.

time decay A term used to describe how the theoretical value of an option "erodes" or reduces with the passage of time. Time decay is specifically quantified by theta. See **theta**.

time spread An option strategy most commonly used by floor traders which involves options with the same strike price, but different expiration dates.

time value The part of an option's total price that exceeds intrinsic value. The price of an out-of-the-money option consists entirely of time value. See **intrinsic value** and **out-of-the-money**.

trader (1) Any investor who makes frequent purchases and sales. (2) A member of an exchange who conducts his buying and selling on the trading floor of the exchange.

trading pit A specific location on the trading floor of an exchange designated for the trading of a specific security.

trading rotation A trading procedure on exchange floors in which bids and offers are made on specific options in a sequential order. Opening trading rotations are conducted to guarantee all entitled public orders an execution. At times of extreme market activity, a closing trading rotation can also be conducted.

traditional expiration cycle See **cycle**.

transaction costs All of the charges associated with executing a trade and maintaining a position. These include brokerage commissions, exchange fees, and margin interest. The spread between bid and ask is sometimes taken into account as a transaction cost.

Treasury bill/call option strategy Same as **90/10 strategy**.

type of options The classification of an option contract as either a put or call.

uncovered option A short option position that is not fully collateralized if notification of assignment is received. A short call position is uncovered if the writer does not have a long stock position to deliver. A short put position is uncovered if the writer does not have the financial resources in his account to buy the stock.

underlying security The asset that can be purchased or sold according to the terms of the option contract.

undervalued An adjective used to describe an option that is trading at a price lower than its theoretical value. It must be remembered that this is a subjective evaluation because theoretical value depends on one subjective input—the volatility estimate.

unit of trading The minimum quantity or amount allowed when trading a security. The normal minimum for common stock is 1 round lot or 100 shares. The normal minimum for options is one contract (which covers 100 shares of stock).

unsystematic risk Same as **nonsystematic risk**.

upstairs trader A professional trader who makes trading decisions away from the exchange floor and communicates his instructions to the floor for execution by the floor broker.

vega A measure of the rate of change in an option's theoretical value for a one-unit change in the volatility assumption.

vertical spread (1) Most commonly used to describe the purchase of one option and sale of another where both are of the same type and same expiration, but have different strike prices. (2) It is also used to describe a delta-neutral spread in which more options are sold than are purchased.

volatility A measure of stock price fluctuation. Mathematically, volatility is the annualized standard deviation of daily returns.

volatility test A procedure in which a multisided options position is evaluated, assuming several different volatilities for the purpose of judging the risk of the position.

wasting asset An investment with a finite life, the value of which decreases over time if there is no price fluctuation in the underlying asset.

write To sell an option. An investor who sells an option is called the writer, regardless of whether the option is covered or uncovered.

INDEX

Note: Bold number indicate illustrations.

D

E

I

Q

R

U

V